SEARCH AND
TECHNOLOGY AND MEDICINE

PRANELA RAMESHWAR
Rutgers University, USA

ELISABETH ENGEL
IBEC, Barcelona, Spain

VERONIQUE SANTRAN
CELTIS Toulouse, France

PERE GASCON
University of Barcelona, Spain

T0313329

ies are submitted to the Web of Science Book Citation
f and to Google Scholar for evaluation and indexing.

ration of areas of basic and applied science with their
es highlights societal benefits, technical and business
emerging and new technologies. In combination, the
us topic cluster analyses of key aspects of each of the
chain.
detailed snapshots of critical issues in biotechnology and
g point in financial investment or industrial deployment,
various specialty areas including pharmaceutical sciences
ology, and biomaterials. Areas of primary interest comprise
gy, molecular biology, stem cells, hematopoiesis, oncology,
polymer science, formulation and drug delivery, renewable
prefineries.
mprehensive review and opinion articles covering all
topic. The editors/authors of each volume are experts in their
s are peer-reviewed.

series, visit www.riverpublishers.com

Basic Cardiovascular Physiology: From Molecules to Translational Medical Science

Pasquale Pagliaro

University of Turin, Torino, Italy

and

National Institute for Cardiovascular Research (INRC)
Italy

Claudia Penna

University of Turin, Torino, Italy

and

National Institute for Cardiovascular Research (INRC)
Italy

Raffaella Rastaldo

University of Turin, Torino, Italy

River Publishers

Routledge
Taylor & Francis Group

LONDON AND NEW YORK

Published 2020 by River Publishers
River Publishers
Alsbjergvej 10, 9260 Gistrup, Denmark
www.riverpublishers.com

Distributed exclusively by Routledge
4 Park Square, Milton Park, Abingdon, Oxon OX14 4RN
605 Third Avenue, New York, NY 10017, USA

Basic Cardiovascular Physiology: From Molecules to Translational Medical Science / by Pasquale Pagliaro, Claudia Penna, Claudia Penna.

Routledge is an imprint of the Taylor & Francis Group, an informa business

ISBN 978-87-7022-200-6 (print)

Contents

Preface

The cardiovascular system is designed to ensure the survival of all cells of the body. Cardiovascular physiology describes the mechanisms by which the appropriate amount of nutrients and oxygen is delivered to the organs in the various conditions during normal life. The cardiovascular system and blood flow ensure the survival of all cells by maintaining the immediate chemical environment of each cell in the body (i.e., the interstitial fluid) at a composition appropriate for that cell's normal function by removing the waste and bringing the nutrients from the interstitial space.

In this book, we will first analyze the conditions necessary for the movement of blood in the cardiovascular system. To better understand the pump function of the heart, we will consider first the functional tissue of the heart describing the structure of the myocardial fibers and then the electric properties of cardiac cells. Cardiac contractility or inotropism is outlined in the details before considering the cardiac cycle, heart sounds, and murmurs. The cardiac output and the venous return to the heart are analyzed in the same chapter to underline that they are two identical concepts. While the former is "seen" from the heart to the arteries, the latter is "seen" from the veins to the heart. Intrinsic and extrinsic regulation of contractile force is considered to explain the adaptation of cardiac output to the needs of the body.

The arterial pressure determinants, regulation, and measurements are considered in Chapter 8 of this book where the mechanisms of control of arterial pressure are described in the details. The fundamental aspects of cardiac work, heart performance, and myocardial metabolism are described. Details of the genesis and interpretation of the electrocardiogram, as well as some electrocardiographic aspects of the conduction disturbances of the main arrhythmias, are reported in this book. Vascular hemodynamics and the microcirculation are treated easily. To the nervous control of the cardiovascular system and the main reflexes of cardiovascular is dedicated a whole chapter. Another chapter is dedicated to the humoral and local control of the cardiovascular system. These last two chapters allow an overview of the cardiovascular system and an integrated approach, before going on to

xiii

analyze in detail the various district circulations. Several district circulations are described, but particular emphasis is given to the coronary, cerebral and pulmonary circulations (a separate chapter is dedicated to the latter). Coordinated cardiovascular adaptations, including physical exercise, alerting response, and responses to hemorrhage are considered to see the fundamental mechanisms at the basis of integrated and coordinated adaptations. To myocardial protection against ischemia-reperfusion injury is dedicated a chapter as a paradigmatic example of pathophysiological mechanisms in response to stressing stimuli in healthy and in pathological conditions.

Finally, lymph circulation is considered a parallel system that cross-talks with the cardiovascular system. The book ends with a chapter dedicated to the functional imaging of the cardiovascular system to study human pathophysiology *in vivo*.

The book is endowed with a rich iconography and various panels (BOXES) in which, on the one hand, aspects connected with physics are treated and, on the other hand, aspects strictly connected with pathophysiology and clinical aspects are considered.

We tried to write the book in a linear and simple language without compromising the scientific rigor of the various topics covered. We hope to find the interest of readers and that they can enjoy using this book during their studies and their working life. We also hope that readers will find what they are looking for in this book, that they will find the texts, boxes, and figures clear and that they will be so generous as to let us know where we can improve this book in the next editions.

Acknowledgments

We are deeply grateful to our dear Prof Gianni Losano who recently passed away. We were lucky enough to get to know him and appreciate his human characteristics and his remarkable knowledge in the physiological and medical history. Prof Losano introduced the authors of this book into the fascinating field of cardiovascular physiology. We also want to thank dr. Amedeo Chiribiri who permitted us to use the material, which served to write much of the chapter "Functional imaging of the cardiovascular system". Finally, we would like to thank the readers who will give us comments and suggestions to improve future editions of this book.

List of Boxes

List of Abbreviations

5-HT	5-hydroxytryptamine or Serotonin
AC	adenylyl cyclase
ACE	Angiotensin-Converting converting Enzymeenzyme
ACE2	Angiotensin-converting enzyme 2
ACh	acetylcholine
ADH	antidiuretic hormone or vasopressin
Akt	(see PKB)
AMI	acute myocardial infarction
Ang I	angiotensin I
Ang II	angiotensin II
ANP	atrial natriuretic peptide
AP	action potential
AQP2	aquaporin 2
ARB	Angiotensin II receptor blockers
ARNi	angiotensin receptor–neprilysin inhibitors
ARVC	arrhythmogenic right ventricular cardiomyopathy
aV	apparent or relative viscosity
AV	atrioventricular
AVN	atrioventricular node
a-vO$_2$	artero-venous difference in O$_2$ concentration
b.p.m	beats per minute
BH$_4$	tetrahydrobiopterin
Bk	bradykinin
BNP	brain natriuretic peptide
BOLD	blood oxygen level dependent
CAM	cellular adhesion molecules
CaMK II	Ca^{2+}/calmodulin-dependent protein kinase II
cAMP (or AMPc)	cyclic adenosine monophosphate
CCS	cardiac conduction systems
cGMP (or GMPc)	cyclic guanosine monophosphate
CICR	calcium-induced calcium release

CM	calmodulin
CMD	coronary microvascular dysfunction
CNP	type C natriuretic peptide
CNS	central nervous system
CO	cardiac output
CO_2	carbonic dioxide
COMTs	catechol-ortho-methyltransferases
COX I	cyclooxygenase I
COX II	cyclooxygenase II
CPVT	Catecholaminergic catecholaminergic polymorphic ventricular tachycardia
CT	computerized axial tomography
CV	conduction velocity
CVLM	caudal ventrolateral medulla
CVP	central venous pressure
DADs	Delayed delayed afterdepolarizations
DAG	diacyl-glycerol
DHP	dihydropyridine channel-receptor (also known as L-type Ca^{2+} channel)
DP	diastolic pressure
dP/dt	derivative of the pressure
EADs	Early early afterdepolarizations
Ec	kinetic energy
ECG	electrocardiogram
EDCF	endothelial derived contractile factor
EDHF	endothelial derived hyperpolarizing factors
EDRF	endothelial derived relaxing factor
EET	epoxy-eicosa-trienoic acids
EF	ejection fraction
EG	excitable gap
EGFR	epidermal growth factor receptor
e-NOS	endothelial NOS
ERK 1/2	extracellular signal regulated kinase 1/2
ESPVR	end-systolic pressure-volume relationship
ESS	end-systolic-slope
ET	endothelin
FFR	fractional flow reserve
fMRI	Functional Magnetic Resonance Imaging
GABA	γ-aminobutyric acid

GC	guanylyl cyclase
GPCR	G-protein-coupled receptors
GTP	guanosine-triphosphate
H_2O_2	hydrogen peroxide
HF	heart failure
HFpEF	HF with preserved EF
HMM	heavy meromyosin
HR	heart rate
HSPs	heat shock proteins
Htc	hematocrit
ICAM-1	intracellular adhesion molecule-1
IMLC	intermediolateral column
i-NOS	inducible NOS
IP3	inositol-triphosphate
IPC	Ischemic preconditioning
IRI	ischemia-reperfusion injury
K_{ACh}	muscarinic potassium channels
Kd	callidin
LDH-H4	lactic dehydrogenase (myocardial isoenzymes H4)
LMM	light meromyosin
L-NAME	L-N-nitro-arginine-methyl ester
L-NMMA	LN-monomethyl-arginine
L-NNA	LN-nitro-arginine
LQTS	long QT syndrome
LVH	left ventricular hypertrophy
MAOs	monoamine oxidases
MAP	mean arterial pressure
MCFP	mean circulator filling pressure (or mean systemic pressure)
MDP	mean diastolic pressure
MEK 1/2	mitogen-activated extracellular regulated kinase 1/2 kinase pathway
MHC	myosin heavy chains
MI	myocardial infarction
mito-KATP	ATP-dependent mitochondrial K^+ channels
MLC	myosin light chains
MLCK	myosin light chain kinase
MLCP	myosin light chain phosphatase
MPLA	monophosphoryl-lipid A

mPTP	mitochondrial permeability transition pores
MSP	mean systolic pressure
MVO_2	myocardial oxygen consumption
NA (or NE)	noradrenaline (or norepinephrine)
NANC	non-adrenergic and non-cholinergic
NCX	$3Na^+/1Ca^{2+}$ exchanger
NEP	Neprilysin
NF-κB	nuclear factor-kappa B
NHX	sodium/hydrogen exchanger
n-NOS	neuronal NOS
NO	nitrogen monoxide or nitric oxide
NOS	nitric oxide synthase
NPR	natriuretic peptide receptor
NSAIDs	non-steroidal anti-inflammatory drugs
NTS	nucleus of the solitary tract
O_2^-	superoxide anion
$ONOO^-$	peroxynitrite
ox-LDL	low density oxidized lipoproteins
P38-MAPK	P38-mitogen-activated protein kinase
PAH	para-aminohippuric acid
PDE	phosphodiesterase
PECAM-1	platelet endothelial cell adhesion molecule-1
PET	Positron emission tomography
PG	Prostaglandin
Pi	inorganic phosphorus
PI3K	phospho-inositol-3-kinase
PIP2	phospho-inositol-diphosphate
PKA	protein-kinase A
PKB	protein kinase B (called also Akt)
PKG	protein kinase G
PLC	phospholipase C
PLN	phospholamban
pO_2	partial pressure of O_2
PP	pulse pressure
PRR	(pro)renin receptor
QTc	corrected QT
RAAS	renin-angiotensin-aldosterone system
RAS	renin-angiotensin system
RBF	renal blood flow

RIC	remote ischemic conditioning
RISK	Reperfusion Injury Salvage Kinases
ROS	reactive oxygen species
RP	refractory period
RPF	renal plasma flow
RVLM	rostral ventrolateral medulla
RyR	ryanodine channel-receptor
SA	senoatrial
SACT	sinoatrial conduction time
SAFE	survivor activating factor enhancement
SAN	senoatrial node
SERCA	Sarco-Endoplasmatic Reticulum Calcium pump (or calcium transport ATPase)
ShR	shear rate
SP	systolic pressure
SPECT	Single photon emission tomography
SQTS	short QT syndrome
SR	sarcoplasmic reticulum
SRV	systolic residual volume
SS	shear stress
STAT-3	signal transducer and activator of transcription-3
STEMI	ST elevation myocardial infarction
SV	stroke volume or systolic ejection
TK	tyrosine kinase
Tn	Troponin
TNFα	tumor necrosis factor α
TPR	total peripheral resistance
VCAM-1	vascular cell adhesion molecule-1
VCS	ventricular conduction system
VDRV	ventricular diastolic return volume
VEDV (or VTDV)	ventricular end-diastolic volume (or ventricular tele-diastolic volume)
VIP	vasoactive intestinal peptide
VSM	vascular smooth muscle
VTDP	ventricular telediastolic pressure
VV	ventricular volume

1

Cardiovascular System

1.1 Overview of the Cardiovascular System

The *cardiovascular or cardio-circulatory system* consists of the heart and the vessels (Figure 1.1). The heart consists of two reservoirs, the right and left atria, and two intermittent pumps, the right ventricle, and left ventricles. The right atrium receives blood from various tissues through the upper and lower *cava* veins, and then blood flows into the right ventricle through the ostium of the *tricuspid valve*. The right ventricle pumps the deoxygenated blood at low pressure (mean pressure $\approx$ 15 mmHg) into the pulmonary artery; the passage of blood from the right ventricle to the pulmonary artery occurs through the *pulmonary valve* ostium. Then blood flows into vessels that form the *pulmonary or small circulation*. This is composed of arteries, arterioles, and alveolar capillaries in which the blood is oxygenated; then the blood flows into the venules and veins ending into the four pulmonary veins, which bring oxygenated blood back to the left atrium. From this reservoir, the blood passes into the left ventricle through the *bicuspid or mitral valve* ostium. Then the left ventricle pumps the oxygenated blood at high pressure (mean pressure $\approx$ 95 mmHg) into the aorta; the passage of blood from the left ventricle to the aorta occurs through the *aortic valve* ostium, from which starts the *systemic or large circulation*, also consisting of arteries, arterioles, capillaries, venules, and veins. Through the various collateral vessels of the aorta, the blood perfuses all organs and tissues. The arterial blood flows in arterioles, capillaries, and venules to be then collected in larger veins that eventually merge into the two cava veins. In passing through the capillaries of the general circulation, the blood returns to be venous, i.e., deoxygenated as it yields a part of the oxygen that carries and is enriched with carbon dioxide produced by the metabolism of the tissues.

From a functional point of view, the blood that reaches the right atrium through two large veins (inferior and superior venae cavae) is *venous (deoxygenated) blood*, that is, relatively poor in oxygen and rich in carbon dioxide.

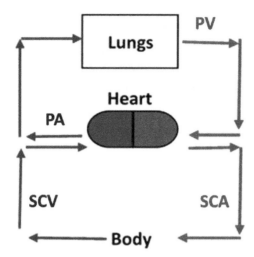

Figure 1.1   Serial arrangement of systemic and pulmonary circulation. PA: pulmonary arteries; PV: pulmonary veins; SCA: systemic circulation arteries; SCV: systemic circulation veins.

As such it passes into the right ventricle and then into the pulmonary artery. As it passes through the alveolar capillaries, the venous blood becomes *arterial (oxygenated) blood*, receiving oxygen from the alveolar air and releasing carbon dioxide. The oxygenated blood then reaches the left atrium, passes into the left ventricle and is then pumped into the aorta and its branches.

It should be noted that "arteries" carry blood from the ventricles to the periphery, whether referring to the pulmonary or systemic circulation, while "veins" are those vessels that carry blood from the periphery to the atria. It is not surprising, therefore, that the vessel that starts from the right ventricle is classified as an artery even if it contains "venous", deoxygenated blood, while the vessels that carry blood in the left atrium are classified as veins even if they contain "arterial" oxygenated blood.

The systemic circulation and the pulmonary circulation are placed in series with each other (Figure 1.1). It follows that all the blood passes through the left heart and the systemic circulation in the unit of time, should also pass through the right heart and the pulmonary circulation. It follows that the amount of blood that the left ventricle pumps in one minute in the aorta should be equal to the amount of blood that the right ventricle pumps in one minute in the pulmonary artery. However, due to some *anatomical shunts* (anastomosis formed by communication, at precapillary or postcapillary level, between the neighboring vessels of the pulmonary and *bronchial circulation*, and some *coronary vessels* that drain blood directly in the left side of the heart), the amount of blood that the left ventricle pumps in one minute in the aorta is a little greater (a few mL) than that pumped by the right ventricle in the pulmonary artery. We can say that **the blood pumped by a ventricle is exactly the same amount it receives as a venous return**, but we cannot say that the amount of blood pumped by the left ventricle is the same amount pumped by the right ventricle.

The term *cardiac output* (CO) refers to the quantity of blood pumped by one ventricle during one minute. It should not be confused with the *stroke volume* (SV) that is the amount of blood ejected per contraction. Since the heart beats a certain number of times per minute and the number of beats per minute (b.p.m.) constitutes the *heart rate* (HR), it is easy to understand how the following relationship between CO, SV, and HR exists:

$$SV \times HR = CO$$

The SV in a resting 70 kg adult subject is about 70 mL. Since the heart normally beats about 70 times per minute, *i.e.,* its frequency is 70/min (70 b.p.m.), the cardiac output is given by 70 mL $\times$ 70/min = 4900 mL/min.

The CO is not fixed, however, and adapts rapidly to the body's needs. Of note, both ventricles are involved in these adaptations, so that if CO of one ventricle increase also CO of the other ventricle inseases after few seconds. This implies that cardiovascular properties and regulatory systems exist to support these mechanisms and adaptations, and these are the main topics of this book.

The cardiovascular system or apparatus is organized in a series of **sections** *and parallel* **districts**.

Sections are in series: we can say that the right heart, the pulmonary circulation, the left heart and the systemic circulation can be considered as *sections* of the cardiovascular system. Sections are also all arteries, the set of arterioles, all capillaries, all venules, and the set of veins for the systemic and the pulmonary circulation. Therefore, the sections are the parts of the cardiovascular system placed in series with each other. In the unit of time sections of the pulmonary circulation and sections of the systemic circulation are crossed by the same amount of blood pumped by the respective ventricles.

Districts are in parallel: the *districts* are different from the sections (Figure 1.2). In Figure 1.2 it is possible to observe how from the aorta depart smaller arteries for the various districts. Vascular districts are constituted by the circulations of organs, tissues, and systems perfused by arteries departing

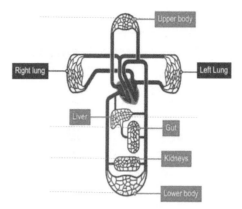

Figure 1.2   Systemic and pulmonary circulations are in series. The dotted lines highlight the parallel organization of the districts within the systemic circulation.

from the aorta. Each of these arteries is subdivided into smaller vessels up to the capillaries, followed by the venules and veins that merge into one of the cava veins. The set of vessels that perfuse a district forms a district circle. Figure 1.2 clearly illustrates how the different circulatory districts are arranged in parallel. Each organ is in parallel with the other, and within an organ, the ramifications are also in parallel. It is obvious that even in each district we can find sections (arterioles, capillaries, etc.) belonging to that specific district and placed between them in series.

1.2 Conditions Necessary for Blood Movements in the Cardiovascular System

To have blood circulation, the cardiac activity must produce a *pressure gradient* within the cardiovascular system by increasing the arterial pressure and lowering the venous pressure. Let us consider the conditions necessary to produce this gradient.

The pumping action of the heart can generate blood movements in the cardiovascular system only if two conditions are present: *a)* the *mean circulator filling pressure* (see below) is greater than the pressure in the ventricles during diastole, and *b)* the vessels are distensible so that in systole the pressure can increase (pressure pulse) and travel along the arterial tree in a finite time (not instantaneously in the entire system, as would occur if the vessels had rigid, non-expansible walls, as predicted by Pascal's law).

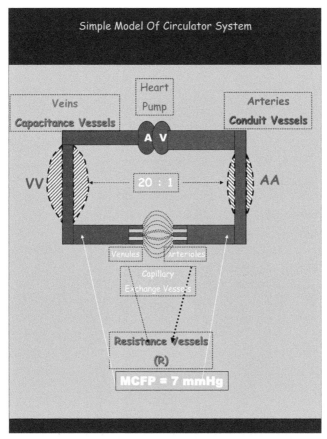

Figure 1.3 Simplified model of the cardiovascular system. The 20: 1 ratio indicates that the venous capacity is 20 times greater than the arterial one. In other words, when the system has an average systemic pressure of 7 mmHg, a volume of blood 20 times greater than in arterial vessels is contained in the venous vessels (capacity vessels). AA: arteries; VV: veins; MCFP: mean circulator filling pressure. For further explanations, see the text.

To understand how the value of the *mean circulator filling pressure* and the distensibility of the vessels can ensure the progression of the blood we can refer to an extremely simplified scheme of the cardiovascular system (Figure 1.3).

The *mean circulator filling pressure* (MCFP) is the pressure exerted by the blood inside the cardiovascular system when the heart is arrested. In this condition, the pressure is the same in all parts of the system whether they are arteries, veins, or capillaries, and into the heart chambers. For the MCFP to be greater than zero, the system must contain an *excess volume* of blood. If we remove 30% *circa* of the blood the MCFP will be zero. Therefore, the *excess volume* is about 30% of the blood usually contained in our cardiovascular system.

In Figure 1.3, the heart is represented as constituted by only one atrium A and only one ventricle V. From the ventricle the arterial tree area AA departs from a single conduit. When the heart is stopped, the wall of this conduit may be in the rest position (continuous lines) when the blood pressure inside the system is equal to zero, or in a tense position (dashed lines) when the pressure is greater than zero (7 mmHg in humans).

The arterial system continues in the venous section VV through a set of resistance vessels R that delay the passage of blood. Also, the venous section, represented by a single duct, can be in rest or tense conditions as indicated by the continuous and dashed lines, respectively. The greater distance between the dashed lines and the continuous lines of the venous system indicates that the amount of blood it can contain, that is the *capacity*, is much higher than the capacity of the arterial system with identical pressure.

While the heart is arrested, if the system is filled with liquid and all vascular walls are in resting position, the pressure in all parts of the system is equal to zero. If under these conditions the ventricle starts to contract (systoles), it can push the blood it contains into the arterial tree. After pumping, the ventricle relaxes (diastole), and the pressure inside it will return to zero. Since the pressure in the venous section is equal to zero, the liquid cannot flow from the veins to the heart, in fact between these two structures there is no pressure gradient. Since the ventricle is not filled after a couple of beats will not produce any SV. So even if the heart continues to beat will not have a CO, *i.e.*, the blood will not move.

If an extra volume (*excess volume*) of liquid is added to the volume that fills the system at zero pressure, moving now the vessel walls to the dotted lines (Figure 1.3), the pressure will rise from zero to 7 mmHg, *i.e.*, the *MCFP or mean circulator filling pressure.* This is the pressure exerted in the circulatory apparatus of the humans with an arrested heart, as could occur with a sufficiently intense and protracted vagal stimulation able to stop the heart.

Now consider that the heart starts pulsing in the presence of a mean *MCFP* of 7 mmHg. At the first beat, it pumps the SV into the aorta. When it relaxes (diastole), the pressure inside the ventricle falls below 7 mmHg. Under these conditions pressure gradient is created between veins and the heart, so the heart can fill and then pump a second SV into the arterial tree.

So many successive heart beat will be possible, and the blood will eventually be in motion. Because the succession of heart beat increases the pressure on the arterial side of the system, it will be able to overcome the resistance R and to push the blood into the veins toward the heart atria by exerting the so-called *vis a tergo*.

Therefore, we can understand that with its activity the heart takes blood from the venous tree and transfers it to the arterial one. The heart can, therefore, be considered a *volume distributor*. Because of this breakdown, we can imagine how in the scheme of Figure 1.3, the increased volume contained in the arterial tree makes their wall to expand beyond the dashed lines, while the reduced volume of the venous tree allows their wall to shrink between the dashed lines and the continuous ones.

In the venous tree, when the volume is reduced, there is a slight decrease in pressure below 7 mmHg, while a significant increase in pressure corresponds to an increase in volume in the arterial tree. Therefore, the heart besides being a *volume distributor* is also a *pressure distributor*. The increase in pressure in the arterial tree is much greater than the decrease in the venous tree due to the lower distensibility of the arteries concerning the veins (Figure 1.4). The veins of the systemic circulation are, in fact, 20 times more distensible than the arteries.

As a consequence, with the continued pulsatile activity of the heart, a situation of equilibrium is established in which the arterial and venous pressures remain almost constant. When the CO is about 5 L/min, the mean

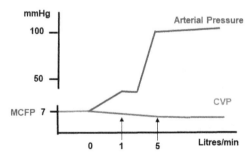

Figure 1.4 Effects of heart-pumping on arterial pressure and central venous pressure (CVP). CVP is the pressure in the big veins and atria. At zero pumping rate, CVP and arterial pressure are equal (that is mean circulator filling pressure (MCFP = 7 mmHg). When the volume transfer-rate increases, the CVP decreases slightly, while the blood pressure increases a lot. This difference is because the arteries are 20 times less compliant than the veins and because the flow from the arteries to the periphery is "hindered" by the resistance vessels (the arterioles).

arterial pressure remains about 90–100 mmHg, and the venous pressure is close to zero. The equilibrium is because when a certain pressure value is established in the arterial section, the blood is pushed through the resistance vessels to pass into the venous section. This step, in addition to preventing the arterial pressure from exceeding certain limits, prevents the venous pressure from falling to zero, allowing it to remain higher than the pressure present in diastole in the heart cavities and thus ensuring the filling of the ventricles (that is *vis a tergo*, see Chapter 6.2).

We have seen that, if the mean circulator filling pressure (MCFP) is below the pressure in the heart cavities in diastole, the blood circulation is impossible even if the heart contracts. We have also seen that MCFP is 7 mmHg (is above 0 mmHg) if an excess volume of blood is present in the cardiovascular system. Generally, this volume represents about 30% of the total blood volume of a normal subject. Therefore, we can infer that a subject can suffer a circulatory arrest by hypotensive shock if during a hemorrhage 30% of the blood is lost. The circulatory arrest, therefore, takes place even if the heart for a certain time continues to beat (ineffective beats). The task of those who intervene to save the life of that subject is therefore to restore excess volume with blood transfusions. If under extreme conditions the blood is not available, the operator can administer plasma, or solutions that contain substances that develop the same colloidosmotic pressure of the plasma proteins, or in extreme cases a simple physiological solution, bearing in mind that the transitory aim is to ensure that the diastolic filling of the heart can be ensured to generate an SV. As soon as possible blood must be used for appropriate tissue oxygenation.

In addition to the existence of a venous pressure higher than diastolic ventricular pressure, the movement of blood is also possible because the vascular system is distensible. Let us imagine that all the sections, arterial, venous, and of resistance are perfectly rigid. The only effect that the contraction of the heart can induce is, according to Pascal's law, a simultaneous increase in pressure throughout the system, without any difference or gradient from one point to another. Therefore, because of the absence of gradient, the blood could not move if the vessels were rigid. Since the vascular system is distensible, the pressure waves travel along the arterial tree with a certain velocity, and the blood can circulate at a lower velocity because beyond the pressure gradient that promotes the blood flow, there are also resistance and impedance to overcome, between sections. The pressure waves travel at about 4–10 m/s in the arterial tree, whereas the blood travels at about 0.2–0.5 m/s in the aorta and even at a lower velocity in the capillaries. The difference between pressure transmission velocity and blood velocity will be considered in other chapters of this book (see for example Chapters 8 and 11). As we will see in Chapter 8 *blood flow is determined by the pressure gradient, divided by the vascular resistance of the considered district.*

1.3 The Sections of the Cardiovascular System

We have already said in Paragraph 1.1 that the various sections of the cardiovascular system are placed in series with each other, while the districts are placed in parallel with each other. Figure 1.5 shows schematically the characteristics of the pressure in the various sections in a series of the systemic circulation and of the pulmonary circulation. In this representation, the heart is considered a pressure generator. It must be said that, even if this is not indicated in Figure 1.5, the pressure generates movement of the blood, *i.e.,* blood flows thanks to the pressure gradient between the subsequent sections. The flow amount is favored by the pressure gradient and is hindered

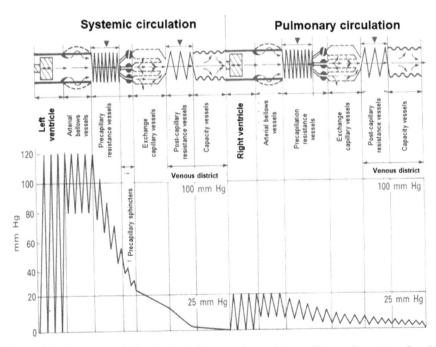

Figure 1.5 Pressure variation in the different sections of the cardiovascular system. See the explanation in the text. (from B. Folkow & E. Neil, Circulation O.U.P, 1961).

by the presence of vascular resistance. The flow direction is determined by pressure gradients and the unidirectional valves located in some parts of the cardiovascular system.

In Figure 1.5, we can see how the pressure presents ample oscillations from a value of about 0 mmHg to a value of about 120 mmHg in the left ventricle. These variations are due to the succession of contractions (systoles) and relaxations (diastoles) of the heart. The large elastic arteries that together with the aorta constitute the so-called *arterial bellow* (for their distensibility), can expand when receiving the SV. In these arteries, the same maximum or systolic pressure value developed by the ventricle is reached, but the arterial elastic recoil, together with the peripheral resistance and the presence of "competent" (properly closed) aortic valve, prevent the fall of diastolic pressure to a value lower than 70–80 mmHg. For instance, when the valve does not close properly (it is incompetent or insufficient) or the resistance is reduced the arterial diastolic pressure can reach lower values. The distensibility or compliance, which is a change in volume per unit change in pressure, and the elastic recoil are responsible for the so-called *windkessel effect, or bellows*, which is the main responsible for the attenuation of the pressure oscillations in the arteries (see below).

In Figure 1.5 we see that when the blood passes through precapillary arterioles or *resistance vessels*, the pressure is reduced both as a mean value and as the amplitude of the oscillations. The reduction in amplitude of the oscillations until their disappearance means that in the capillaries, or *vessels of exchange*, the pressure ceases to be pulsating and becomes continuous. Along the capillaries, there is a modest degree of resistance, thereby the pressure still has a certain dampening, even if it less than that observed in the resistance vessels.

While the reduction of the blood pressure along the precapillary arterioles is due to the resistances present in them, the progressive reduction of the pulsatility until its disappearance in the capillaries is due to the combined effect of the resistances and elastic distensibility or compliance of the arteries. Although on this topic we will return in other chapters (*e.g.,* Chapter 11) of this book, where we will consider arterial pressure and flow waveforms in terms of *wave propagation and reflection*, it is opportune to briefly consider how the distensibility and elasticity of the large arterial vessels, including the aorta, attenuates the systolic pressure and raises the diastolic pressure, compared to a rigid (no elastic) tube. Indeed, the arterial vessels expand into systole during ventricular ejection and return to the resting position during diastole. The expansion during ventricular ejection causes the arterial walls

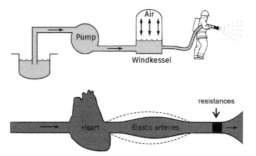

Figure 1.6 "The leveling" of pressure and flow over the entire cardiac cycle is somehow similar to the maintenance of water pressure in the fire engines by a dome-shaped, air-filled compression chamber (*windkessel* in German translation). Indeed, the volume of blood that is ejected into the circulation during systole is "stored" by the stretching (elasticity) of the vessel walls due to their elasticity, and during diastole, flow and, consequently, the pressure is maintained by the elastic recoil of these same vessel walls. Also, the resistance vessels play a role in the maintenance of the diastolic pressure. On the contrary, in rigid tubes, flow and pressure oscillate more despite a similar stroke volume.

to absorb and accumulate part of the energy otherwise destined to increase the systolic pressure. This accumulated energy is returned to the blood in the form of diastolic pressure just when the arteries return to the resting position due to the elastic recovery. This effect can reduce the amplitude of the pressure oscillations and it is called the *windkessel effect* (Figure 1.6). The combined action of the *windkessel effect* with that of the resistances placed downstream of this causes the pressure pulsations to disappear and the flow from pulsatile to be continuous in capillaries.

*Actually, in peripheral arteries, the **pressure oscillation** can be greater than in aorta. This is due to the summing of the wave-pressure propagation and wave-pressure reflection (see Chapter 11). On the other hand, the **flow oscillation** is maximum in the ascending aorta where it oscillates from zero velocity in diastole to maximum velocity in systole, thereafter it has less and fewer oscillations in the distal arteries and become continuous in terms of pressure and flow in the systemic capillaries, but blood velocity is never zero in the distal sections (see also Chapter 11).*

Once we have passed the capillaries of the systemic circulation, we find the venous district formed by venules and veins. In these vessels, in a lying subject, the pressure is a few mmHg above zero. Because of their high distensibility, veins and venules are considered *capacity vessels*. Their blood content increases with the transition from the clinostatic to the orthostatic position. Indeed, in the upright position, the hydrostatic pressure increases in the arteries and veins located below the heart in proportion to the distance between the heart and the considered district. The blood volume increases appreciably only in the distensible veins (*pooling effect, see Chapters 6.2*).

Through the right atrium, the venous blood passes into the right ventricle from which the pulmonary circulation begins. *Because the resistances in the arterioles of the pulmonary circulation are much lower than those of the systemic circulation, the pulmonary pressure values are also lower.* Therefore, the value reached by the pressure in the right ventricle and the pulmonary artery is about one-sixth of the value reached in the left ventricle and the aorta.

In Figure 1.5, we can also see how the pressure oscillations, absent in the capillaries of the systemic circulation, persist instead in the capillaries of the pulmonary circulation. Persistence of pulsatility is because in the pulmonary circulation the precapillary resistance is very low. It must be said, however, that the pulsations observed in the capillaries of the pulmonary circulation are not only those originating in the right ventricle and transmitted in *via* anterograde but also those originating in the left atrium and transmitted retrogradely via the low post-capillary resistance (see the *wedge pressure* Box 1.1).

The amount of blood that per unit of time flows through each section of the cardiovascular system is the *cardiac output* that we have already seen to be the product of the SV for the HR. Referred to the individual sections or district the cardiac output can also be called *flow rate*. As we have already said, the flow rate is the same amount for all the sections of the cardiovascular system. If instead of the sections we consider the districts, we see that the amount of blood that in the unit of time flows through a district can be indicated with the term of *blood flow* for that district. Each district's blood flow is a fraction of the cardiac output (Figure 1.2).

Whether we are referring to cardiac output or blood flow, values are expressed in units of volume per time unit, usually as mL/min. Since the current flow measurement instruments are sensitive to the velocity of the blood, knowing this value the flow can be obtained by *multiplying the velocity for the transverse section area* of the explored vessel, i.e. the section area of the ascending aorta if one wants to measure the cardiac output or the section area of any district artery (e.g., renal artery) if you want to measure a district blood flow. If the speed is expressed in *cm/min* and the vessel section in cm^2, the flow will be *cm/min* $\times$ cm² or in cm^3/min, namely, *mL/min*. Therefore, we must keep in mind that blood velocity and flow are different concepts, as only the latter informs about the quantity of blood that passes through a vessel or circulatory district per time unit.

We have said that the flow rate is virtually the same in all sections of the cardiovascular system. Not all sections have the same extension. Indeed, the section area of the *ascending aorta* is about 4 cm², while the area of the equivalent section of all the *precapillary arterioles* taken together is 60 cm², i.e., 15 times higher. The section area increases further to become 60 times greater than that of the arterioles in *capillaries* in resting conditions when only a quarter of these vessels are perfused. In this situation the total area of the systemic capillaries is about 3000–4600 cm². If all the capillaries are open, the total cross-sectional area increases further by about four times, reaching the value of 11000–12000 cm² with an obvious reduction in the blood velocity. Let us now consider that the amount of blood pumped from the left ventricle in one minute in a single vessel (the aorta) returns to the heart in one minute through two big vessels (the two cave veins), it is clear that in each of these two vessels the flow velocity is about a half of that in the aorta (Figure 1.7).

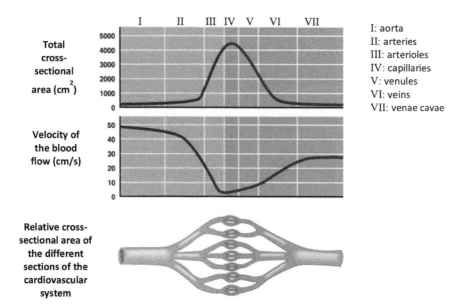

Figure 1.7 The velocity of blood flow varies inversely with the total cross-sectional area of the different sections (indicated with Roman numerals) of the cardiovascular system. As the total cross-sectional area of the section vessels increases, the velocity of flow decreases. Blood flow velocity is slowest in the capillaries. This is functionally important as it allows time for the exchange of nutrients and gases.

1.4 The Blood Contenied in the Various Sections of the Cardiovascular System

We have already seen that in the systemic circulation the venous section is 20 times more distensible than the arterial side. As a consequence of this greater distensibility, the veins should contain a proportionally larger quantity of blood than the arteries, at equal pressure, and with a stopped heart. However, since the pulsating heart acts as a distributor of volume and pressure between veins and arteries, the distribution of blood in the various parts or sections of the cardiovascular system at rest is that shown in Figure 1.8, from which it appears that the quantity contained in the venous side is only 7–8 times higher than the quantity contained in the arteries. "The heart job" consists of taking the blood from the veins to pump it into the arteries.

Starting from the situation of arrested heart with an MCFP of 7 mmHg, when the cardiac activity starts, the heart removes blood from the veins with a consequent slight decrease in venous pressure and pumps blood in the arteries where the pressure increases (Figure 1.4), because of the lower compliance and the presence of arteriolar resistance. Therefore, we can understand how in the presence of heart failure the distribution of blood will be compromised. After a reduction in systolic energy, the heart will eject a reduced amount of blood into the arterial tree causing an increase in the amount of blood

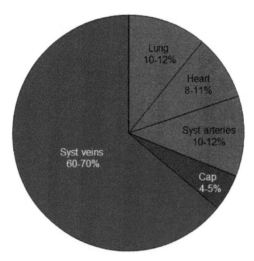

Figure 1.8   Distribution of blood volume in the circulatory system of a subject at rest.

contained into the venous tree. Moreover, the pressure will also tend to decrease in the arterial tree and to increase in the venous bed.

When left ventricle insufficiency occurs, pressure and blood volume are reduced in the arteries of the systemic circulation and increased in the pulmonary circulation. When the contractile insufficiency concerns the right ventricle, pressure and volume decrease in the pulmonary artery and increase in the veins of the systemic ciruculation. Clinically this increase in venous pressure is revealed by a turgor of the neck veins. The increase in pressure and volume in the veins of the systemic circulation is not necessarily due to a primary insufficiency of the right ventricle: it can increase when a left ventricular insufficiency exerts its effect on the right ventricle through the short pulmonary circulation which, as previously described, has low vascular resistance (in pre- and post-capillary section). This process involves a "retrograde" increase in pressure in the pulmonary capillary (an increase of wedge pressure) and circulation. The right ventricle cannot withstand a pressure overload so that in a relatively short time its pump effectiveness decreases, *i.e.,* it also becomes insufficient.

Box 1.1 Wedge Pressure and Pulmonary Edema

In the pulmonary circulation, the retrograde propagation of pressure oscillations generated in the left atrium allows the registration of left atrial pressure with a minimally invasive method. A light and flexible catheter, carrying an inflatable balloon around the apex and connected with a manometer, is introduced into a peripheral vein (*e.g.,* in a femoral vein) and pushed into the pulmonary artery through the inferior cave vein, the atrium, and the right ventricle. From the pulmonary artery, the tip of the catheter and the balloon are pushed in one of the two branches of the artery and pushed until the balloon is *wedged* in a peripheral branch. In this way, the tip of the catheter receives the pressure that propagates backward from the *alveolar capillaries*, whose pressure is greatly *influenced by left atrium pressure*. The pressure recorded with this catheter wedging procedure is called *wedge pressure*.

The great influence of capillary pressure by left atrial pressure is possible because, in addition to the low precapillary resistance, also the postcapillary resistance is low in the pulmonary circulation. The low value of the post-capillary resistance is important not only because it makes possible to register the pressure in the left atrium, but also because it normally limits the pressure inside the alveolar capillaries. Note that if the postcapillary resistance of the pulmonary circulation were equal to those of the systemic circulation, they would hinder the passage of blood from the capillaries to the veins while the low precapillary resistance would favor entry with an obvious increase in capillary pressure, leakage of fluid in the interstices and consequent pulmonary edema (see Chapter 15). When the edema occurs in the pulmonary tissue, it may first be interstitial with a strong limitation of the gaseous exchanges and then it can become *"frank pulmonary edema"* with extravasation of fluid in the alveoli. The second case is a condition comparable to drowning. It is said that in the pulmonary edema the patient drowns in his liquid. Therefore, because of the low post-capillary resistance, pulmonary edema may occur when the outflow of blood from the alveolar capillaries is limited by the pressure increase in the left atrium due to the insufficiency of the left ventricle or a defect of the mitral valve.

2

Structure and Function of the Myocardial Fiber

2.1 Myocardial Fiber

It may seem strange, but cardiomyocytes represent only 3% of all cardiac cells. The heart is also composed of many different cell types, including connective tissue cells, fibroblasts, neurons, epicardial and endocardial cells, and vessel cells (endothelial and smooth muscle cells), as well as functional tissue cells of the heart (cells of the conduction system). However, being the cardiomyocytes large (about 20 μm in diameter) they compose the majority of the heart volume.

Myocardial fibers, or cardiomyocytes, are striated muscle fibers similarly to skeletal muscle fibers, but, unlike these, each possesses a single nucleus and a large number of mitochondria. Indeed, the mitochondria occupy 35% of the volume of cardiomyocytes. Myocardial fibers are surrounded by an abundant capillary network (about 1 capillary per fibers).

The cardiomyocyte, composed of *myofibrils*, has a membrane called *sarcolemma*. In the sarcolemma the **T** *tubules* open and sink transversely into the sarcoplasm between the **I** and the **A** band of sarcomeres (Figure 2.1). The sarcoplasm is also crossed by the sarcoplasmic reticulum (SR) formed by tubules, which runs along the fiber perpendicular to the **T**-tubules and according to the longitudinal axis of the fiber and are therefore called *longitudinal tubules*. At the crossing with the T-tubes, the longitudinal tubules have expansions called *cisterns (corbular SR)*. At this point, the SR takes the name of *junctional SR*, while the **T** tubule and cistern constitutes the *dyad*. From the inside of the fiber, the longitudinal tubule sometimes approaches the sarcolemma, where it presents expansions similar to those that participate in the dyad also called *junctional SR*. Figure 2.1 clearly illustrates the presence of T tubules and SR in a cardiomyocyte. As we will see, the main function of the SR is to store calcium ions (Ca^{2+}). Moreover, the dyad plays a central role

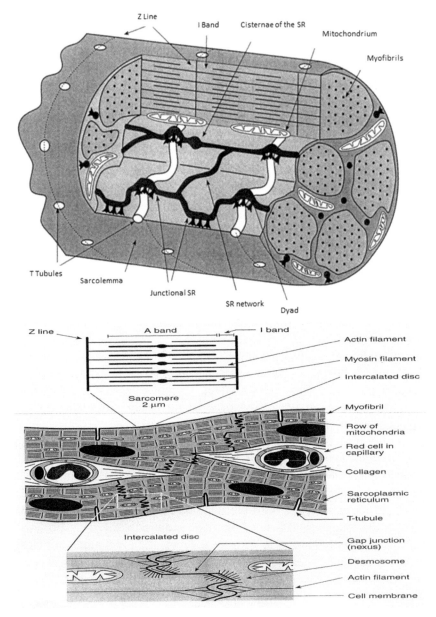

Figure 2.1 Top: schematic cross-section and longitudinal section of a cardiomyocyte. Bottom: schematic organization of the sarcomere and its location in the myocardial fiber. See the explanation in the text.

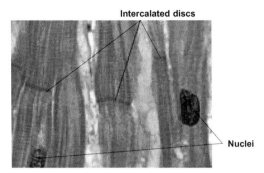

Figure 2.2   Microscope slide showing a section of cardiac muscle stained with hematoxylin to show the intercalated discs.

in the processes of *electromechanical coupling* during cardiac contraction (see Chapter 4.3).

Within the myocardial fiber, we find the *myofibrils* arranged along the long axis of the fiber and parallel to each other. The myofibrils are made up of *sarcomeres* that represent the contractile units of the fiber and therefore of the myocardium. The sarcomere has a length of about 2 μm. Several sarcomeres are arranged in series and separated from each other by the **Z** lines, which therefore represent the external limits of each sarcomere. The sarcomeres belonging to different myofibrils of the same myocardial fiber are parallel to each other.

The cardiomyocytes have ramifications that connect them with other cardiomyocytes through *intercalated discs* (Figures 2.1 and 2.2). These are formed by the interdigitation of the two adjacent cardiomyocytes connected by desmosomes rich in glycoproteins. The presence of the desmosomes ensures the adherence between the two cardiomyocytes through what is called the *adherens junction*. The passage of the *impulse* (the traveling action potential) from one cell to another occurs through the presence of a *gap junction* in the *intercalated disc*. The gap junctions are made of two *hemi channels* or *connexons* each located on one of two neighboring connected cells. Each connexon is made up of six *connexins* and regulates the passage of small biological molecules and ions. Through the connexons, the ions can pass from one cardiomyocyte to the other allowing the transmission of the impulse.

Figures 2.1 and 2.2 illustrate the organization and structure of cardiomyocytes. As can be seen in the figures, the sarcomeric organization confers a typical striated appearance to the myocardial fiber, which can be observed under an optical microscope. The striations are first of all constituted by the alternation of light bands and dark bands. In Figure 2.1, the light bands are indicated as *I* bands, as they are isotropic in polarized light, while the dark ones are indicated as *A* bands, as they are anisotropic in polarized light. Within an **I** band is visible a darker transverse line, the **Z** line, which divides it into two regions called *half I-band*. The central part of the **A** band is lighter and called the **H** zone, which contains in the middle a thin dark line called **M** line.

Each sarcomere is included between two consecutive **Z** *lines* and is formed of a *half I-band*, an *A* band, and a second-*half I-band*. During the contraction the two *Z lines* approach each other, the *half I-band* becomes shorter, while the length of the *A* band remains unchanged. However, within the *A* band, the *H zone* is shortened.

The structure of the sarcomere and the sliding of the filaments explain both the nature of the striation and the variations that it undergoes during contraction. As can be seen in Figure 2.1, the thick filaments, composed of myosin, are located in the *A* band, while the thin filaments, consisting of actin, troponin, and tropomyosin, are located in the sarcomere periphery and anchored to **Z** *lines*.

The *half I-bands* are those parts of the sarcomere in which the thin filaments are not juxtaposed to thick filaments, while the darker *A* band is the part where the thick filaments are juxtaposed with the thin filaments. As said, the *A* band is clearer in the middle (*H zone*) where the juxtaposition of thick and thin filaments is not present.

Since the contraction is due to the sliding of the thin filaments towards the median part of the sarcomere, we understand how the **Z** *lines* are approaching during sarcomere shortening. Moreover, during sarcomere shortening the *half I-bands* and the *H zone* reduce their length, while the length of the *A* band remains unchanged as it corresponds anatomically to the thick filaments.

The cardiac sarcomere is made up of 3 classes of proteins: myofibrillar or contractile proteins, regulatory proteins, and structural proteins.

2.2 Myofibrillar or Contractile Proteins

Myofibrillar proteins are myosin and actin. The molecular weight of myosin is 470,000 D, while that of actin is 43,000 D only.

Thick filaments: many parallel myosin filaments form the *thick filament* of the sarcomere, while the actin protein, together with the tropomyosin and troponin regulatory proteins, forms the *thin filament* (Figure 2.3). A thick

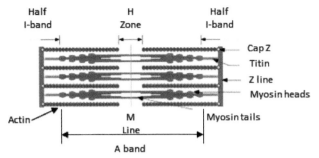

Figure 2.3 A schematic organization of the sarcomere components of the myocardial fiber. See the explanation in the text.

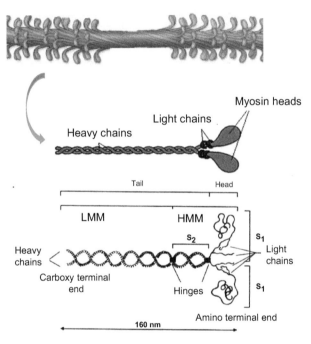

Figure 2.4 The sarcomeric thick filament and the myosin molecule structures. Note that light (LMM) and heavy meromyosine (HMM) do not correspond to light and heavy chains. Indeed, the HMM includes the two light chains S1 and the S2 neck of the heavy chain. The remaining of the two heavy chains represents the LMM. See the text for further explanation.

filament has a diameter of about 100 Å and a length of 1.5–2.0 μm, while a thin filament has a diameter of about 45 Å and a length a little shorter than the thick filament.

Each thick filament is made up of a large number of myosin molecules (Figure 2.4 upper part). Each myosin molecule is formed by an elongated

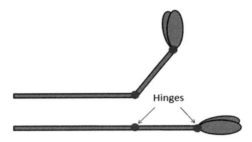

Figure 2.5   Hinges of the myosin molecules. See the explanation in the text. (modified from Caldarera C.M., 1995).

portion, called the *tail*, which ends at one end with 2 globular portions that together constitute the *head* (Figure 2.4 lower part). In particular, it consists of 2 polypeptide chains (*myosin heavy chains, MHC*) wrapped around each other to form a double spiral α-helix. Initially, the two MHC run straight to form the tail of myosin. At the end of the tail, the two MHC continue in the neck of myosin, indicated in Figure 2.4 by the symbol S2. This is separated from the tail and the head by two hinges which allow the head to flex (Figure 2.5).

Once the neck is over, the two heavy chains separate, fold and associate each with 2 short polypeptide chains (*myosin light chains, MLC*) forming the myosin head, indicated in Figure 2.4 by the symbol S1.

The head is the location of *transverse bridges* because it has a specific binding site for actin. This allows myosin to bind to the actin of the thin filament. The myosin head has a slight ATPase activity and acquires a stronger ATPase activity when binds to actin.

In the myosin molecule, we can also distinguish 2 portions: heavy meromyosin (HMM) and light meromyosin (LMM). Figure 2.4 illustrates the arrangement of these two components of myosin. LMM forms the tail, while the HMM forms both the neck and the head.

Heavy chains can also be α or β types. The different presence of these chains in the heavy meromyosin allowed us to recognize three different functional forms of myosin, characterized, according to the decreasing speed of their ATPase activity, in V1, V2, and V3.

We have already said that each thick filament is made up of a large number of myosin molecules. Therefore, the tails of many myosin molecules are placed parallel to each other, while the globular heads stick out from the surface of the thick filament with a regular helical arrangement, where each head is 60 degrees rotated respect to the adjacent one and 43 nm far

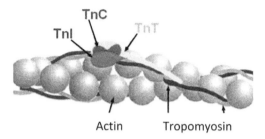

Figure 2.6 Structure of thin filaments. See the explanation in the text.

away from the head on the same plane since the myosin molecules are offset slightly in the longitudinal plane. As shown in the upper part of Figure 2.4 the disposition of the myosin molecules is symmetrical. The heads are located only at the edges of the thick filament, while the middle region there is a bare region formed only by tails.

Thin filaments: *Actin*, a constituent of *thin filaments*, can be found in a globular (G-actin) or fibrous (F-actin) form (Figure 2.6). G-actin is in the form of globose monomers, each weighing 43,000 D. Each monomer has a specific binding site for the myosin head, but it is not available at rest. The F-actin is instead a polymer formed by many monomers arranged as a chain of pearls.

In the thin filaments, two chains of F-actin are wrapped around each other to form a double right-hand helix with a pitch of about 360 Å (Figures 2.4 and 2.6). In the two solid angles (furrows or grooves) of the double helix are located, one per corner, two filamentous molecules of *tropomyosin*, at the end of which there is a complex consisting of *three troponin subunits*. *Tropomyosin* and *troponins* are considered regulatory proteins and their role are explained in the following paragraph. In brief, tropomyosin/troponin regulation by Ca^{2+} makes actin specific binding site available for the myosin head where the *transverse bridge* of the myosin head is inserted during contraction. Following the bond of myosin to actin, actomyosin is formed which has an ATPase power 200 times higher than that of myosin alone. This ATPase activity is indispensable for the *excitation-contraction coupling* (see Chapter 4.3).

If we observe Figure 2.7 we can see how, in the cross-section of a sarcomere at the extremity of the *A band*, each thin filament is surrounded by 3 thick filaments arranged at the corner of a triangle. While each thick filament is surrounded by 6 thin filaments arranged at the corners of a hexagon

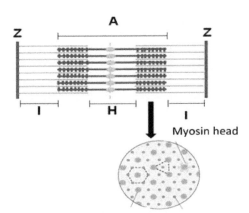

Figure 2.7   Arrangement of thick (pink) and thin (blue) filaments in the transverse section of the sarcomere. See the explanation in the text.

where the heads protrude towards the thin filaments on which the *transverse bridges* (also called actomyosin complex) are inserted during the contraction.

2.3 Regulatory Proteins

Tropomyosin and troponin are the regulatory proteins of the sarcomere and therefore of contraction. Their function will be also considered in the excitation-contraction coupling description (Chapter 4.3).

Tropomyosin has a molecular weight of 68,000 D and is made up of two α-helix spiral chains (Figure 2.6). It extends along the two grooves of the double helix formed by the two chains of F-actin and ensures the stability of the thin filaments. Therefore, the disposition of this protein masks the actin-binding site, which is specific for the myosin head. The protein length covers only 7 monomers of actin, but tropomyosins attached each other form a continuous double helix that is wrapped around along the entire length of the thin filament.

Troponin (Tn) has a molecular weight of 76,000 D. It is made up of *3 subunits: TnC, TnI, TnT*, arranged, concerning the actin of the thin filament, as indicated in Figure 2.6.

From a functional point of view, TnC binds Ca^{2+} at the end of the *excitation-contraction coupling* which represents the beginning of the *cross-bridge cycle*; TnI covers and inhibits the actin-binding site for myosin head during diastole and, finally, TnT is bound to tropomyosin. When Ca^{2+} binds

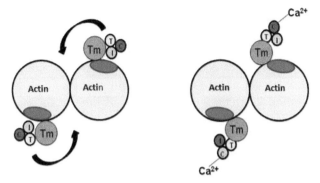

Figure 2.8   Cross-section of the thin filament with exposure of the actin-binding site for the transverse bridge following the increase in the calcium concentration in the sarcoplasm. Tm: tropomyosin; I, C, and T: troponin subunits. See the explanation in the text. (Modified from Potter JD and Gergely, 1974).

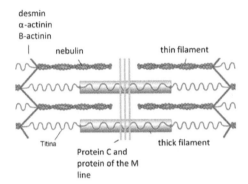

Figure 2.9   Structural proteins of the sarcomere. See the explanation in the text.

to TnC, TnI leaves the actin-binding site uncovered with the formation of a transverse myosin bridge (Figure 2.8). Thus, allowing the cross-bridge cycle and contraction.

2.4 Structural Proteins

Structural proteins of the sarcomere are *α- and β-actinin, desmin, protein C, proteins of the M-line, nebulin,* and *titin* (Figure 2.9).

The *α- and β-actinin* are, together with the *desmin*, components of the Z line. In particular, *α*-actinin is necessary for the attachment of actin filaments to the Z lines.

Protein C, whose function has not yet been sufficiently clarified, is linked to the tail of the myosin filament.

The *nebulin* helps to maintain the alignment of the actin.

The *proteins of the M line* are arranged transversely along multiple juxtaposed sarcomeres helping to keep thick myosin filaments organized (Figure 2.9), while the *filaments of titin* join the Z lines at the two ends of the thick filaments. The protein of the M line and titin prevent the sarcomere from drying out in case of excessive stretching. In particular, the proteins of the M line prevent the removal of the various thick filaments from the central part of the sarcomere.

3

Cardiac Electrophysiology

3.1 Cardiac Electrophysiology: Overview

As said, the heart is composed of many types of cells, but we can distinguish three different types of *cardiac tissue*, which allow the heart to function continuously as an intermittent pump. They are called *functional tissues of the heart,* which consist of:

(1) atrial and ventricular *myocardium*, composed by cardiomyocytes, - specialized muscle cells able to contract - which make up the muscular layers of the heart wall;

(2) the *nodal tissue (sinoatrial and atrio ventricular nodes)*, composed by groups of specialized cells found in the right atrium, which can *auto-generate impulses* that travel and trigger the myocytes to contract;

(3) the *internodal, interatrial, and atrioventricular conduction systems.* This tissue of the cardiac conduction systems (CCS) consists of cardiac cells and conducting fibers that are specialized to conduct the impulse rapidly through the heart.

Physiologically, the cardiac impulse is generated at the *sinoatrial node (SAN)*, but in some pathological circumstances, all other cardiac cells (especially those of nodal and CCS) can initiate the impulses to generate arrhythmias. Therefore, we have a main pacemaker, the SAN, and several latent pacemakers in the nodal and CCS.

Like nerve and other muscle cell membranes, the membranes of the cardiac cells are also polarized, *i.e.,* in resting conditions they have positive charges outside and negative charges inside of the membrane (Figure 3.1a). On the contrary, when they are excited, they undergo depolarization, so that the exciting portion of the membrane has negative charges outside and positive inside (Figure 3.1b).

The degree of polarization of a membrane in rest conditions is said *resting membrane potential.* In this condition, when an appropriate stimulus

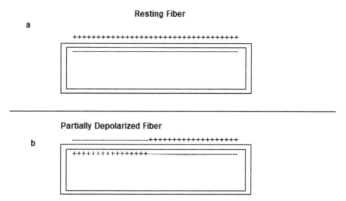

Figure 3.1   Scheme of the distribution of the charges on the myocardial fiber membrane in different conditions: *(a)* at rest; *(b)* partially excited. See the explanation in the text.

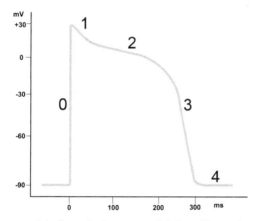

Figure 3.2   Action potential of ventricular myocardial fiber. The resting membrane potential of −90 mV is stable (phase 4). The depolarization stage (phase 0) is so rapid (500 mV/s) that it is the shortest phase of the action potential. It is followed by a transient attempt of repolarization (phase 1) a plateau (phase 2) in which the membrane potential is stable around 0 mV and a repolarization phase 3. When the fiber is completely repolarized, it is again in phase 4. See the text for further explanation.

is applied the membrane can depolarize and can generate an ***action potential (AP)***, *i.e.*, an inversion of membrane polarity that can propagate at distance. *Repolarization* is the process by which a depolarized membrane returns to the resting state (Figure 3.2).

In the case of a *ventricular cardiomyocyte*, the resting membrane potential is stable and it has an amplitude of about −90 mV. In other words, the

Table 3.1 Concentrations of anions and cations present inside and outside the plasma membrane and relative relationship between the two sides of the membrane. Concentrations are expressed in milliequivalents per liter (mEq/L). A^- = protein anions

CATIONS	EXTERN	INTERN	Ratio
Na^+	145	14,5	10:1
K^+	4,6	140	1:30 *circa*
Ca^{2+}	3	1	3:1
Mg^{2+}	2	40	1:20
Total cations	**154,6**	**195,5**	
ANIONS	EXTERN	INTERN	Ratio
Cl^-	117	3	39:1
HCO_3^-	31	7	4:1 *circa*
A^-	–	140,5	–
Altri	6,6	45	1:7 *circa*
Total anions	**154,6**	**195,5**	

inner side of the membrane is negative by about 90 mV compared to the outside. When depolarization occurs, the inner side becomes rapidly positive compared to the outside, reaching +30 mV at the apex of an AP. From Figure 3.2 we can see that the amplitude of the AP is about 120 mV (from −90 to +30 mV).

After the action potential has reached its maximum amplitude, repolarization begins, which determines the slow return to the resting membrane potential.

To understand the mechanisms responsible for *resting membrane potential*, it is necessary to remember the distribution of ions on the outside (interstitial fluid) and inside (sarcoplasm) the membrane and consider some properties of the membrane (Table 3.1). The concentrations are expressed in milliequivalents per liter (mEq/L).

The actual values of the concentrations may be different from those shown in Table 3.1, as various determinations can give different results.

Note that both the outside and the inside of the membrane the sum of the cations is equal to the sum of the anions. However, the gradient of the single ions across the membrane assures the diffusion of ions along specific channels, making possible a difference of potential.

Because of the membrane and ion characteristics, in determining the resting membrane potential and action potential, a pivotal role is played by the following ions: Na^+, K^+, protein anions A^- and Cl^-.

As regards the characteristics of the membrane, it is necessary to remember that in resting conditions it is freely permeable to K^+ and Cl^-, very poorly

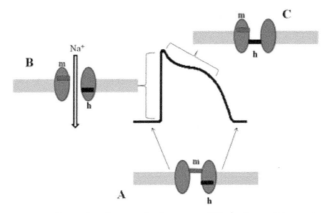

Figure 3.3   Three conformational state of voltage-gated Na^+ channel due to the opening and closing of the m (in red) and the h (in black) gates. **A**: Na^+ channel is closed but excitable; **B**: Na^+ channel is open; **C**: Na^+ channel is closed and unexcitable. See the explanation in the text.

permeable to Na^+ and, always, totally impermeable to A^-. It follows that in resting conditions the Na^+ cannot enter the cardiomyocyte and the A^- cannot get out of it.

During the initial development of the AP, the membrane becomes transiently permeable to Na^+, and becomes transiently less permeable to K^+, while it remains impermeable to A^-. As above mentioned, the passive movement of ions through the membrane takes place along specific channels and it is driven by the electrochemical gradient.

The Na^+ channels are closed in concomitance with the resting potential, while they open during the initial development of the AP to close immediately after.

Voltage-gated Na^+ channels have three main conformational states: closed but excitable, open, and close but unexcitable (or inactivated). These three states of the Na^+ channels are ensured by the opening or closing of two different gates: the *m* and *h gates* (Figure 3.3). The *m gate* controls Na^+ channel opening, while the *h gate* controls Na^+ channel closing.

During the resting potential, the *m gate* is closed and keeps the channels closed but excitable (Figure 3.3A). When an excitement occurs, the *m gate* opens thus allowing Na^+ to flow into the cell (Figure 3.3B). This state of opening of the channels, characterized by both gates opened, lasts very little, less than 2 msec. The *h gate* closes, when the membrane depolarization has reached its maximum level, thus "blocking" the incoming sodium current (Figure 3.3C). When the *h gates* have closed the channels and the cells are unexcitable. Afterward, the *m gates* close, while *h gates* re-open gradually during the repolarization. Until the *h gates* are closed the channels cannot be activated no matter how strong is the stimulus, thus maintaining the cardiac cell into a *refractory period*. Once the repolarization is over, the channel is back in the closed but again excitable state (Figure 3.3A) and it remains in this conformational state (*m gates* closed and *h gates* opened) until the arrival of a next impulse.

The passage through these three different conformational states is responsible for (1) the excitability of the membrane during the resting phase (Figure 3.3A), (2) the excitation at the beginning of an AP (Figure 3.3B), (3) the refractory period during an AP (Figure 3.3C), and again the recovery of excitability only after the completion of the AP (Figure 3.3A).

The membrane has also calcium channels that allow the entry of Ca^{2+} into the cell during action potential which, as will be seen further on, slows down the repolarization.

The cell membrane has also *ion pumps* and *ion exchangers*. The pumps are active transports that consume ATP to move ions across the membrane even against the ionic gradient, whereas the *exchangers* operate passively according to ionic gradients usually created by the pump activity (they are secondary active transports).

The *pumps* of interests in cardiac electrophysiology are the Na^+/K^+ pumps or Na^+/K^+ ATPses, and the calcium pump. Na^+/K^+ pumps are membrane proteins that continuously extrude from the cell $3Na^+$ and enters $2K^+$. Calcium pumps expel Ca^{2+}, thus maintaining a low cytoplasmic level of these ions that allows the complete relaxation of the cardiomyocytes (see below).

The *exchanger* that is interested in this context is the $3Na^+/1Ca^{2+}$ exchanger, which is an antiporter that exchanges these ions between the outside and the inside of the membrane depending on the concentration of the two ions. This exchanger generally extrudes $1Ca^{2+}$ and enters $3Na^+$, but it can act in *reverse mode* when the concentration of Na^+ in the inner side of the membrane is enhanced. This exchanger can work in *reverse mode* in pathological conditions characterized by intracellular Na^+ overload, thus supporting Ca^{2+} overload, or transiently, in physiological condition, for example, at the apex of the phase 0 of the AP (see below).

3.2 Genesis of Resting Membrane Potential

To explain the genesis of resting membrane potential are necessary for some preliminary information.

Let us first consider the *Gibbs-Donnan equilibrium*: imagine a tank divided into two compartments, X and Y, by a semipermeable membrane that allows the passage of water and electrolytes (K^+ and Cl^-) from one compartment to another, but not proteins (non-diffusible anions, A^-).

As long as *only* K^+ and Cl^- are present, their concentration will be equal to the two sides of the membrane.

$$[K^+]_x = [K^+]_y$$

and

$$[Cl^-]_x = [Cl^-]_y$$

If A^- are added to the compartment X, the excess of negative charges on this side will induce the passage of Cl^- into the compartment Y, to have the same concentration of negative charges on the two compartments:

$$[Cl^-]_x + [A^-]_x = [Cl^-]_y.$$

Yet, the K^+ will remain equally concentrated on the two sides of the membrane.

Being Cl^- now more concentrated in the Y compartment, some (not so many) will pass in the X compartment followed by some K^+ for reasons of electric balance. Therefore, will we have

$$[K^+]_x > [K^+]_y \quad \text{and} \quad [Cl^-]_y > [Cl^-]_x.$$

On the X compartment, there will, therefore, be a greater concentration of ions than on the Y side.

$$[K^+]_x + [Cl^-]_x + [A^-]_x > [K^+]_y^+ [Cl^-]_y.$$

The only ions that can cross the membrane by concentration gradient are the K^+, while the A^- cannot cross the membrane and the Cl^- cannot move against the electrochemical gradient.

The final result of the infinitesimal passage of K^+ is the separation of charges between the two compartments until a *balance for diffusible ions* is reached, in which

$$[K^+]_x/[K^+]_y = [Cl^-]_y/[Cl^-]_x,$$

or

$$[K^+]_y \times [Cl^-]_y = [K^+]_x \times [Cl^-]_x$$

The balance that has been achieved is the *Gibbs-Donnan equilibrium.*

The potential difference due to the K^+ micro-diffusion, at the Gibbs-Donnan equilibrium, is about 30 mV.

In the cells, the amplitude of the resting membrane potential (-90 mV) is achieved thanks to the Na^+/K^+ pump that expel 3 Na^+ and import 2 K^+ for each hydrolyzed ATP. Because of this pump activity, there is a high concentration of K^+ within the cardiomyocyte, which creates a concentration gradient that causes the exit of a small amount of these ions, bringing the value of the membrane potential to about -90 mV.

Moreover, the presence of ion pumps in cell membranes avoids the explosion of the cells due to excess osmolarity by preventing the *Gibbs-Donnan equilibrium.* Indeed, when the Gibbs-Donnan equilibrium is established, the different amounts of total ions from the two sides of the membrane generate an osmotic pressure that could explode the cells, the pumps avoid this explosion.

In summary, the Na^+/K^+ pump creates the ion gradients, is a little electrogenic, and avoid the cell explosion. The gradient for K^+ is fundamental for the exit of potassium and the genesis of the resting membrane potential; instead, the gradient for Na^+ is fundamental for the entry of Na^+ during the genesis of the AP when the Na^+ channels will open.

Due to the poor Na permeability of the membrane at rest, very little Na^+ penetrates the cardiomyocyte and therefore this ion remains 10 times more concentrated outside than inside. Conversely, the K^+ ions are concentrated into the cardiomyocyte, because are actively pushed by the Na^+/K^+ pumps. Even if the membrane is highly permeable to them, the K^+ remains more concentrated inside the cells (few K^+ exits) due to continuous activity of the pump and the electrical attraction exerted by the A^- protein anions, which limits the K^+ exit.

At this point, we have understood that in the absence of pumps, ions reach an equilibrium (concentration gradient is balanced by electrical force). A potential exists across the membrane at the equilibrium, which is the *ion equilibrium potential.* In general, the ion equilibrium potential can be defined as the force that the ion exerts on a membrane because of its concentration gradient between the two sides of the same membrane and that is balanced by the electrical gradient. In other words, the ion equilibrium potential is the translation in terms of the voltage of the force that the ion exerts on a membrane with its concentration gradient. If *i* is the ion, its equilibrium potential E*i* is calculated with the *Nernst equation:*

$$E_i = RT/nF \times \log_n [i]_e/[i]_i,$$

where R is the gas constant (8.316 joules per °C of absolute temperature), T the absolute temperature (273 + 37 = 310°C), n the valence of the ion, F the Faraday number (96.500 Coulombs per mole) and $[i]_e/[i]_i$ indicate the concentration ratio of ion outside and inside the membrane respectively. Log_n indicates the natural logarithm that corresponds to 2.3 times the decimal log.

Using the Nernst equation, we can calculate the equilibrium potential for Na^+, for K^+ and for any diffusible ion of which we know the concentrations.

In the case of Na^+ we will have:

$$E_{Na} = RT/nF \times 2,3 \log[10]_e/[1]_i. = RT/nF \times 2,3 \log 10$$

Considering that both the n value of Na^+ and the value of log 10 is equal to 1, we have

$$E_{Na} = RT/nF \times 2,3$$

Replacing RT and F with the respective values, we obtain 0.061 V, which is 61 mV.

We can say that, due to its difference in concentration, Na^+ exerts from the outside on the membrane a force equal to 61 mV. If the membranes were permeable to them, this force would push the Na^+ towards the inside of the cardiomyocyte until the membrane potential will be positive inside of 61 mV. However, since at rest the membrane is very little permeable to Na^+, they do not penetrate the cardiomyocyte.

In the case of K^+, being 1/30 the concentration ratio between potassium ions outside and inside the cell, the membrane equilibrium potential for K^+ is indeed

$$E_K = RT/nF \times 2,3 \log 1/30.$$

Since the logarithm of 1/30 is equal to $-\log 30$, the equation will become

$$E_K = RT/nF \times 2,3(-\log 30).$$

Since the value of $-\log 30$ is -1.47, and n is equal to 1, we will have

$$E_K = 61 \times (-1,47) = -90 \text{ mV}.$$

The *Nernst equation* reveals that the K^+ concentrations exert a force of -90 mV on the membrane. This force is exerted from the inside out, *i.e.*, it tends to *push* K^+ out of the cardiomyocyte. Note that in this case the sign – indicates that the ions act from the inside to the outside.

Being the membranes permeable to K^+, this force would *push* the K^+ towards the outside of the cardiomyocyte until the membrane potential will be negative inside of 90 mV. Since at rest the membrane is highly permeable to K^+, the resting membrane potential is very close to this value.

From the above explanation, we have understood that *the membrane potential is closer to the equilibrium potential of that ion for whom the membrane is more permeable.* Also, the ions are subject to two forces: the electrical and the concentration gradient which can push the ion in the same or in the opposite direction across the membrane that is permeable to them.

Because of high intracellular K^+ concentration and their permeability membrane, a small number of K^+ can escape from the cardiomyocyte without a measurable change in their outside concentration. However, it results in a charge separation between the two sides of the membrane, since they are not followed by the A^- that cannot cross the membrane. The outflow of the K^+ ceases when the voltage difference across the cell membrane has reached 90 mV, which occurs when the balance between the concentration gradient (that tries to get them out) and the electric gradient (that tries to keep them into the cardiomyocytes) has been achieved. At this stage, it is possible to determine the membrane potential at rest, which is therefore due to an electrical potential difference across the cell membrane resulting from K^+ diffusion.

The real resting membrane potential is slightly less negative than the K^+ equilibrium potential. This slight difference is due to the pumping back of K^+ by Na^+/K^+ ATPase which avoids the achievement of the equilibrium and to the very low permeability of the cell membrane to other positive ions, among which the Na^+, which could enter the cardiomyocyte.

3.3 The Action Potential

Action potentials are self-regenerating waves of voltage that travel along the cell membranes. Following stimulation, the cell becomes depolarized, i.e., it becomes positive of about +30 mV inside compared to the outside, which is therefore negative. The depolarization of the cell membrane and the subsequent repolarization constitute the AP.

In the heart, the action potential can have a different morphology depending on the functional tissue in which it occurs. As a reference, we take the AP of a ventricular cardiomyocyte, with which we will compare the potentials of the other parts of the heart (Chapter 4.1).

The action potential of the ventricular myocardium is reported in Figure 3.4. It is caused by the variation of the membrane conductance to the Na^+, K^+, and Ca^{2+} ions. As can be seen, the AP has duration (300– 350 ms)

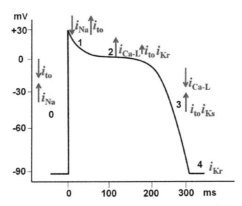

Figure 3.4 Action potential of ventricular myocardial fiber. In red the ionic currents responsible for the various phases of myocardial potential. i Na: sodium current; i Ca-L: calcium current; i_{to}, i_{kr}, and i_{ks}: potassium currents. See the explanation in the text.

much longer than the potentials of the neuron and skeletal muscle. The longer duration is due to rather slow repolarization that assures the myocardium a long state of *refractoriness* (non-excitability) that lasts almost the same time of the mechanical contraction. The long refractory period prevents the cell from undergoing a new contraction before it has been released. We can then understand how, unlike skeletal muscle that has a short refractory period confined to the latency phase after a stimulus, *the heart cannot undergo tetanization.*

From Figure 3.4 we can see that the action potential of a ventricular cardiomyocyte can be divided into four phases plus a resting phase. Starting from the resting condition (*phase 4*), we observe a sudden and rapid depolarization (*phase 0*) at the end of which the inside is 30 mV positive compared to the outside. The positive part of the AP is called *overshoot*. Immediately after phase 0, transient repolarization occurs (*phase 1*). This phase lasts very little because, as soon as the membrane potential is close to zero, there is a slowing down of the repolarization that constitutes *phase 2* or *plateau*. As can be easily understood, the *plateau* is the cause of the long duration of the action potential of the ventricular myocardium and therefore of the consequent long refractory period. After phase 2, repolarization reaccelerates (*phase 3*) bringing back the myocardium to the resting condition or phase 4. Now let us see the ionic movements that determine the features of this AP.

When a stimulus is applied or an impulse reaches a cardiomyocyte, the voltage-dependent rapid Na^+ channels open, thus leading to an increase in the Na^+ conductance (conductivity for Na^+ is indicated as gNa). The rapid Na^+

channels are called voltage-dependent as they are activated by a reduction in the amplitude of the membrane potential, as occurs at the arrival of an impulse. The Na^+ ion crosses rapidly the membrane due to its strong electrochemical gradient, thus the opening of these channels produces a sodium current indicated as iNa.

The ionic current can be directed toward the inside and it is said *inward current*, or out of the cell membrane, that is *outward current*.

The iNa$^+$ is an inward current responsible for phase 0 depolarization of the action potential in non-pacemaker or working cardiac cells. Indeed, the flow of positive charges inside to the cardiomyocyte reduces the amplitude of the membrane potential up to reverse the polarity, so that the inside of the membrane becomes positive of 30 mV compared to the outside. The duration of phase 0 is a few milliseconds. In case the resting potential is −90 mV, the amplitude of the AP is 120 mV.

During phase 0, in addition to an increase in conductance to Na^+, there is a decrease in conductance to K^+. It means that the entry of Na^+ is not accompanied by the exit of K^+, thus causing a rapid depolarization.

Once phase 0 is completed, repolarization begins. This is the short phase 1, which is due to a rapid decrease in conductance to Na^+ accompanied by an initial resumption of the conductance to K^+ for the opening of specific transient K^+ channels (called $i$to, transient outward current)

After phase 1, the potential presents phase 2 or *plateau*, which is responsible for the long duration of the cardiac action potential. The plateau is due to three factors: *a)* an increase in conductance to Ca^{2+}, *b)* a slowdown of conductance to K^+ and *c)* the persistence of a certain, albeit small, degree of conductance to Na^+ (late sodium current that is not reported in the figure, see Chapter 10).

The increase in Ca^{2+} conductance is due to the opening of L-type Ca^{2+} channels (also known as the dihydropyridine channels, or DHP channels). These are voltage-dependent channels, that start to open slowly when in phase 0 the membrane potential has reached −45 mV. The resulting movement of calcium ions is called L-type calcium current and is referred to as iCa-L. Therefore, positive charges enter the cardiomyocyte, slowing down the repolarization process. These channels play an important role in excitation-contraction coupling. The entry of Ca^{2+} into the cardiomyocyte during phase 2 of the AP explains why, as we shall see later, the increase in heart rate is accompanied by a certain increase in contractility. As said, during the plateau, the slowing of the resumption of the conductance to K^+ limits the leakage of positive charges necessary for repolarization.

The persistence of a certain degree of conductance to Na$^+$ is due in little part to incomplete closure of a small number of channels (*late sodium current*), and in greater part to the activation of the passive $3Na^+/2Ca^{2+}$ exchanger induced by the increase in Ca^{2+} concentration within the cardiomyocytes, as a consequence of the entry of this ion. The activation of this exchanger causes a prevalence of positive charge entry which prolongs the plateau.

At the end of phase 2, the L type Ca^{2+}channels are closed and the activity of the exchanger slows, while the conductance for K^+ starts to rise again by the opening of specific K^+ channels. The opening of these K^+ channels allowing the spillage of these ions from the cell completes the repolarization (phase 3). Once this phase is over, the cell membrane is once again at the resting potential or phase 4. We can say that the repolarization is due to outward K^+ currents.

As reported in Figure 3.4, the K^+ current in phase 1 is called transient outward current and is referred to as *i*to. Cardiac repolarization is controlled by the rapidly (*i*Kr) and slowly (*i*Ks) activating delayed rectifier potassium channels. The currents present at the end of phase 2 and in phase 3 are mainly the two delayed rectifying currents indicated respectively as *i*Kr and *i*Ks. Each current for K^+ corresponds to a specific channel.

After repolarization, in phase 4 there is an outward current due to the so-called inward rectifying channels. Inward-rectifier potassium channels (Kir, IRK) are a specific subset of potassium channels. The term "inward rectifier" should not be confused with the direction of the flux of these ions, because the net K^+ current in the heart cells is always outwards, *i.e.,* K^+ flux out of the cardiomyocytes. The phenomenon of inward rectification of Kir channels is the result of complex mechanisms that "plug" the channel pore at positive potentials, resulting in a decrease in outward currents. This voltage-dependent "block" results in more efficient conduction of current only in the inward direction. In other words, the inward rectifying channels pass easier a current in the inward direction than the outward one. Nevertheless, in cardiomyocytes, the net result is a current outward which is greater in phase 4 and much smaller at positive potentials.

During an action potential, we have two inward currents responsible for phase 0: the Na^+ current and the Ca^{2+}current, which persists during phase 2. Figure 3.5 illustrates the variations that undergo an AP following the combined treatment of the cardiomyocyte with *adrenaline,* which facilitates the opening of the L-Type Ca^{2+} channels, and with *tetrodotoxin,* which closes the Na^+channels. The administration of adrenaline causes a more

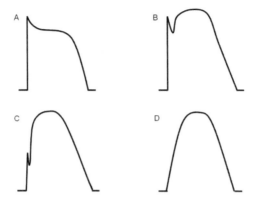

Figure 3.5 Effect of adrenaline and tetrodotoxin on the myocardial action potential. A: potential action in the absence of treatment; B: adrenaline treatment; C: addition of tetrodotoxin; D: combined effect of the two substances (this AP resembles a sinus node AP). See the explanation in the text. (from Carmeliet E and Vereeke J, 1969).

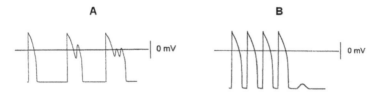

Figure 3.6 Early after depolarizations (A) and delayed after depolarizations (B). See the explanation in the text.

positive phase 2 which is mainly due to an iCa^{2+}, while the administration of tetrodotoxin reduces the amplitude of phase 0 due to the rapid iNa^+. As can be seen in Figure 3.5D, the combined effect of these two substances changes the appearance of the AP making it bell-shaped. This is due to the slowdown of phase 0, together with the potentiation of phase 2. This experiment shows that action potentials supported only by the opening of L-Type Ca^{2+} channels exist. As we will see in Chapter 4 they are typical of nodal cells, the sinus node, or atrioventricular node.

3.4 After Depolarizations

The mechanisms of arrhythmias will be considered in Chapter 10. Here we consider only the fact that an action potential can be responsible for an "induced activity" capable of provoking the appearance of membrane re-excitation and, thus, *extrasystoles* (*premature beats*). This induced activity is referred to as early and late or delayed after depolarizations. As can be seen from Figure 3.6, *early after depolarizations* occur along with phase 3 of the action potential, usually when repolarization has reached -60 V, whereas *delayed after depolarizations* occur mainly at the end of repolarization or phase 4. Early and delayed after depolarizations can be attributed to imbalances in Ca^{2+} currents. In particular, the early ones seem to be caused by the early reactivation of the Ca^{2+} channels, while the delayed ones are due to an intracellular overload of Ca^{2+} which activates the $3Na^+/1Ca^{2+}$ exchanger, which is electrogenic. After depolarizations can generate extrasystoles and sometimes give rise to ventricular tachycardia, which may result in fibrillation (see also Chapter 10).

4

Functional Tissue of the Heart

4.1 The Properties of Functional Tissues of the Heart

In Chapter 3.1, we have seen that the functional tissues of the heart are: the myocardium composed by cardiomyocytes, the nodal tissue (sinoatrial and atrioventricular nodes), and the tissue of the conduction systems, composed by specialized cells or fibers.

While the working myocardium can be divided into atrial and ventricular myocardium, the *cardiac conduction systems* (CCS) can be divided into impulse generating but slowly conducting *sinoatrial node* (SAN) and *atrioventricular node* (AVN), and the rapidly conducting *internode* (IN), *interatrial* (IA) and *ventricular conduction system* (VCS). Of note, inside the SAN and AVN the velocity of the impulse is the lowest we can find in the heart (about 0.01 m/s). The VCS, which consists of the *His bundle*, *bundle branches*, and the *Purkinje fiber network*, conducts rapidly (about 4 m/s) and allows a coordinate and effective ventricular contraction.

4.1.1 Automatism or Chronotropism

To understand the meaning of automatism it is necessary to define two terms that are substantially different but often confused by students: *stimulus* and *excitement*.

A *stimulus* is something that we can apply to cause a physiological response in a tissue that can give that response. In the case of excitable cells, when we apply an adequate stimulus (*e.g.,* an electrical stimulus), as a response, we can get an excitation. For all excitable tissues, the presence of *excitement* is revealed by the insurgence of an *action potential* (AP) which, in the case of muscles, is followed by contraction (see Excitation-contraction coupling, Paragraph 4.3). Then there is a third term, the *impulse*, which mainly indicates the excitement that moves from one point to another of the excited tissue. An impulse represents a stimulus for neighboring cells.

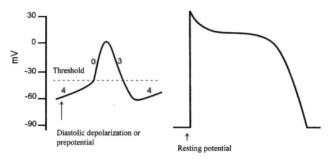

Figure 4.1   Action potential of the sinus node (left) compared with a ventricular myocardial action potential (right). While the ventricular myocardial fiber has a stable *resting potential*, the sinus node cell presents a slow depolarization called *diastolic depolarization or prepotential*.

Automatism, or chronotropism, is the ability of some parts of the heart to self-excited with a certain frequency, *i.e.* this part of the heart does not require the application of an external stimulus to generate an AP. This capacity is very high in the SAN, somewhat less elevated in the AVN and progressively lower in the other parts of the atrioventricular conduction system, such as the His bundle, the right and left branches, and the subendocardial Purkinje network. Therefore, physiologically many sites can potentially generate excitement and then an impulse. However, the only active center is usually the SAN which, imposes its frequency and controls the activity of the whole heart because it has a higher frequency. Therefore, the SAN is the natural heart pacemaker.

The activity of another pacemaker can be appreciated only if it has a faster frequency than the SAN. Besides the structures above mentioned, also the adult cardiomyocytes may be the site of automatic activity both at the level of the atrial and the ventricular myocardium under certain pathological conditions (see Chapters 10 on arrhythmias).

The self-excitation of the SAN can be explained by the morphology of the phase that precedes the AP development. As can be seen in Figure 4.1, the AP is not preceded by a constant voltage resting potential. In fact, at the end of the previous AP, the membrane potential is not −90 mV as in the ventricular myocardium but is only about −60/−70 mV. Importantly, this value does not remain constant during diastole but decreases progressively to about −40 mV. At this point, the AP that has a bell-like shape is triggered.

The progressive reduction in the amplitude of the membrane potential (from −70 to −40 mV) is called *prepotential* or *diastolic depolarization*. In

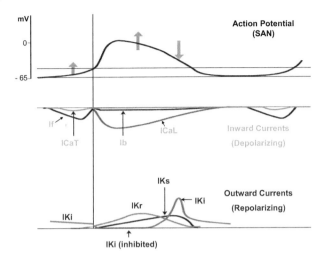

Figure 4.2 Action potential of the SAN (upper part); depolarizing (middle part) and repolarizing (lower part) currents. The arrows on the action potential of the SAN indicate the prevalence of inward currents (orange arrows) and outward currents (blue arrow) in the different stages of the AP. *if*: funny current for sodium; *i* b: basal sodium current; *i* Ca-T: transient current of transient calcium channels; *i* Ca-L: current of long-lasting calcium channels; *i* Ki, *i* Ks, and *i* Kr: currents for potassium. *See the explanation in the text.*

brief, the *prepotential* is due to an increase in the conductance of the cell membrane to Na^+, followed by an increase in conductance to Ca^{2+} with a simultaneous decrease in conductance to K^+. The increase in conductance to Na^+ and Ca^{2+} involves an inward current of positive ions. Furthermore, the decrease in K^+ conductance results in a reduced outwards current of this ion. Therefore, the membrane potential becomes progressively less negative and can, therefore, reach the threshold value of about -40 mV. Indeed, $-40/-45$ mV is considered a *threshold* value at which an AP takes place.

During *prepotential*, the initial increase in the conductance to Na^+ occurs as a consequence of the opening of channels that for their singular behavior are called *funny channels* (Figure 4.2). The funny channels are cationic channels that allow Na^+ to enter and K^+ to exit. However, since the iNa^+ is greater than the iK^+, the net current consists of the entry of positive charges within the cell *via* the funny channels. This inward positive funny current is referred to as *If*. The HCN channels (hyperpolarization-activated cyclic nucleotid-gated channels) mediate the funny current.

The concentration of cyclic adenosine monophosphate (cAMP or AMPc) can influence this current: if the cAMP increases the current in turn increases and *vice versa*. If the funny current intensity increases, the potential pacemaker will take less time to reach the action potential threshold, increasing the heart rate. The concentration of cAMP is increased by the stimulation of norepinephrine and epinephrine while it is reduced by acetylcholine.

The channels are called *funny* because they are opened at the end of repolarization and not during depolarization as usually Na^+ channels do. The decreasing conductance for K^+, which contributes to the development of the prepotential, occurs with certain slowness (Figure 4.2). Moreover, it seems that a weak inward current, which is part of the basal Na^+ current referred to as *i*b, contributes to the *prepotential*. When the membrane potential is brought to -55 mV, transient voltage-dependent Ca^{2+} channels (T type) opens and an inwards Ca^{2+} current occurs, thus hastening the diastolic depolarization. It seems that also the $3Na^+/1Ca^{2+}$ exchanger contributes to this hastening. In summary, *the prepotential* is mainly due to funny current, with contribution of K^+ conductance reduction and T type Ca^{2+} current increase, and to $3Na^+/1Ca^{2+}$ exchanger intervention.

When the prepotential reaches the *threshold* (about -40 mV), the voltage-dependent long-lasting Ca^{2+} channels (L type) are opened and the AP develops. The *SAN action potential*, therefore, has the characteristic not only of being preceded by the prepotential, but also of developing by the entry of Ca^{2+} instead of Na^+ in phase 0. The Ca^{2+} entry occurs slowly, thus explaining the slowness of phase 0 depolarization and the bell-shaped APs. During AP, K^+ conductance is low.

The *repolarization* of the SAN occurs by the closure of L-type channels and the recovery of the K^+ conductance due to the opening of delayed rectifier K^+ channels, which allow an outward iK^+. Repolarization is always due to the prevalence of an outward K^+ current.

As said, the SAN is not the only structure with automatism. The AVN and all the other parts of the atrioventricular conduction system are indeed provided with automatic activity, *i.e.*, they have a few funny channels. Since their discharge frequency is progressively lower in the structures between the SAN and the ventricular apex, they are normally silent, hence the name "latent pacemakers". Indeed, before the automatic depolarization can reach the threshold, a sinus impulse generated by the SAN arrives. Of note, these latent pacemakers have both a more negative resting potential and a slower development of the prepotential, which takes more time to reach the threshold, thus explaining the lower discharge frequency. While there is little difference in AP between SAN and NAV, the differences between the nodal tissue and the His bundle or Purkinje's network are much more

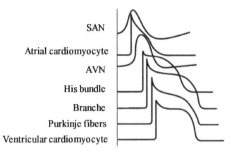

SAN
Atrial cardiomyocyte
AVN
His bundle
Branche
Purkinje fibers
Ventricular cardiomyocyte

Figure 4.3 Morphology of various Action Potential of the different cardiac cells: from the main pacemaker cells (SAN) to the ventricular myocardium. SAN: sinoatrial node; AVN: atrioventricular node. See the explanation in the text.

pronounced. In fact, in the fibers of these last structures, the amplitude of the resting potential is about -90 mV and the diastolic depolarization is very slow. In the case of complete blockage of the conduction between atria and ventricles, one of these centers of the conduction systems can give to the heart a discharge frequency of 20–30 beats per minute (b.p.m.). Another clear difference between the nodal tissue (SAN and AVN) and the other parts of the conduction system is in the overall morphology of the AP. While in the nodal tissues the majority of cells have a bell-shaped AP, the APs of the fibers of His bundle and Purkinje network have almost the same shape of those of ventricular cardiomyocytes (Figure 4.3). While genesis and development of the AP of the majority of cells of nodal tissues are due to the opening of L-Type Ca^{2+} channels, the AP of the cells of His bundle and Purkinje network, once a threshold value has been reached, is characterized by the opening of Na^+ channels and other ionic currents seen for cardiomyocytes AP.

Therefore, if the SAN is inactivated, a junctional rhythm may appear from the AVN. The discharge frequency can go from the typical sinus rhythm (about 70 b.p.m.) to the typical discharge frequency of the AVN (about 50 b.p.m.). Moreover, if an impulse arising in the SAN cannot reach the ventricles, as in the case of the complete atrio ventricular block (see below), the ventricular myocardium, the His bundle or one of its branches becomes the pacemaker and the HR will be about 20–40 b.p.m., that is the *escape rhythm or idioventricular rhythm*.

4.1.2 Excitability or Bathmotropism

Excitability or bathmotropism is the property of the heart to be excited in response to external stimuli. The excitability must not be confused with automatism, because, as previously seen, while the latter consists in the capacity of the cells to auto-generate impulses, the former requires that a stimulus comes from the outside. In other words, we can say that many cardiac cells can be excited if they receive an impulse from another fiber, while a few cells can be the place of the automatic activity. The latter is naturally located in the SAN and generates the impulse.

To be effective, a *stimulus* must have a minimum intensity sufficient to depolarize the cell membrane to reach the threshold value and thus trigger the action potential (AP). A stimulus below this intensity is ineffective and is said *subliminal*, while if it has intensity equal or higher the minimum intensity it is called *supraliminal*. According to *all-or-none law*, the response to a subliminal stimulus is null, while it is maximal when the stimulus has the intensity equal to or greater than the minimum required. This means that the strength of a response, i.e., the "magnitude/amplitude" of the AP is not proportional to the intensity of a stimulus. Therefore, it will not be a graded response. In simpler terms, when a stimulus is applied the AP will occur or it will not. If it occurs, it will have the maximum amplitude possible in that particular moment. Whatever is the intensity of supraliminal stimuli the AP will be the same as it depends on the properties of ion channels present on the cells. Therefore, each cell type has a typical AP.

During an AP, the cell is unexcitable, i.e., unable to respond to a stimulus. The period in which a cell cannot be easily excited is called the *refractory period*. This can be absolute or relative. During the *absolute refractory period,* the cell cannot respond to any stimulus whatever is its intensity. In the *relative refractory period*, however, the cell does not respond to regular stimuli but can be excited by greater intensity stimuli.

It said that in the case of *skeletal muscle*, the AP and the relative refractory period occur in the *latent period*. This is the time interval occurring between the instant in which a stimulus is applied and the beginning of the contraction. It follows that, when it starts to contract, the muscle is excitable again. Therefore, it can respond to the stimulus with a new AP and contraction, and the latter merges and sums with the previous contraction, with a consequent increase of the developed tension. If the stimuli follow each other at a sufficiently high frequency they induce the fusion and the sum of several individual twitches, which is *muscular tetanus*.

Unlike skeletal muscle, the *cardiac muscle cannot be tetanized*. Indeed, the AP has a duration that keeps the myocardium in a state of refractoriness for the whole time it is contracting. Figure 4.4 shows the contraction of a ventricular cardiomyocyte and the AP that caused it. From Figure 4.4, we can

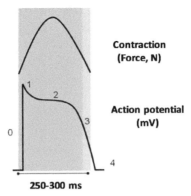

Figure 4.4 Absolute refractory period (in yellow) and relative (in blue) referring to the contraction and action potential. When the cardiac fiber is in the refractory period it is contracting. *See the explanation in the text.*

see how the contraction begins at the end of phase 0 and it is over when the membrane potential is again in resting conditions (phase 4). A blue vertical band indicates the *relative refractory period* (a stronger-than-usual stimulus is required to produce another AP), while the yellow one indicates the *absolute refractory period* (the cell cannot produce another AP) that extends to most of the contraction coinciding with phases 1 and 2 and with the great part of phase 3 of AP. In the heart, the absolute refractory period lasts about 230–270 ms, while the relative refractory period lasts about 20–30 ms when HR is about 60–80 b.p.m.

The variation in the cardiomyocyte excitability is attributed to the fact that Na^+ channels have two different types of gates, m and h gates (Chapter 3.1; Figure 3.3). While the closing state of the m gates (activation gates) prevents the entry of Na^+ but allows the cell to be excited when a stimulus brings the membrane potential to the threshold, the closure of h gates (inactivation gates) makes the fiber in an unexcitable state.

In resting conditions (phase 4 of the membrane potential) the m gates are closed while h gates are open. The cell is therefore excitable. When a *supraliminal* stimulus is applied m gates of the voltage-dependent Na^+ channels rapidly open, the Na^+ enters the cell and a phase 0 occurs. As soon as phase 0 has ended, the m gates remain open but the channels are closed by h gates. From this moment the fiber has become unexcitable. Therefore, it is in an absolute refractory period during phases 1, 2, and initial phase 3. During the late phase 3, several h gates are reopened, while several m gates are closed again. The majority of Na^+ channels are closed, but the fiber becomes progressively less refractory. When the AP has returned to phase 4, only the *m* gates are closed and the fiber is again normally excitable.

At the end of phase 3 (at end of repolarization) there is a period of increased excitability, said *supernormal period*. This is an extremely limited increase in excitability since the stimulus required to induce the response is just 0.01–0.05 mA below the minimum intensity required. The mechanism of this supernormal period is not known, but it has been suggested that the supernormal period results from the slow kinetics of the currents and does not depend on sodium current activation or inactivation or the after-depolarization.

4.1.3 Conductivity or Dromotropism: Origin and Diffusion of the Cardiac Impulse

Conductivity or dromotropism is the property of all the cardiac tissues to conduct the impulse allowing its diffusion to all parts of the heart. The impulse is a propagating AP, which is the formation of an AP in a fiber adjacent to the already active one. Conduction takes place due to local electrotonic currents between the active point and the resting adjacent fibers. Normally the impulse originates in the SAN and diffuses to the rest of cardiac tissues *via* electrotonic currents. The *gap-junctions* favor impulse diffusion. Gap junctions are a type of cell junction in which adjacent cells are connected through protein channels, which make cells chemically and electrically coupled.

An *electrotonic current* is a flow of charges. Outside the fiber, the flowing charges go from a polarized resting point (positive outside) to the adjacent depolarized point (negative outside). Inside the fiber, the flowing charges also go from the depolarized point (positive inside) to the adjacent polarized resting point (negative inside; Figure 4.5). The presence of these flowing charges causes a reduction in the amplitude of the potential in the resting point. When the reduction is about 15 mV the threshold is reached and an AP occurs. In this way, the excitation travels (propagates) to the point previously at rest.

The impulse originated in the SAN propagates rapidly to all heart *via* specific tissues which constitute the conduction systems of the heart, including the internode bundles, the interatrial bundle and the atrioventricular system (Figure 4.6). The *internode bundles* consist of an anterior, a mid, and a posterior bundle. These bundles carry excitation from the SAN to the AVN with which they are connected using the so-called transitional fibers. The

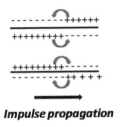

Impulse propagation

Figure 4.5 Propagation of excitation through electrotonic currents. See the explanation in the text.

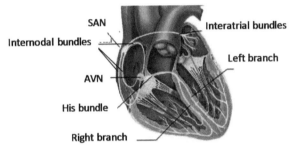

Figure 4.6 Schematic representation of the cardiac conduction system (CCS). CCS is formed by the sinoatrial node (SAN), atrioventricular node (AVN), and internodal and interatrial bundles, as well as by ventricular conduction system formed by His bundle with the two (right and left) branches, which terminate in the subendocardial Purkinje network.

interatrial bundle connects the SAN with the left atrium passing along the front wall of the two atria. The *atrio ventricular system* consists of the AVN situated in the atrial septum. The AVN has posterior (slow pathway, SP) and anterior (fast pathway, FP) fibers, which have the unique characteristics of *decremental conduction,* for which the more frequently the AVN is stimulated the slower it conducts. This is the property of the AVN that prevents rapid conduction to the ventricle in cases of rapid atrial rhythms, such as atrial fibrillation or atrial flutter. The conduction system continues with the *atrio ventricular bundle of His,* which penetrates, with the piercing portion, the atrioventricular fibrous tissue to reach the ventricles, where it divides in the *right and left branches* of the His bundle. These branches run respectively from the left and right sides of the septum. The left branch is then divided into the *anterior, middle and posterior fascicles*: the anterior and posterior bundles are respectively carried to the anterior and posterior papillary muscles of the left ventricle, while the center extends over the septal part of the same ventricle. The right branch, which runs on the right surface of the septum, reaches the so-called *trabeculae of Leonardo.*

From branches and fascicles originates the *subendocardial network of Purkinje* which is distributed to the inner surface of the heart from the apex to the base of the ventricles. The basis of the ventricles is, therefore, the last part of the heart to be excited. The *Purkinje fibers* do not cover the entire thickness of the ventricular wall but stop at the limit between the two inner thirds and the outer third of the wall. For reasons that will be clarified describing the *electrocardiogram* (ECG, see Chapter 10), in the myocardial wall the imaginary surface that separates the two inner thirds from the external part of the ventricular wall is considered as the *electrical myocardium.*

As we have seen, in the SAN is the diastolic depolarization or prepotential that generates the AP, i.e. the excitement. When the prepotential reduces the amplitude of the membrane potential from about -60 to about -45 mV, the opening of the L type Ca^{2+} channels determines the development of the AP. Once formed in the SAN, the AP is propagating to the surrounding tissues, including atrial muscles, and internodal and interatrial bundles. In the atrial myocardium, the impulse spreads to about 0.3 m/s, while in the bundles to about 1 m/s.

Within the SAN, which cannot be considered conductive tissue, the conduction velocity is only 0.01–0.02 m/s. The values of conduction velocity within the SAN and atrial conduction system are very important, as this is the time taken to excite the SAN and to activate the surrounding myocardial tissue.

The time taken by the sinus impulse to activate the atrial myocardium or *sinoatrial conduction time* (SACT) is about 75 ms. This is a time that can be measured by intracavitary electrography: with two electrodes inserted in the right atrium, it is possible to evaluate when the AP of the SAN is present and when the AP appears in the adjacent myocardium.

In SAN pathologies called *sick sinus syndrome, or* sinus dysfunction, or sinoatrial node disease or in its variant called *bradycardia-tachycardia syndrome,* SACT is much higher than 75 ms. In these patients very low SAN frequency (*e.g.,* 50 b.p.m.) episodes of very high tachycardia or paroxysmal supraventricular tachycardia have been described. These patients can suffer from Stokes-Adams attacks: fainting due to the absence of ventricular systole (asystole or ventricular fibrillation).

After propagating through the internode bundles, the excitement reaches the AVN. Within the AVN the impulse slows down. The time between its exit from the SAN and the arrival at the AVN is about 30 ms. In the AVN the conduction velocity is very slow the impulse takes another 90 ms before the excitement reaches the first or penetrating portion of the His bundle. The passage of the excitation through the penetrating portion requires 40 ms. The pulse thus undergoes a 160 ms delay when passing from SAN to the ventricles. This delay has considerable functional importance because it prevents premature beats arising in the atria from spreading too quickly to the ventricles.

The His bundle is surrounded by fibrous connective tissue rather than myocardium, and then enters the muscular septum and divides to form the right and left bundles. His bundle fibers, the two branches, the fascicles, and the Purkinje network are made of Purkinje fibers. With the sole exception of the penetrating portion, these fibers are rather large having a diameter even higher than those of the working myocardium. The conduction velocity of the excitation along these fibers is very high and can reach up to 4 m/s. The fibers of the subendocardial Purkinje network deepen in the thickness of the myocardial wall; transmit an impulse at the same velocity of His bundle. From the subendocardial Purkinje network, the impulse spreads to the myocardium (working cardiomyocytes) through which it propagates at a velocity ranging from 0.4 to 1 m/s.

Due to the distribution of the conduction system the activation of atria goes from right atrium to left atrium and the sequence of activation of the ventricles is as follows: septum/apex, free wall (non-septal), and base of the ventricles. Due to the rapid conduction in His bundle and branches, the apex of the heart is activated almost simultaneously with the septum (see ECG description in Chapter 10).

4.1.4 Contractility or Inotropism

Contractility or inotropism concerns the ability of the heart to contract and develop strength.

4.1.4.1 The cross-bridge muscle contraction cycle

At the end of the excitation-contraction coupling process (see Paragraph 4.3), the thin filaments slide on the thick filaments with a shortening of the sarcomere. The sliding occurs because the *cross-bridges* of the myosin filaments are inserted on the actin of the thin filaments. Passing from a position perpendicular to the sarcomere's axis (Figure 4.7A) to a position inclined towards the center of the sarcomere (Figure 4.7C), the cross-bridges exert a pull on the thin filaments in the direction of the central part of the sarcomere.

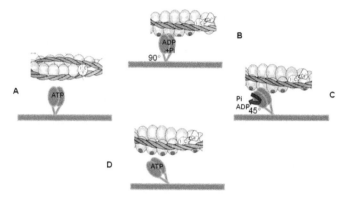

Figure 4.7   Cycle of cross-bridges. A: initial and final "rearmament" situation; B: hydrolysis of ATP into ADP and inorganic phosphate (Pi) with the formation of the actomyosin complex; C: the release of Pi and ADP from the myosin head and "power stroke"; D: detachment of myosin head from the thin filament induced by binding of a new ATP molecule.

Being anchored to the lines Z, the thin filaments exert on these traction, thus determining a shortening of the sarcomere. With each *cycle of cross-bridges* the thin filament slips by 5–10 nm. The cycle of the cross-bridges describes the insertion of the myosin heads on the globule of an actin filament, their flexion with relative traction on the thin filaments, the subsequent detachment, and the return of the myosin in the initial resting position.

In Figure 4.7 the cycle is described with some details. In the resting position, ATP is bound to the myosin head (Figure 4.7A). In this form the affinity of myosin head for actin is low. After excitation-contraction coupling, Ca^{2+} ions are bound to troponin C, induces a conformational change in the troponin–tropomyosin complex, with consequent unmask of the active sites on the thin filament. Moreover, ATP is split into ADP and inorganic phosphorus (Pi), but both products of hydrolysis remain linked to myosin head (Figure 4.7B). In this form, myosin enhances the affinity for actin and binds it tightly.

The myosin head forms an angle of 90° with the tails. The link between myosin and actin causes the rapid detachment of ADP and Pi from the active site of myosin, which is the cause of the increase in the hydrolysis rate of ATP. The free energy made available induces a conformational change in the myosin head, such that its inclination concerning the actin passes from 90° to 45° (Figure 4.7C). It is said that, after insertion, the myosin head flexes until the angle is reduced to about 50°. At this stage, Pi is detached from myosin. In an instant immediately following the ADP is also detached and the flexion is accentuated by bringing the angle to 45°. The flexion of the myosin head is called the *"power stroke"*, as it is the step at which force is produced. Then immediately after power stroke, a new molecule of ATP attaches to the myosin head causing its detachment from actin, thus myosin reassumes again the resting position (Figure 4.7 D-A). The return of myosin in the rest position is called *rearming*, as it puts the cross-bridges in the condition of performing a new cycle. Cross-bridge cycles continue until calcium level remains elevated.

The force that the thick filaments exert on the thin ones depends on the number and velocity of cross-bridges that are inserted during the electro-mechanical coupling process that in turn depend on the amount of calcium that binds to troponin C and the affinity of calcium for filaments. *When this force varies for the variation of calcium concentration and/or variations of calcium affinity for filaments and the consequent variations of **number and velocity of the cross-bridge cycles**, it is said that a variation in **myocardial contractility** occurred with a consequent variation of contractile force.* Note that this variation of contractile force may occur without variations of the initial length of the muscle. The different modalities of the variations and regulation of myocardial contractile force will be discussed in Chapter 7.

4.1.4.2 Isometric and isotonic contraction

The contraction of a muscle can occur at least with two different modalities when we consider force and length of the muscle: i*sometric and isotonic contraction.* In the *isotonic contraction,* the shortening of the sarcomeres determines the shortening of the muscle and the tension remains constant for the whole duration of the muscle shortening. In the *isometric contraction,* the whole muscle is not allowed to shorten, and the length of the muscle does not change, while it is developing force due to the sliding of filaments and tensioning of the elastic structures of the muscle.

Indeed, the muscle has a contractile and an elastic component. The *contractile component* consists of the set of sarcomeres, while the *elastic component* consists of elements in series represented by the so-called skeleton of the heart and by the tendinous strings of the papillary muscles.

For convenience, imagine that a papillary muscle behaves like a skeletal muscle. If the muscle is fixed to the two ends and cannot be shortened in its entirety when it is stimulated it only shortens the contractile compo-nent, while the elastic component in series (tendons and other structures) is extended by the same length of contractile component shortening. The distension of the elastic component in the series is responsible for the increase in tension that the muscle develops. We can say that an isometric contraction occurs when an excess load to be lifted is applied to the muscle (see below).

The concept of isotonic contraction applies to the fibers of the skeletal muscles and the myocardium, to the papillary muscles, and the pieces of the myocardium, but not to the heart as a whole. As we will see when we talk about the cardiac cycle, the heart should speak of a *contraction against a variable afterload*. Indeed, during cardiac ejection, the heart muscle exerts a sort of *auxotonic contraction* which is an increase in strength when the muscle is shortening, like when we push a spring whose resistance to being pushed increases from moment to moment.

Despite this clarification, it is useful to describe again the various modalities of an isotonic contraction using the papillary muscle as an experimental model. Indeed, an isotonic contraction can occur with preload, with afterload, or with both. ***Preload is the load the muscle "feels/bears" when it is relaxed and the afterload is the load the muscle "feels/bears" during contraction.***

When the same object is used as preload and afterload we can have:

4.1.4.3 Isotonic contraction with a suspended load

If, as shown in Figure 4.8, the muscle is fixed to one end while at the other end is attached to a suspended load (not resting on support), the relaxed muscle is subjected to a preload, which is the weight of the object attached to the muscle (Figure 4.8A). This preload determines in the muscle a stretch and a tension that we can define *passive tension* due to the distension of the elastic components of the muscle, which is in series with the contractile component constituted by sarcomeres. If the muscle is stimulated to contract, the sarcomeres are shortened by lifting the tendon to which the load is applied (Figure 4.8B). In this way, the muscle as a whole is shortened, but its tension, given by the distension of the tendons already before the contraction began, remains unchanged. *Isotonic contraction with suspended preload* is therefore isotonic throughout its duration.

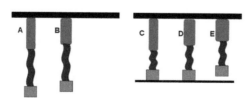

Figure 4.8   Isotonic contraction with preload (left) and with afterload (right). The contractile component (in red) in series with the elastic component (in blue). See the explanation in the text.

4.1.4.4 Mixed contraction: isometric and isotonic contraction with a sustained load

Let us consider now that the load applied to the tendon of the relaxed muscle is resting on a support. Under these conditions, the elastic component is not stretched and the muscle does not develop any passive tension, *i.e.* the load is not felt by the muscle (Figure 4.8C). When the muscle is stimulated, the contractile component is shortened by initially stretching out the elastic component which begins to develop tension. The muscle as a whole does not shorten because the elastic component lengthens as much as the contractile component is shortened (Figure 4.8D). When this developed force (tension) equals the weight of the load, the further shortening of the sarcomeres lifts the load, without causing a further increase in tension (Figure 4.8E). Having been felt by the muscle after the beginning of the contraction, the load is now called an *afterload*. In the case in which the muscle lifts an afterload, we speak of *isotonic contraction with afterload*. The contraction is properly isotonic only in the phase in which the whole muscle is shortening in its contractile and elastic components, while in the first phase, in which only the shortening of the contractile component and a simultaneous distention of the same amount of the elastic component, the contraction was *isometric*. Therefore, this is a *mixed contraction*: isometric first and isotonic later. This resembles the cardiac contraction which is first isometric and then isotonic (although it is better to say *auxotonic*, see below).

Isometric contraction: it is clear that the contraction becomes isometric from beginning to end if the load is so heavy that it cannot be lifted, whether it is sustained or not during muscle relaxation.

When different objects are used as preload and afterload we can have a mixed contraction whether or not the load is sustained and this condition may better resemble the heart contraction (*preload:* diastolic ventricular pressure; *afterload:* aortic systolic pressure).

4.1.4.5 Mixed contraction with different preload and afterloads

A said preload is the load that the muscle "bears" at rest. The afterload is the load that the muscle "bears" during contraction. Here we consider when the contraction of the muscle can be simultaneous with a preload and with an afterload. *During muscle relaxation,* the elastic component can be *completely stretched* (the passive tension is equal to the weight of the object not sustained and the muscle lengthening is an index of preload), *partially stretched* (the

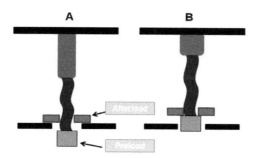

Figure 4.9　Isotonic contraction with preload and afterload. In **A**, the contraction begins with preload only, while in **B**, the afterload is added to the preload. The contractile component (in red) in series with the elastic component (in blue). See the explanation in the text.

passive tension is lower than the sustained weight) or *not stretched* at all by the completely sustained preload.

As can be seen in Figure 4.9A, a preload is applied to the muscle at rest, which is stretched and assumes a certain value of passive tension. It can be equal or not to the weight of the object, depending on if sustained, partially sustained, or not sustained. Starting from the resting value of tension, when the muscle is stimulated to contract in the first phase of the contraction, it contracts until the tension equals the weight of the preload, then the weight of the second object is added (is felt) to the contracting muscle; now the afterload is equal to the weight of the two objects and the muscle continues to contract isometrically. Only after that, the muscle tension equals the weight of the objects the muscle starts to shorten keeping the tension constant (*isotonic*) and equal to afterload (Figure 4.9B). Therefore, the contraction of a muscle can then be simultaneous with preload and afterload if the load is increased during contraction.

If the lifted load is increasing during the shortening, adding more and more objects, the tension developed by the muscle increases for the remaining part of the contraction. In this case, the contraction is *auxotonic*. This occurs, for example, pushing a spring and it is somehow similar to the ejection phase of heart contraction, in which we can imagine that weight is first added to the afterload and then removed while the muscle is shortening. Indeed, in the first part of systole the aortic pressure increases, and the second part decreases (see the cardiac cycle, Chapter 5).

In brief, the cardiac contraction happens with preload and with afterload. The preload is represented by the ventricular diastolic pressure and the afterload by the systolic pressure present in the aorta during ventricular ejection. As said, for the heart we cannot speak of an isotonic phase referring to the period in which the myocardial fibers are shortening. In fact, during ventricular ejection, the pressure in the aorta is not constant but varies continuously as a result of both the amount of blood ejected in the elastic aorta moment by moment and the passage of blood from the arterial vessels to the microcirculation regulated by peripheral resistance of the arterioles. For this reason, instead of defining the systolic ejection as an isotonic contraction of systole, it is better to describe it as the *systolic ejection or efflux phase at a variable afterload*.

In the heart also the preload differentiates in some respects from the preload schematically illustrated for the skeletal muscle. While in the skeletal muscle the passive tension is given by the stretching of the elastic component in series, in the case of the ventricles it is given above all by the distension that blood exerts on the myocardial fibers of the ventricular wall. Therefore, being the heart a hollow sphere, the Laplace's low help to better define the load as *ventricular wall stress*, S, which is proportional to ventricular pressure P; and ventricular radius, *r*, and inversely proportional to the wall thickness:

$$S P \times r / W.$$

Therefore, wall stress (load) is wall tension (P $\times$ *r*) divided by wall thickness (W). In diastole, P is the diastolic pressure, and *r* is the radius of the filled ventricle. Of course, we can consider this formula also in the ejecting heart, in which S is the afterload of cardiomyocytes in a given moment, as in systole P and *r* changes appreciably moment by moment.

An athlete's heart can be hypertrophic with a thickened ventricular wall. This ventricle has less wall stress that reduces the pre and afterload. Hypertrophy, in this non pathological case, can be seen as a mechanism that allows more parallel muscle fibers and more sarcomere units to "share" the wall tension that is determined by pressure and radius. The more the ventricular wall is thick, the less tension undergoes every sarcomere unit.

4.1.4.6 Contractility and the strength of contraction in the heart

Now we have the elements to understand that every time contractility increases, the energy (velocity and/or strength of contraction) developed by the heart increases. We can now discuss a different way to increase the force of contraction without increasing contractility. When we treat the *Frank-Starling's law*, we will see that by stretching the myocardial fibers, within certain limits, the heart can contract to develop a greater force without leading to any appreciable increase in inotropism or contractile velocity.

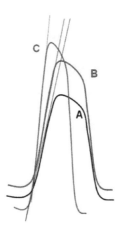

Figure 4.10 Ventricular pressure curves. A: basal condition; B: Increased preload; C: increase in contractility. See the explanation in the text.

Figure 4.10 illustrates how the rate of increase in ventricular pressure can help to understand if an increase in the contraction force may be due to an increase in contractility or to a greater stretching of the fibers.

The rate of increase of the ventricular pressure is given by the *derivative of the ventricular pressure* (dP/dt), where dP is the variation of pressure and dt the variation of time. Each increase in the derivative indicates an increase in contractility (dP/dt is a good index of contractility). The curve A of Figure 4.10 represents the ventricular pressure in the basal situation; the curve B represents instead the effect of stretching of the fibers and the curve C the effect of an increase in contractility.

In the passage from A to B, it is seen how a greater stretching of the fibers corresponds to an increase of the ventricular pressure in diastole, *i.e.* an increase in preload. Stretching is due to a greater degree of filling of the ventricle, which corresponds to an increase in the ventricular end- diastolic volume or ventricular tele–diastolic volume (VEDV or VTDV).

In B it is seen how in systole the pressure reaches a higher level than in the control conditions but using the same time it used in A to reach its maximum value. The dP/dt derivative is therefore increased a little bit by an increase in dP, while it remains unchanged dt (an index of the velocity of contraction). We can say that this is an increase in developed force, without significant changes in contractility.

Contractility is increased in curve C, for example, following the administration of a substance with an inotropic action such as adrenaline. In systole the pressure reaches a higher value in C than in A, but with an increase in dP and a decrease of dt, *i.e.* in a shorter time than in the two previous situations. In this case, a significant increase in contractility is responsible for the increase in developed force. It is an increase in force and velocity of contraction.

In brief, every time the contractile force increases, the dP/dt derivative is also increased. However, it depends on better contractility if the increase of the derivative is due to the increase of dP and to the decrease of dt. The increase in contractility is an *omeometric* regulation of the force of contraction. We do not need to stretch the fibers to increase the developed force.

If the force increase depends on the preload, the increase in the derivative is minimal and is due to an increase of dP with an unchanged dt. This is an *eterometric* regulation of force of contraction. Note that dP/dt variations are indicative of variations in contractility when there are appreciable variations of dt and not only in dP.

In addition to variations in the velocity of contraction, there may also be variations in the relaxation rate, that is the negative derivative, indicated as – dP/dt. The property that regulates the velocity of relaxation is called *lusitropy*. This will be discussed in more detail in the next paragraph.

We can see that there are huge differences in increasing the force of contraction by changing the contractility or the preload, analyzing the initial velocity of shortening of a piece of the myocardium (e.g. papillary muscle) at different initial lengths and at different afterloads to be lifted. Figure 4.11 shows the value of the afterload applied to a papillary muscle on the abscissa and the ordinate the initial velocity of shortening.

The initial length is adjusted by changing the position of the support on which the load rests. Once a certain length at rest has been established (index of preload), it can be observed that at each increase of the afterload decreases the velocity of shortening, until for a too-high afterload the contraction becomes completely isometric (curve 1 of Figure 4.11). If the resting length (preload) of the papillary muscle is increased (curve 2 of Figure 4.11), we see that for each afterload value the velocity of shortening is greater and the load necessary to render the contraction isometric is also greater. If the resting length of the papillary muscle is decreased (curve 3 of Figure 4.11), it is seen that for each value of afterload the velocity of shortening decreases and that a lower afterload is necessary to make the contraction isometric. Of note, for this family of curves *the maximum velocity of shortening* (A) at zero afterloads is identical for all, regardless of the values of length at rest (preload).

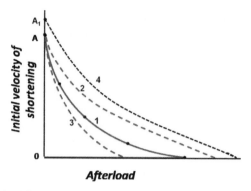

Figure 4.11   Afterload-Shortening Velocity Curve. Curve 1: basal condition; curve 2: an increase in the initial length (preload); curve 3: reduction of the preload; curve 4: an increase in contractility due to adrenaline delivery at an initial length equal to curve 1. A: initial velocity at zero loads equal for all initial lengths in the absence of adrenaline. A_1: demonstrates the increase in the initial velocity of shortening at zero loads when contractility is increased.

On the contrary, in curve 4 of Figure 4.11 the maximum velocity of shortening (A_1) is increased, regardless that this is the same muscle with identical rest length of curve 1.

While curve 1 refers to a papillary muscle in basal conditions, curve 4 refers to the same muscle treated with adrenaline which increases the contractility. In this case, it can be observed that *the increase in contractility increases all the velocities and even the maximal velocity of shortening at zero loads.* Moreover, a greater afterload is necessary to make the contraction isometric. We can say that *the increase in contractility increases the velocity and the force of contraction, independently from pre-load and afterload.*

4.2 Cardiac Contractility and Heavy Meromyosin Isoforms

Above we have seen that contractility depends on calcium quantity and filament affinity to calcium. Another factor that can affect contractility is the type of meromyosin expressed by cardiac muscle. In Chapter 2 we have seen how the heavy meromyosin presents ATPase activity in the head of the thick filaments. Meromyosin can be constituted by á or â chains. In the rat heart, these chains are combined to form three heavy meromyosin isoforms, meromyosin V1, V2, and V3.

The *meromyosin V1* is formed by two chains á and has a high velocity of the ATPase activity, the *V2* is formed by a á chain and a â chain and shows an enzymatic activity with intermediate velocity, while the V3 is formed by two â chains and it is characterized by a lower velocity. The greater or lesser velocity of the ATPase action leads to a greater or lesser velocity of shortening of the fiber and therefore a greater or lesser myocardial contractility.

The expression of these meromyosins may change in physiological (exercise training) and pathological conditions (hypertiroidism and malnutrition). As said, the preponderance of one or another of the heavy meromyosin isoforms is not the only structural factor affecting the velocity of shortening. In fact, with the same isoforms, the velocity of shortening is greater in the atria than in the ventricles.

4.3 Excitation-Contraction Coupling

The *excitation-contraction coupling* or *electro-mechanical coupling* is a sequence of events through which the excitation determines the contraction of the myocardial fiber.

As can be seen from Figure 4.12, when the fiber is excited, an AP travels through the sarcolemma and enters the T tubules which form the dyads with the cisterns of the sarcoplasmic reticulum (SR). Channels for Ca^{2+} are present in the sarcolemma, in the membrane of the T tubule, and the cisterns of the SR. The channels of sarcolemma and tubule T are called dihydropyridinic (DHP) channels because they are blocked by drugs of the

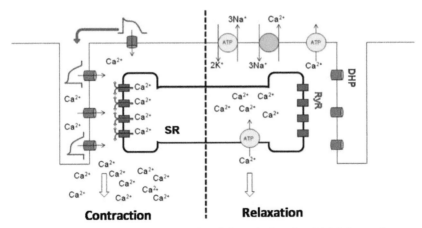

Figure 4.12 Excitation-contraction coupling (left) and relaxation (right) due to the removal of calcium from the sarcoplasm. SR: sarcoplasmic reticulum; RyR: ryanodine channel; DHP: dihydropyridinic channel. See the explanation in the text.

dihydropyridine class (also known as L-type calcium channel blockers). The channels of the cisterns are called ryanodine (RyR) channels or receptors because this class of intracellular calcium channels has a high affinity for a plant alkaloid named ryanodine.

While the dihydropyridinic calcium channels, called also L-type channels, are opened by the passage of the excitation, the RyR channels are opened by the Ca^{2+} entering into the fiber through the DHP channels. The release of Ca^{2+} from the RyR channels is called *calcium-induced calcium release* (CICR). As a consequence of the opening of these two types of channels, the cell concentration of Ca^{2+} increases from 10^{-7} M to 10^{-5} M. 25–40% of Ca^{2+} entering the cytosol comes from the outside and 60–75% from the SR.

Following the binding of Ca^{2+} with troponin C, the whole troponin complex undergoes a conformational modification that also concerns the tropomyosin which uncovers the myosin-binding sites on actin. Conformational change allows the activation of the ATPase function of myosin and, of actomyosin, after actin and myosin are linked. Due to this ATPase function, the ATP splits into ADP and inorganic phosphate, developing the energy that is necessary for filaments sliding.

The number of RyR channels that opens depends mainly on the Ca^{2+} concentration in their proximity, namely the subsarcolemmal calcium, which depends on the L-type channels that have been opened. Therefore, explaining why all factors, including *catecholamines (adrenaline and noradrenaline)*, that affect the opening of L-type channels, affect also contractility. Also, an increase in HR increases *per se* a little bit the contractility because the greater number of action potentials in the time unit corresponds to a greater number of phases 2 during which the L-type channels are opened increasing subsarcolemmal calcium (see BOX 5.2 on Bowditch scale phenomenon).

As already mentioned in Chapters 2.2 and Paragraph 4.1, contraction occurs by inserting the cross-bridges (i.e.: head projections from the thick filaments that attach to certain sites of the thin filaments). Insertion is followed by a change in the orientation of the myosin heads (cross-bridges) which, from a position perpendicular to the longitudinal axis of the fiber, are brought into an inclined position facing the center of the sarcomere. The cross-bridges repeatedly detach and reattach, thus determining the sliding of the thin filaments on the thick ones. This movement induces the approach of the two zeta lines with a shortening of the sarcomere at ascertaining velocity. Note that the developed force of contraction depends on the velocity of sliding and the number of cross-bridges that are established between thick filaments and thin filaments in a unit of time, which in turn depend on the concentration of Ca^{2+} in the sarcoplasm, which depends for the most part on the number of open RyR channels.

Once the contraction is over, relaxation occurs (Figure 4.12, right part). This is not a passive but active phenomenon, as it requires ATP for the active mechanism of the expulsion of Ca^{2+} from the cytoplasm both outside (25–40%) and in SR (60–75%) by specific pumps and exchanger. The decrease of the Ca^{2+} concentration in the sarcoplasm involves a detachment of the ion from the troponin C with consequent re-masking of the sites for the myosin head bridging leading to relaxation.

The expulsion of Ca^{2+} through the sarcolemma takes place, as shown in Figure 4.12, for the intervention of a $3Na^+/1Ca^{2+}$ exchanger and the activity of a Ca^{2+}-ATPase pump. While the pump uses the energy released by ATP hydrolysis, the exchanger allows the removal of the Ca^{2+} facilitated by the entry of Na^+ due to the high concentration gradient created by the $3Na^+/2K^+$ pump.

A Ca^{2+}-ATPase pump is also present on the SR membrane. It is called the Sarco-Endoplasmatic Reticulum CAlcium pump (SERCA). The isoform present in the heart is SERCA2; it is activated by the increase of free Ca^{2+} in the cytosol which is then pumped into the SR. Its activity is also inhibited by the phospholamban. When threonine and/or serine residues of phospholamban protein are phosphorylated by a protein-kinase (usually from protein-kinase A or PKA), the inhibitory power of phospholamban on SERCA2 is dampened. The Ca^{2+} is then pumped more quickly into the SR and the fiber release rate increases that is an increase in *lusitropy*, which is an increase in the rate of myocardial relaxation.

If the amount of Ca^{2+} returned to the SR increases, the Ca^{2+} also becomes available for a subsequent release through the RyR channels. This explains why an increase in lusitropy is generally associated with an increase in inotropism. *Catecholamines*, in addition to increasing the number of L-type channels and RyR channels that open, inactivate also the phospholamban *via* PKA activation and subsequently phospholamban phosphorylation: the overall result is an increase in inotropism and lusitropism. The heart contracts and relaxes quicker thus allowing a greater increase in heart rate, which accompanies catecholamine effects on the SAN.

Of note, the amount of Ca^{2+} that passes from the RyR channels into the cytoplasm may also increase due to the sudden increase in the duration of diastole. In this case, the time allows the ions to move from the point where they have returned to the SR or *intake compartment* to the terminal cistern or *release compartment*. This fact may explain, at least in part, the increase in the force of contraction after a prolonged diastolic pause, as it can occur after an extrasystole. On the other hand, the greater quantity of Ca^{2+} entering from L-type channels increases the CICR mechanism and this may explain also the small increase in contractility that accompanies the increase in HR (see BOX 5.2 on Bowditch scale phenomenon).

5

The Cardiac Cycle

5.1 The Heart as a Pressure Gradient Generator

In Chapter 1.3, it has been said that the heart generates a pressure gradient that generates flow. The heart must, therefore, be considered as a pressure generator. By raising the pressure of the ventricle to a value equal or greater than the pressure present in the large artery that starts from it, at each beat, the ventricle introduces the blood of *stroke volume* (SV). The heart, taking blood from the veins and expelling it into the arteries, increases the pressure in the latter and decreases it in the former.

If we refer to the left ventricle, we see how at each beat a certain quantity of blood is introduced into the aorta whose pressure is much higher than what we find in the following sections, particularly in the atrium and in the right ventricle during diastole. Because of this difference or pressure gradient, the blood moves from the aorta to the right atrium and ventricle, passing through the whole systemic circle. A similar description is valid for the pulmonary circle which, starting from the right ventricle ends in the left ventricle due to an analogous, albeit minor, pressure gradient.

Even the atria contracting generate pressure. This is a fairly modest pressure, which serves to complete the filling of the ventricles at the end of the diastole, when the ventricles are almost full of blood, as we shall see shortly.

The two atria contract almost simultaneously and while they contract the ventricles relax, and *viceversa* when two ventricles contract, almost simultaneously, the two atria relax. This sequence of contraction (systole) and relaxation (diastole) determine pressure and volume variations that can be measured with different methods. Looking at these changes in ventricle volume and pressures the cardiac cycle and its phase can be studied. Note that the cardiac cycle is described as events occurring in the ventricles and that, if not specified, when referring to systole or diastole we refer to ventricular events.

5.2 The Phases of the Cardiac Cycle

Here we will describe the *four phases of the cardiac cycle*, two systolic: *isovolumic contraction* and *ejection*, and two diastolic: *isovolumic relaxation* and *ventricular filling*. Between the end of the systole and the beginning of diastole, some authors describe *protodiastole* as the shortest period in the cardiac cycle coinciding with the closure of the semilunar valves.

The cardiac cycle of the left and the right heart are constituted by the same phases and, therefore, by the same sequence of events. However, there are little timing differences, but a significant difference in pressure values developed by the two ventricles. Indeed, the right ventricle pressure is much lower (1/6) than that developed by the left ventricle.

For a comprehensive study of the left heart cycle at least ventricular pressure, aortic pressure, and left atrial pressure curves should be taken into consideration (Figure 5.1). It is also useful to examine the ventricular volume curve and the aortic flow curve.

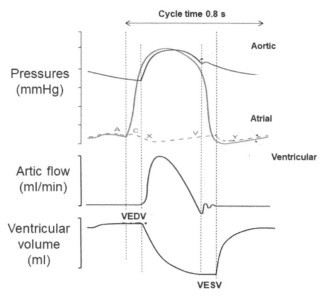

Figure 5.1   Top: Curves of aortic (black solid line), atrial (black dashed line), and ventricular (red line) pressures. Middle: the aortic flow curve. Bottom: ventricular volume curve. VEDV: ventricular end-diastolic volume (called also VTDV, ventricular tele-diastolic volume); VESV: ventricular end-systolic volume. The dotted vertical lines mark the isovolumic systole and the isovolumic relaxation. See the explanation in the text.

The phases of the cardiac cycle can be listed from the moment in which, at the end of ventricular filling, the ventricles begin to contract. The sequence of the various phases is then the following:

Systole:

> *Isovolumic systole*
> *Ventricular ejection phase*, divided into two efflux phases:
> - *The phase of rapid ejection or ejection with acceleration*
> - *The phase of slow ejection or ejection with deceleration*

Protodiastole;

Diastole:

> *Isovolumic relaxation*
> *Ventricular filling,* divided into three filling phases:
> - *Rapid filling phase*
> - *Slow filling phase (diastasis)*
> - *Atrial systole*

Please note that, if not differently specified, we consider the arterial and venous pressures as if they had been recorded when the subject was lying in a *clinostatic resting condition* so that the pressure sensors are at the same hydrostatic level. These avoid physics and therefore descriptive complications caused by the fact of having to take the force of gravity into account in an orthostatic position. Indeed, we will see that gravity affects appreciably the venous return of the blood to the heart (see Chapter 6.2). If in a hydraulic system the pressure sensor is positioned 1.3 cm above or below a reference level, 1 mmHg is subtracted or added, respectively, to the recorded pressure.

Since the cardiac cycle is studied as referring to the ventricles, the first phase considered is the one in which the increase of ventricular pressure or isovolumic systole begins, *i.e.,* immediately after the ventricular filling has been completed.

When the ventricular filling has been completed, the mitral valve is open while the aortic valve is closed. Since the mitral valve is open the atrium and ventricle form a single cavity in which the pressure under normal conditions is approximatively close to zero mmHg. This pressure, which in some cases may be higher or, rarely, lower than zero, is an index of the ventricular preload. Since during ventricular filling the blood proceeds from the atrium to the ventricle, in Figure 5.1 atrial pressure is reported as being slightly higher than the ventricular pressure. Throughout the filling of the ventricle, the aortic pressure, although progressively decreasing, is much greater than the ventricular pressure. The net difference between aortic pressure and ventricular pressure keeps the aortic valve closed.

5.2.1 The Isovolumic Systole

Aortic valve: closed
Ventricular pressure: increasing
Mitral valve: closes
Duration: 0.05–0.06 s
As the ventricle begins to contract, the pressure in its cavity suddenly becomes greater than that in the atrium, and immediately the atrio ventricular (AV) valve closes. It is the inversion of the pressure gradient between the atrium and the ventricle that causes the mitral valve to close. From this moment until the end of this phase of the cardiac cycle the ventricular muscles contract on a closed cavity in which the blood cannot enter and from which it cannot "escape", thus the ventricle contracts without varying its volume.

The contraction of the ventricle on a closed cavity causes the intracavitary pressure to continue to increase until it exceeds that present in the aorta. At this moment, under normal conditions, the aortic pressure is equal to 70–80 mmHg. This is the value called *tele-diastolic aortic pressure*. As soon as the ventricular pressure exceeds the tele-diastolic aortic pressure, the aortic valve opens and ventricular ejection begins. The time interval between "closing of the mitral valve" and " opening of the aortic valve" during which the ventricle contracts is the isovolumic systole phase. Under normal conditions, the duration of the isovolumic systole is about 0.05–0.06 s. This duration may change with ventricular contractility and aortic diastolic pressure variations. Indeed, an increase in ventricular contractility can result in a more rapid increase in intracavitary pressure, while an increase in aortic diastolic pressure may result in an *increase in the time* necessary for the intracavitary pressure to increase and to open the aortic valve during the subsequent systole.

Since the isovolumic systole occurs after the ventricular filling has been completed, during this first phase, the *ventricular volume* remains fixed at the highest value that can be observed throughout the cardiac cycle (about 130–135 mL, that is the VEDV) as can be seen in Figure 5.1.

5.2.2 The Ventricular Ejection Phase

Mitral valve: closed
Aortic valve: opens
Ventricular pressure: increasing and decreasing
Duration: 0.28–0.32 s
Ventricular ejection consists of two *sub-phases: ejection with acceleration (also called rapid ejection)* and *ejection with deceleration (also called slow ejection).*

The ventricular ejection phase is also improperly called isotonic systole to distinguish it from the previous isovolumic systole. Since the afterload is related to the pressure present in the aorta and since it varies throughout the ejection phase, we should speak of *variable load contraction* instead of isotonic contraction.

5.2.2.1 The ejection with acceleration

With the opening of the aortic valve, the efflux phase with acceleration begins. As can be seen from the systolic phase of aortic flow curve (called also systolic flow curve or SV curve) of Figure 5.1, during the first part of ejection the blood is introduced into the aorta at increasing velocity. At the same time, the aortic and ventricular pressures increase almost parallel to each other, with the ventricular pressure having at each instant a slightly higher value than the aortic pressure.

The value reached by the pressure at the end of this phase corresponds to the aortic systolic pressure (120–130 mmHg).

At this stage, the ventricular volume curve (Figure 5.1) reveals how the size of the ventricle is rapidly decreasing. The duration of this phase is about 0.12–0.13 s.

5.2.2.2 The ejection with deceleration

On the aortic flow curve, you can see that the rate of blood introduced into the aorta begins to decrease after it has reached its maximum value during the acceleration phase.

The ventricular pressure now faces a gradual reduction during which the ventricular pressure is no longer slightly superior, but slightly less to the aortic one. Nevertheless, the blood is continuing to move from the ventricle to the aorta (see Box 5.4: The analysis of the aortic flow curve during ejection).

The decelerating efflux lasts about 0.16 s. During this phase, the ventricular volume decreases with a lower velocity than that seen in the ejection with the acceleration phase. At the end of ejection with deceleration, the ventricular volume reaches the minimum value of the entire cardiac cycle (55–65 mL).

Once the ejection with the deceleration phase is completed, ventricular relaxation begins. It is noteworthy that the ventricular contraction ends when the ventricular pressure is already decreasing. As can be seen in Figure 5.1, at the end of the contraction (end of the ejection phase with deceleration), the ventricular pressure is lower than the aortic pressure and the ventricular outflow rate (*systolic flow curve*) has dropped to below zero (*backflow*).

In the subsequent phase, the onset of ventricular relaxation causes a further decrease in ventricular pressure, while the aortic pressure displays the *dicrotic notch or "incisura"* and subsequently the *dicrotic wave* (see below).

Box 5.1 The Ventricular and Aortic Pressures Increase in the Ejection Phase with Acceleration and Decrease in the Deceleration Phase

From the description of the two ventricular ejection phases, one question arises: *why the ventricular and aortic pressures increase in the ejection phase with acceleration and decrease in the deceleration phase?* The answer to this question is the following: the blood present at the root of the ascending aorta does not move (zero velocity) until the ventricular ejection intervenes. This blood has an inertial mass that is opposed to the pumping force of the heart. The latter tends to push the mass of blood towards the peripheral vessels, through the pressure imparted by the stroke volume.

As a consequence of the inertia of the blood present in the aorta, when the ventricle starts to pump its systolic flow at a rising rate, *i.e., during the ejection phase with acceleration, the amount of blood that enters the aorta from the ventricle is greater than the amount of blood that goes from the aorta to the smaller vessels.* Because of this difference between the amounts of blood that enters the aorta and the quantity that at the same time from the arterial tree moves toward the venous system, *the pressure in the aorta can only increase*.

During the ejection phase with deceleration, the speed at which blood is ejected into the aorta is decreasing, while the inertia now tends to keep blood in that state of motion that was imposed in the previous phase. It follows that *during the ejection phase with deceleration, the amount of blood that enters the aorta from the ventricle is less than*

---

the amount of blood flowing from the aorta toward the smaller vessels, as a consequence the pressure decreases. In other words, since the blood that entering is less than the blood leaving the aorta, the aortic pressure gradually decreases during this deceleration ejection phase. Of course, *since the aorta and ventricular cavity during ejection can be considered a single structure, the pressure changes that occur in the aorta and ventricle are consensual.* In other words, the pressure in the aorta (*i.e.*, the afterload) is continuously changing and the ventricle force of contraction adapts to it to develop a ventricular pressure similar to the afterload.

As can be seen from Figure 5.1, during the deceleration ejection phase, the aortic pressure remains above the ventricular pressure, so it seems that blood is flowing from a lower pressure compartment towards a higher pressure compartment. The last statement raises a second question: *why blood can flow from a lower pressure compartment to a higher pressure compartment?* The answer to this question is more complex. Also, in this case, the explanation must take into account the inertia of the blood and the total energy of a fluid.

Since the blood moves from one compartment to another with oscillatory rather than stationary motion, the total pressure difference between ventricle and aorta consists of a resistive component, an inertial component and a gravitational component, that is:

$$\Delta Ptot = \Delta Pres + \Delta Pi + \Delta Pg$$

where $\Delta Ptot$ represents the total pressure gradient between ventricle and aorta, $\Delta Pres$ the resistive component of the gradient, ΔPi the inertial component and ΔPg the gravitational component.

Note that here the resistive component is not given by the peripheral resistance, but it is due to the aortic ostium that the blood must cross to move from one cavity to another. This is a very modest resistance; thus we may not consider it. Moreover, in the clinostatic position of the subject, we can also ignore the gravitational component.

Since in the aorta there is a considerable mass of blood with inertia, it is clear that the inertial component prevails over the resistive one in the aorta. It is known that, when in a pressure gradient the inertial component prevails over the resistive one, the total difference $\Delta Ptot$ does not determine the direction of the movement of the fluid, but it determines the velocity variations (acceleration or deceleration) of blood flow. Therefore, *when the ventricular pressure is higher than the aortic one during the first phase of efflux, the ejection accelerates, whereas when the pressure gradient is reversed during the second phase of efflux, the ejection decelerates. The direction of flow remains the same until the ventricle contracts, giving energy to the system.* In other words, the total fluid energy is greater in the ventricle than in the aorta until the ventricle contracts (the *outward momentum*). When it relaxes, the energy and the momentum immediately reverse with consequent closure of aortic valve (protodiastole): the isovolumic relaxation has begun.

5.2.3 The Protodiastole

Mitral valve: closed
Ventricular pressure: decreasing
Aortic valve: closing
Duration: 0.01–0.015 s

The protodiastole is when the aortic valve closes. As mentioned above, in protodiastole ventricular relaxation starts and accentuation of the difference between the aortic and ventricular pressure occurs (2–3 mmHg difference), thus the blood reverses the sense of its movement and tends to return from the aorta into the ventricle. This inversion of the direction of movement of the blood (backflow) is however immediately prevented by the closing of the semilunar leaflets of the aortic valve. In its closing, the aortic valve protrudes slightly towards the ventricle. Note that the backflow is only in the aorta! The flow meter is positioned a little higher than the level of the valves and can record the backflow. It is the aortic backflow that closes the aortic valve. Normally no appreciable amount of blood returns to the ventricle. Only if the valve is leaky (aortic insufficiency or incompetence) the backflow increases and appreciable amount of blood goes back into the ventricle.

Therefore, the short time in which the aortic valve is closing is the protodiastole. It is the shorter phase of the cardiac cycle because it lasts only 0.01–0.015 s.

The blood that moves from the aorta to the ventricle has kinetic energy. The sudden arrest of this movement by closing the aortic valve causes the kinetic energy to be transformed into pressure energy. This pressure energy determines the appearance of a small pressure wave, called the *dicrotic wave*, which is observed in the diastolic phase of the aortic pressure curve. The phenomenon for which kinetic energy is transformed into pressure energy by an obstacle (the closed aortic valve) is called the "water hammer". The intervening dip between the main systolic wave and the dicrotic wave is called the *incisura* or *dicrotic notch* and corresponds almost exactly to the protodiastole. In brief, in protodiastole we see the incisura in the aortic pressure curve there is the backflow at the end of the systolic flow curve. While in the aortic pressure the dicrotic wave occurs in the isovolumic relaxation phase.

5.2.4 The Isovolumic Relaxation Phase

Mitral valve: closed
Aortic valve: closed
Ventricular pressure: decreasing
Duration: 0.06–0.07 s

After the protodiastole, aorta and ventricle form two separate cavities. In particular, the ventricle is now a closed cavity, as both the aortic and the mitral valves are closed. Starting from the dicrotic wave, the aortic pressure slowly declines, while ventricular pressure falls very rapidly due to the release of the myocardium and the mechanical recoil of the elastic component (collagen fibers) within it.

Due to the closure of the two valves, the ventricle relaxes without varying its volume, which remains fixed at the lowest value of the cardiac cycle (55–65 mL), so that this phase of the cardiac cycle is called *isovolumic relaxation phase*, which lasts about 0.06–0.07 s. This diastolic phase ends when ventricular pressure has fallen just below atrial pressure with the consequent opening of the mitral valve, so that blood flows from atrium to ventricle.

From the time the mitral valve closes, *i.e.* from the beginning of the isovolumic systole, atrium and ventricle form two separate cavities. During all this time, blood continues to enter the atrium from the pulmonary veins but cannot pass into the ventricle. The blood that accumulates in the atrium increases the pressure in this cavity: from the atrial pressure curve from Figure 5.1 we see a progressive increase in atrial pressure that rises to about 10 mmHg. The rise of atrial pressure ceases when the ventricular pressure becomes lower than the atrial pressure. At this moment the isovolumic relaxation phase ends and the rapid filling phase of the ventricle begins.

The duration of the isovolumic relaxation phase can vary in part depending on the level of pressure at which protodiastole occurred, which, of course, depends on the aortic pressure level and partly also about the velocity of myocardial relaxation or *lusitropy*.

5.2.5 The Rapid Ventricular Filling Phase

Aortic valve: closed
Mitral valve: opening
Ventricular pressure: decreasing
Duration: 0.11–0.12 s

As the ventricular pressure falls below the atrium pressure, where the blood has accumulated during ventricular systole, protodiastole, and isovolumic relaxation, the mitral valve opens and blood floods into the ventricle from the atrium. Atrial and ventricular pressures quickly fall to a level close to zero. During this rapid ventricular filling phase, while ventricular continues to relax and pressure continues to fall, about 70–75% of the volume of blood that will form the "stroke volume" in the ventricular systole of the next beat enters into the ventricle. In this phase, as well as in the following phases of slow filling and atrial systole, the atrial and ventricular pressure run parallel to each other while the atrial pressure is slightly higher than the ventricular one. The duration of the rapid filling phase is 0.11–0.12 s.

5.2.6 The Slow Ventricular Filling Phase

Mitral valve: opened
Aortic valve: closed
Ventricular pressure: increasing a little
Duration: 0.17–0.20 s

At the usual HR (60–90 b.p.m.), this phase is the longest of the whole cardiac cycle (*0.17–0.20 s*). All phases of the cardiac cycle shorten to a different degree their duration or last a little longer when HR increases or decreases, respectively. However, the duration of the slow ventricular filling phase can vary greatly, from 0 to 0.30 s or more, when HR changes. As seen in Figure 5.1, it begins when the atrial and ventricular pressure curves become horizontal or begin to rise very slowly, i.e., immediately after it had reached the lowest value during the rapid filling phase. In other words, the transition from the fast filling phase to the slow filling phase coincides with the instant in which the two pressure curves suddenly change direction, first decreasing (rapid ventricular filling), then increasing a little (slow ventricular filling).

During the slow ventricular filling phase, about 25% of the volume of blood that will form the "stroke volume" in the ventricular systole of the next beat passes from pulmonary veins into the atrium and then into the left ventricle.

At the end of the slow filling phase, the ventricular volume is very close to its maximum value. The slow filling phase ends when the atrial systole occurs.

5.2.7 The Atrial Systole Phase

Mitral valve: opened
Aortic valve: closed
Ventricular pressure: increasing a little
Duration: 0.07–0.10 s

This phase represents a modest, albeit active, the contribution of the atrium to ventricular filling. Atrial systole duration can reach 0.10 s. During all previous phases of the cardiac cycle, the atrium was relaxed, and the blood accumulated in it from the beginning of the isovolumic systole until the end of the isovolumic relaxation. During rapid and slow ventricular filling, the majority of blood had already left the atrium. When the atrium contracts, it forces some additional blood into the ventricle, which enhances ventricular filling by about 10–20% in a resting healthy subject. The percent contribution of the atrial systole to ventricular filling is elevated at high HR s and progressively decreases as the frequency decreases. For more details, see Paragraph 5.5 and 5.6.

Another function of the atrial systole is also to avoid the regurgitation of blood from the ventricle to the atrium when the mitral valve will close at the beginning of the upcoming isovolumic systole. Indeed, at the end of the atrial systole, the ostium of the AV valve is a little narrower and the valve leaflets are closer. The leaflets are closer because of blood vortices caused by atrial contraction. Because of this position of semi-closing, as soon as the ventricular pressure increases with the beginning of the isovolumic systole, the AV valve closes rapidly, preventing blood regurgitation from the ventricle to the atrium. We can say isovolumic contraction begins and the atrioventricular valve closes almost simultaneously.

In the presence of a prolongation of AV conduction, the isovolumic systole initiates when vortices are finished, the AV valve is completely open, and regurgitation occurs (some blood goes back to the atria).

During the two phases of slow ventricular filling and atrial systole, in conditions of HR of 70–80 b.p.m., only 25–30% of the blood that will form the next stroke volume returns to the ventricle. The majority of blood fills the heart during rapid ventricular filling. The absence of an atrial contraction as it can occur in patients suffering *atrial fibrillation* (chaotic depolarizations without efficient contraction, see Chapter 10) often makes little difference to resting stroke volume and cardiac output. When HR increase or when the ventricular function decrease, as in aged people with the stiff ventricle and those suffering heart failure, the atrial contraction becomes important because the passive filling may be compromised. Also during exercise, the atrial contraction at high HR may contribute significantly to ventricular filling.

5.3 The Atrial Pressure Curve

As can be seen in Figure 5.1, during the cardiac cycle the atrial pressure curve has three upward waves (*a, c* and *v*) and two downward waves or depressions/descents (*x* and *y*).

The *wave* corresponds to the atrial systole and precedes the ventricular isovolumic systole. The *wave c* is due to a protrusion of the mitral valve leaflets in the atrial cavity during isovolumic systole. The wave *c* is followed by the *x descent* due to the lowering of the atrioventricular fibrous disk, with an enlargement of atria with a consequent slight decrease in the pressure within them. Indeed, while the ventricular volume is rapidly decreasing because of rapid ejection: the atrium floor is pulled towards the ventricles.

Due to the closed mitral valve, the blood accumulation in the atrium causes the pressure to rise slowly up to the apex of the *v wave*, which marks the beginning of rapid ventricular filling because at the apex moment the ventricular pressure is falling below the atrial pressure. In this phase, the rapid passage of blood from the atrium to the ventricle causes a further fall in atrial and ventricular pressure, which in atrial pressure is called *y descent*. B etween the atria and the big veins draining, there are no valves; therefore, the cycle of pressure changes in the atria producing a similar cycle of events in the veins.

Particularly interesting are the external jugular veins as in these neck veins the waves and descents can be easily appreciated in a recumbent lean subject enabling the physician to appreciate the pathological variation of these events during a visual inspection. For example, atrial contraction at closed tricuspid (dysrhythmias, nodal rhythm or heart block) or stenotic tricuspid may produce big *a*-waves called *"cannon waves"*, while tricuspid insufficiency can produce exaggeratedly *"wide v waves" or "giant v waves"*.

5.4 The Cycle of the Right Heart

As said *a)* the right heart cycle has the same phases of the left heart cycle; *b)* the right ventricle develops a mean pressure of about 1/6 only compared to the left ventricle; *c)* this low pressure is due to the low vascular resistance that the blood flow meets in the pulmonary circulation.

The lower pressure and the lower afterload that must be overcome to open the pulmonary valve lead to a shorter duration of the isovolumic systole of the right ventricle. This is about 0.02 s only, instead of 0.06–0.07 s for the left ventricle. Since it starts from a lower value, the isovolumic relaxation phase also is very short. However, the ejection period of the right ventricle lasts longer than that of the left ventricle, so that the pulmonary valve closes a little later than the aortic valve and the second cardiac sound can be appreciated as two separated components. During inspiration, there is an increase in venous return to the right ventricle and thus increased flow through the pulmonary valve delaying its closure, *vice versa* in expiration. Therefore, the delay of the pulmonary component of the second sound can vary with respiration (seen in Paragraph 5.8).

It is worthwhile to point out here, that the lower thickness of the walls of the right ventricle (1/3 of the left ventricular wall) is a consequence of the lower pressure load that it has to face during systole. Moreover, despite a similar stroke volume compared to the left ventricle, the right ventricle is characterized by a lower end-systolic volume and tele-diastolic volume. As will be seen in Paragraph 5.9, these volume differences are accompanied by corresponding higher values of the *ejection fraction* of the right ventricle. Indeed, the ejection fraction is the ratio between the stroke volume and the tele-diastolic volume, expressed as a percentage.

5.5 Changes in the Heart Rate and Duration of the Phases of the Cardiac Cycle

Of course, when HR increases the duration of the cardiac cycle has to shorten, and *vice versa* when the HR is reduced the cardiac cycle lasts longer. However, the various phases do not shorten or prolong to an equal degree. As can be seen from Figure 5.2, in the increase of the HR from 60 to 180 b.p.m., *to a rather modest shortening of the duration of the systole is accompanied by a more pronounced shortening of the duration of the diastole. This is because an action potential and a ventricular contraction hardly shorten more than a certain degree and rarely last less than 160–180 ms.* Therefore, an increase in the frequency can be considered as approaching two successive beats with a reduction of the time interval that separates them, which is a shortening of diastole and in particular of the slow ventricular filling phase, which usually is the longest phase of the cardiac cycle.

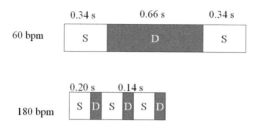

Figure 5.2 Changes in the duration of cardiac cycle phases due to changes in heart rate. With a heart rate of 60 beat per minute (b.p.m.), the diastole lasts about 0.66 seconds. When the heart rate increases three times (180 b.p.m.), the diastole shortens by almost 5 times (from 0.66 to about 0.14 s), while the systole shortens by 1.5 times only (from 0.34 to about 0.20). S: systole; D: diastole. See the explanation in the text.

An appreciable shortening of the duration of the systole usually takes place when a sympathetic stimulation increases HR and myocardial contractility, with a consequent increase in the rate at which the ventricular pressure increases (*positive inotropic effect*) and decreases (*positive lusitropic effect*).

Box 5.2 Bowditch Scale Phenomenon

When the HR increases for artificial pacing with a pacemaker device, in the absence of sympathetic stimulation, the shortening of the systole duration and the increase in developed pressure, *albeit* limited, depend on the *Bowditch scale phenomenon*. This phenomenon consists of a progressive increase in contractility occurring when a heart is stimulated to beat at higher frequency (Figure 5.3). After the transition from a low to a higher frequency, the developed pressure progressively increases until a stationary situation is reached in which the pressure is higher than that developed at low frequency.

This phenomenon is also *Scale/Treppe Phenomenon* and can be explained by *a)* the progressive increase in the intracellular Ca^{2+} concentration following the increase in the rate of action potentials, each with its plateau phase (phase 2) during which the Ca^{2+} enters the fiber and CICR phenomenon occurs; *b)* the reduction of the duration of intervals between beats and, therefore, of the time necessary to remove Ca^{2+} from the cytosol. In brief, the intracellular concentration of Ca^{2+} increases, and therefore the myocardial contractility increases.

An initial reduction in contractility and developed pressure is often observed in the early beats immediately after an increase in frequency

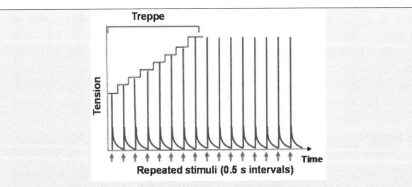

Figure 5.3 The beginning of pacing or the increase in heart rate causes an increased step by step of the developed tension until a steady-state is reached.

by external pacing. This is due to a lower release of Ca^{2+} from the sarcoplasmic reticulum because the sudden reduction in diastole time has limited the passage of the ions from the compartment of intake or longitudinal tubule, to the release compartment, or terminal tank, of the same sarcoplasmic reticulum.

During an increase in HR, the *shortening of the diastole* is much more pronounced than that of the systole and can involve important hemodynamic modifications. For modest increases in HR, the phase of the cycle that is curtailed is that of slow ventricular filling, during which usually about 30% of the blood that will form the next SV returns to the ventricles. Therefore, if an increase in HR cuts or even cancels the slow ventricular filling phase, a correspondent SV reduction can occur. The reduction of stroke volume is compensated by the increase in HR. It is also possible that the reduction of the SV is initially lower than the reduction of blood returning to the ventricle in diastole because of greater use of the systolic residue determined by the increased contractility that accompanies the increased HR. Therefore, a hemodynamic equilibrium that is established between increased HR and reduction of SV explains why the cardiac output (CO = SV × HR) may increase a little or remain constant. In anesthetized dog subjected to atrial pacing an HR increase from 100 to 160 b.p.m. does not vary significantly the CO.

For further increases in HR, it is possible that the shortening of the diastole, after having completely canceled the slow ventricular filling phase, also "cuts" the rapid filling phase. Since now, less than 70% of the usual SV returns to the ventricles in this phase, such a shortening causes a drastic reduction of the next stroke volumes so that a substantial increase in HR can be accompanied by a reduction in cardiac output. It must be said that *initially,* the SV reduction may be in part due to the higher diastolic aortic pressure (there is less time to decrease to the usual level the initial afterload). In humans, a sudden decrease in CO and aortic pressure occurs when the HR by external pacing exceeds 180 b.p.m. This is the so-called *upper critical limit of Wenkebach.* Ventricular filling time of 0.11–0.12 s is the minimum interval that allows an adequate filling of ventricle despite the "help" of atrial systole. Higher heart rates can be achieved in athletes in which the increase in inotropy and lusitropy shortens the systole and isovolumic relaxation below 0.2 seconds. So that, for example, with systole of about 0.17 s and ventricular filling time of 0.11 s *circa* an HR of about 210 b.p.m can be sustained for a while, during an exhausting physical exercise. In this case, the ventricular filling is also helped by skeletal muscle massage and sympathetic venoconstriction (see Chapter 6 on Venous Return)

When the HR is reduced, of course, the diastole prolongs, and the cardiac output is initially unchanged since the lower HR is accompanied by an increase in SV. The SV increase is caused by both the greater time of ventricular filling and by the reduction of the initial afterload, determined by the longer time of diastole which allows an appreciable reduction of the telediastolic aortic pressure (Figure 5.4). However, if the frequency falls below 40 b.p.m. *circa*, the further lengthening of the diastole does not guarantee a greater filling of the already full ventricle. At this point, the reduction in HR prevails over the increase in stroke volume, and therefore the cardiac output begins to decrease: the *lower critical limit of Alella* has been reached, as in the case of complete AV block (see Chapters 10 on Arrhythmias).

5.6 The Hemodynamic Role of Atrial Systole

The hemodynamic role of atrial systole varies with frequency. The contribution of atrial systole at a normal HR is modest, whereas it pumps from atria to ventricles an increasing percentage of the next SV as the HR increase and the diastolic interval is reduced. In dogs, at a frequency of 150 b.p.m., the atrial systole has been seen to contribute approximately by about 50% to the next SV.

When the HR is so high that the slow ventricular filling phase disappears completely and the rapid filling phase is shortened, the atrial systole contributes relatively more to the ventricular filling to sustain the next stroke volume.

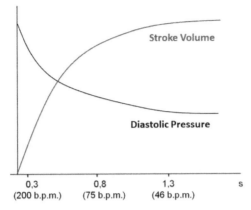

Figure 5.4 Changes in diastolic pressure and stroke volume following a change in heart rate. When the cardiac cycle duration increases (that is a decrease in b.p.m.) the diastolic pressure decreases and the SV increases and *vice versa*. See further explanation in the text.

Box 5.3 Atrial Systole in the Presence of Arrhythmias

The comparison between *atrial flutter* and *atrial fibrillation* helps to understand the contribution that atrial systole (when present) does to ventricular filling at high HR. In both arrhythmias, HR may be very high and the interval between two heart beats may be less than 300 ms (systole and diastole time about 0.18 and 0.12 s, respectively), corresponding to a frequency of 200 b.p.m. In the case of *atrial flutter* each beat is preceded by atrial systole and the stroke volume, although reduced, is present. In the case of *atrial fibrillation* (desynchronized depolarization/contraction of atrial fibers), however, the atrial systole is absent: in this case, a ventricular beat that appears after an interval of 300 ms from the beginning of the previous beat, can be devoid of SV, that is an ineffective beat. This suggests that at very high frequencies the presence of atrial systole can be the *conditio sine qua non* to have an SV. Nevertheless, as said, even with the help of atrial systole, 0.11–0.12 s is about the minim filling interval that allows an adequate refilling of the human ventricle to have an adequate SV.

The contribution of the atrial systole to the subsequent SV also depends on the time in which it occurs. In the case of ventricular extrasystoles, the atrial systole can occur at the same time as the ventricular systole. Since in this case the AV valves are closed when the atrium contracts, the contribution of the latter at the ventricular filling is null, regardless of the HR. A similar, but not identical, the situation may occur in the case of sinus rhythm, when the interval between the onset of atrial and ventricular systole is greatly reduced as in the syndrome of ventricular pre-excitation or *Wolf-Parkinson-White syndrome*. In this syndrome, the PR interval of the electrocardiogram is abnormally shortened and the atrial systole takes place almost simultaneously to the ventricular systole. Pathological tachycardia (>250 b.p.m.) may occur causing a decline in cardiac output because of inadequate refilling during the short diastole. As mentioned, in aged people with ventricular failing and ventricular stiffness, atrial systole contributes substantially to ventricular filling, and its absence (fibrillation) may compromise stroke volume, even at low HR. Indeed, atrial fibrillation is characterized by complete arrhythmias of ventricle which will beat at variable HR between 40 and 220 b.p.m. (see Chapters 10 on Arrhythmias).

5.7 The Aortic Flow Curve

The analysis of the flow curve allows understanding of how the ventricular ejection phase is divided into an efflux phase with acceleration and an efflux phase with deceleration.

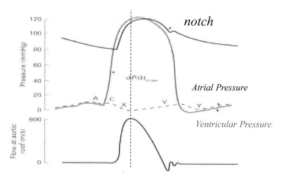

Figure 5.5 Aortic, ventricular and atrial pressures (top) and aortic flow (bottom). Note that the amount of blood ejected in the first part of the systolic flow is less than that ejected in the second part (the area under the curve is proportional to the amount of blood ejected). See the explanation in the text.

Figure 5.5 represents a blood flow curve recorded at the root of the aorta during a cardiac cycle. Each point in the trace indicates a velocity. To transform it in an instantaneous flow value expressed in mL/min, we need to know the internal vessel radius and, therefore, the sectional area. Of note, in Figure 5.5 the aortic flow curve is synchronized with the aortic and ventricular pressure curves. It can be seen that during the diastole and the isovolumic systole, the aortic flow is equal to zero. When the ejection begins, the flow increases together with the pressure reaches its maximum value a little before the maximum pressure value. It ends when the pressure *incisura* appears on the pressure curve that corresponds precisely at the end of the ejection and the closure moment of the aortic valve (protodiastole). At the instant corresponding to the *incisura*, the flow curve has a very small downward deflection, determined by the movement of blood which at this moment occurs from the ascending aorta towards the aortic valves (*backflow*). This movement is immediately stopped by the closing of the valve. From this time until the beginning of the next ejection phase, the aortic flow is again equal to zero.

If we draw a vertical line from the highest point of the aortic flow curve, we divide the area underlies into two parts. As you can easily see from Figure 5.5 the left side has a smaller area than the right side, meaning that in the first part of the ventricular ejection a smaller quantity of blood is introduced into the aorta than that introduced in the second phase.

Box 5.4 The Analysis of the Aortic Flow Curve During Ejection

The analysis of the aortic flow curve during ejection (Figure 5.5) renders inadequate the terms maximum and reduced ejection phase as well as rapid and slow ejection. Indeed, during ejection, the blood velocity varies continuously during each of the two ejection phases. Since the only parameter that distinguishes each of these phases is the increase or decrease of the velocity of the blood, it is appropriate that the first phase is defined as *the ejection with acceleration* and the second phase as the *ejection with deceleration*.

It should, however, be kept in mind that, when the peripheral resistances are sufficiently high, the value of the aortic systolic pressure can be reached when the ventricular ejection speed has already begun to decrease. We can ask: *why the maximum value of the aortic flow does not coincide with the maximum value of the pressure, but the second is delayed compared to the first?* The most likely hypothesis is that on the curve of the pressure wave a *reflected wave component* is summed to the *forward wave*. The pressure wave travels forwards and backward (reflected wave) and it travels faster than blood. The forward wave is reflected at the level of bifurcation and resistance vessels so that a reflected wave can be inserted and summed at various levels of the original forwarding wave. In this case, a reflected component results in a further increase in the pressure at the moment of the ejection phase with deceleration.

The SV can be recorded with the use of eco-Doppler systems or electromagnetic flowmeters placed around the ascending aorta. In both cases what is recorded is the instantaneous blood velocity expressed in cm/min. Since the instantaneous velocity varies during ejection, the graphical integration of an aortic flow curve during the cardiac cycle (including the diastole in which the velocity is zero!) allows obtaining the average velocity value during an entire cardiac cycle.

Multiplying the velocity value for the sectional area of the aorta expressed in cm^2 we obtain the flow expressed in cm^3/min or mL/min. Therefore, we can express the mean flow of the single beat with the same units of cardiac output (CO). If this CO (mL/min) is divided by HR, the value of a single SV (mL) is obtained. In other words, after conversion, each point of the curve represents a flow (mL/s or mL/min). If for absurd, the flow would be for the entire cycle equal to the level reached at the apex of aortic flow curve, we would have a huge CO (600 mL/s = 36 L/min) at rest. However, the aortic flow oscillates between zero and a maximum, and for this reason, in steady-state conditions, we can consider the mean flow of the single beat (diastole and systole) as the actual CO (if at each beat the same SV is ejected CO = SV × HR).

5.7.1 The Sounds and the Heart Murmurs

Placing a stethoscope on the anterior surface of the chest, the rhythmic repetition of two *heart sounds or heart tones* can be heard. The *first tone* or *first sound* is due to the *closure* of the atrio ventricular valves while the *second sound* is generated by the *closure* of the aortic and pulmonary valves. Consequently, while the first sound coincides with the isovolumic systole and is called the systolic tone, the second sound coincides with the protodiastole and is called the diastolic tone. The first sound lasts 0.10–0.17 s and has a frequency of 25–50 cycles per second (c.p.s.), while the second has duration of 0.10 s and a frequency of 50–60 c.p.s (Figure 5.6).

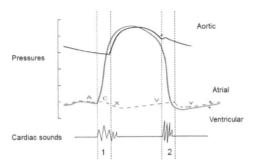

Figure 5.6 First and second heart sounds and their relationship with the phases of the cardiac cycle.

As a consequence of the *valve closure*, vibrations of the structures on which their leaflets are inserted, *i.e.,* the ventricular myocardium for the AV valves as well as the walls of the aorta and pulmonary artery for the respective valves generate the sounds.

For this reason, the first sound is well audible at the apex of the heart, where the left ventricle comes in contact with the chest wall, while the aortic and pulmonary components of the second sound are perceived along the aorta and the pulmonary artery for the respective components. Finally, it must be said that the first sound also contains a small muscular component, due to the vibrations of the contracting myocardium. Therefore, the sites of better perception of cardiac sounds on the chest are obtained in those points that are called *auscultation points (see below)*. These do not correspond to the points of anatomical projection of the valves, but to the points in which vibration is better transmitted. The first and second sound is usually represented as lubb-dupp onomatopoeic sounds followed by a pause, in a waltz-like time. The *lubb* (first sound) is heard after the long pause (diastole) and the dupp (second sound) is heard after the shorter pause (systole). In Italian, the two sounds are described as dum-dau.

With a device called a *phonocardiogram (a microphone placed on the thorax precordium)*, in addition to the first and second sounds, it is also possible to record a third and a fourth sound, which are sometimes audible. The *third sound* is produced during the rapid ventricular filling phase by the vibrations generated in the ventricular wall by the fall of blood from the atria. The *fourth sound* is attributed to the vibrations generated in the atrial systole both by the tensioning of the muscular walls and by the acceleration and vortexes of the blood that is forcefully pushed into the ventricles. Therefore, the fourth sound immediately precedes the first sound of the next cardiac cycle.

During auscultation, the perception of the third and fourth sounds is rare. In young and fragile subjects, it is sometimes possible to perceive one or, much more rarely, two of these sound. However, even when the fourth sound has an intensity that can be perceived, it can be difficult to recognize it because it is fused with the first sound that follows it immediately. In this case, it is said that the first sound also consists of an atrial component.

In patients, all four sounds can be detected at the auscultation in conditions characterized also by a very high HR. The rhythm that is perceived is the so-called *gallop rhythm*, which is often considered to be a sign of the dangerous poor myocardial condition.

As mentioned above the second heart sound is due to the closure in a sequence of the aortic and pulmonic valves. The former, in physiological situations, occur before the pulmonic component due to the rapid closure of the aortic valve caused by high aortic pressure. When these sounds are distinguishable from each other a split second sound can be heard. The trend of the splitting of this sound varies physiologically during the respiratory cycle, being the pulmonic component a little delayed in exhalation and a little more delayed in inspiration compared to the aortic component. Under pathological conditions, we can observe an *enlarged splitting* (*e.g.,* in pulmonary hypertension or blockade of the right branch) or a *paradoxical splitting* (*e.g.,* in case of the blockade of the left branch the aortic component occurs after the pulmonic one), and *fixed splitting* (*e.g.,* atrial septal defect). A *fixed splitting* very often indicates an atrial septal defect. In this condition, there is always a delay in the closure of the pulmonary valve and there is no further delay with inspiration. This occurs because during inspiration, as usual, there is an increase in venous return to the right ventricle and thus increased stroke volume through the pulmonic valve delaying its closure. As the person expires, there is less venous return in the right atrium and increased venous return in the left atrium, therefore the pressure gradient between the left and right atria increases. The increased gradient allows more blood to flow through the atrial defect from the left atrium to the right atrium resulting again in increased inflow into the right ventricle and increased stroke volume through the pulmonary valve, again delaying its closure as in inspiration.

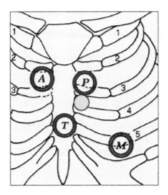

Figure 5.7 *Auscultation points.* M: mitral component of the first tone; T: tricuspid component of the first tone; A: aortic component of the second tone; P: pulmonary component of the second tone. See the explanation in the text.

5.7.2 Auscultation Points (Figure 5.7)

(a) for the mitral component of the first sound, the better auscultation point is at the left V intercostal space about 1 cm from the hemiclavic line. In this point, the apex of the ventricles vibrates following valve closure; for the tricuspid component the area above the xiphoid process is the better auscultation point;

(b) for the aortic component of the second sound, the better auscultation point is the second intercostal space immediately to the right of the sternum, *i.e.,* at the point where the ascending aorta approaches the chest wall, as it initially heads to the right;

(c) for the pulmonary component of the second sound, the better auscultation point is the second intercostal space immediately to the left of the sternum, where the pulmonary artery comes into contact with the chest wall, as it heads to the left almost wrapping the origin of the aorta.

5.7.3 Murmurs (Figure 5.8)

In pathological conditions, it may be possible to hear the heart murmurs. They are systolic murmurs if they fall between the first and second sounds, while they are diastolic murmurs if they fall after the second sound.

Murmurs are audible when vortexes are produced in the passage of blood through the damaged heart valves. Audible vortices can be generated both in the passage of blood through stenotic valve orifices and in regurgitation through valves that fail to close completely (incompetent valves).

The *systolic murmurs* can be due to:

(a) incompetence or insufficiency of AV valves (mitral or tricuspidal incompetence), the murmur appears in early systole when the contraction of the ventricle causes a certain amount of blood to return to the atrium. The murmur can be holosystolic (*i.e.,* can be heard for the entire systole). Murmurs due to defects of the tricuspid valve are very rare.

(b) stenosis of semilunar valves (aortic or pulmonary stenosis), the murmur appears in meso-systole (the meddle part of systole) when intraventricular pressure reaches a peak and the ejection of the blood generates vortexes passing through the stenotic valve. Pulmonary stenosis is often due to congenital heart disease and a meso-systolic (midsystolic) murmur can be appreciated in newborns.

The *diastolic murmurs* can be due to:

(a) semilunar valve insufficiency (aortic or pulmonary incompetence), the murmur appears early in diastole for regurgitation in the ventricle of part of the just ejected blood.

(b) stenosis of AV valves (mitral or tricuspidal stenosis), the murmur (rumble) appears in diastole and is reinforced at the end of the diastole (telediastolic murmur), i.e., when the atrial systole pushes blood forcefully through the stenotic valve.

5.8 Cardiac Volumes and Ejection Fraction

The percentage of blood that leaves the ventricles at each systole is called ejection fraction (EF), it is calculated by stroke volume (SV) times the quantity of blood contained in the ventricles at the end of the filling phase, which is the ventricular end-diastolic volume (VEDV) or ventricular tele-diastolic volume (VTDV).

$$EF = SV/VEDV \times 100$$

Therefore, only part of VEDV is ejected to form the SV. The VEDV is composed of the systolic residual volume (SRV) plus the ventricular diastolic return volume (VDRV). In the presence of a constant hemodynamic situation,

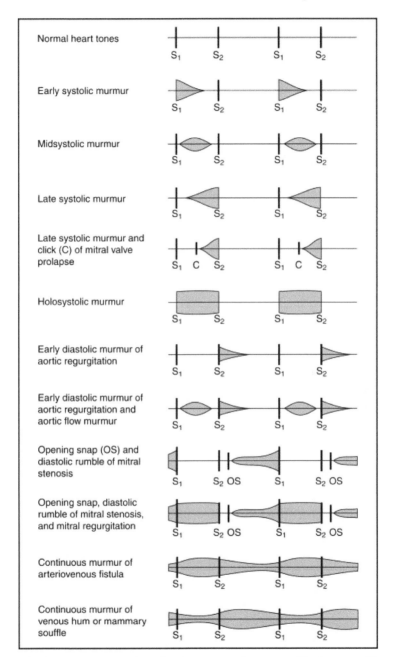

Figure 5.8   Graphic representation of various types of murmurs. S1: first tone; S2: second tone.

the VDRV corresponds exactly to the SV. Therefore,

$$VEDV = SRV + VDRV = SRV + SV$$
$$VDRV = SV$$

The ejection fraction is not the same for the two ventricles. While for the left ventricle it is equal to 50-60% of the VEDV, for the right ventricle it has a greater value (55–65%) due to a lower VEDV, in the presence of an almost identical SV.

The ejection fraction is a good index of contractility. Indeed, EF generally increases and decreases respectively with the increase and decrease in contractility.

In the clinic the EF of the left ventricle is carefully studied, the knowledge of which allows the diagnosis and the quantitative evaluation of heart failure. Often a reduction in contractility with a reduced EF may have a normal SV at rest. In this case, the VEDV has increased. The mechanism by which this compensated heart failure is established begins with a reduction in SV from loss of contractility. Then the blood that is not pumped with the systolic flow remains in the ventricle where it determines an increase of the SRV and, consequently, of the VEDV if VDRV is unchanged. As we saw in Chapter 4.2, an increase in VEDV involves stretching of myocardial fibers which allows them to develop greater contraction force. It can then achieve a return to the norm of the SV in the presence of an increase in the VEDV. If the systolic function is then calculated as the percentage of VEDV ejected, namely EF, we see that EF is decreased. We must say that EF is a good, and not the perfect index of contractility (see below). Moreover, in Chapter 6 we will see that heart failure with preserved EF exists.

5.9 The Ventricular Pressure-Volume Loop

The *ventricular pressure-volume loop* can be constructed using a Cartesian coordinate system in which variations of the ventricular volume are reported on the abscissa axis and those of the pressure on the ordinate axis (Figure 5.9). Point A of the loop represents pressure and volume values after a ventricular filling is completed. This volume is the VTDV or VEDV.

During the isovolumic systole, the pressure rises from A to B, without any variation in volume. Point B represents the end of isovolumic systole and the beginning of ventricular ejection. During this phase, the volume decreases

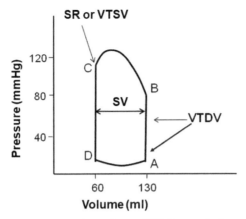

Figure 5.9 Ventricular pressure-volume loop. VTDV: ventricular tele-diastolic volume; VTSV: ventricular tele-systolic volume or SR: systolic residue; SV: Stroke volume. See the explanation in the text.

from B to C with a curved shape of the pressure trend during the ejection phase.

At point C the ventricular volume reached the minimum value of the whole cardiac cycle. From this moment and throughout the subsequent isovolumic relaxation phase, the ventricle contains only the SRV also called VESV or VTSV. During the isovolumic relaxation, the pressure falls to point D.

The isovolumic relaxation is followed by the three phases of the ventricular diastolic filling (rapid filling, slow filling, and atrial systole) that bring the loop back to point A, i.e., to the VTDV. As we will see later, the area included within the loop represents the work that the ventricle performs in moving the SV into the aorta, giving it the pressure that is reached during ejection. Practically, the heart's potential work ($P_x V$) can be obtained by multiplying the volume (V) of blood ejected, namely SV, with the aortic mean pressure (P) during the ejection phase. This is the only phase in which the heart performs external work since it sets in motion a mass of blood at a certain speed. Therefore, the kinetic component ($1/2mv^2$) must also be added to the potential component of the work. However, as we will explain in Chapter 9, for the left ventricle, the kinetic component (≈ 2 percent) is usually negligible at rest and becomes more important during physical activity. For the right ventricle, the kinetic component is ≈ 12 percent at rest.

The end-systolic pressure-volume relationship (ESPVR), which can be obtained connecting the end-systolic point of different loops, allows calculating an optimal contractility index, namely the end-systolic-slope (ESS, the

PRESSURE/VOLUME RELATIONSHIPS UNDER DIFFERENT CONDITIONS

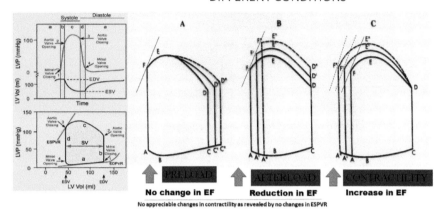

Figure 5.10   Left ventricular pressure-volume loops can be derived from pressure and volume traces recorded during the cardiac cycle. The numbers (1, 2, 3, and 4) in the left diagrams indicate the same moments in the pressure and PV loop traces. The end-diastolic pressure-volume relationship (EDPVR) is an index of ventricular stiffness. An upward shift of EDPVR is observed in stiff ventricles.

slope of ESPVR). ESPVR delimits the maximal energy that can be generated at a given inotropic condition whatever is the preload and afterload. Being independent of pre- and after-load, ESS is a better contractility index than EF. Although contractility is unchanged, EF is reduced at higher afterload (Figures 5.10 and 5.11).

In brief, the pressure-volume loop allows us to graphically highlight and compare the various values of cardiac volumes and their variation during the cycle, as well as to calculate EF and to evaluate the contractility of the heart.

PV loops can be studied changing preload, afterload, or contractility. Only variations of contractility are accompanied by shifts of the ESPVR to the left (increased contractility) or the right (decreased contractility). In the PV loops A: correspond to mitral valve opening, B: the beginning of the slow filling phase, C: mitral valve closing, D: aortic valve opening, E: peak systolic pressure, F: end-systole and aortic valve closure. End-systolic pressure-volume relationship (ESPVR) can be obtained connecting the end-systolic points of the beat with the same contractility. ESPVR shifts to the left and upward when contractility increases or to the right and downward when it decreases.

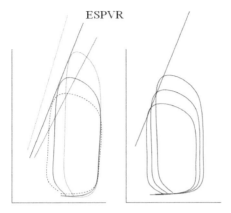

Short-term changes to the afterload (or preload)
can be used to generate a series of pressure-volume loops.
The slope of the ESPVR is an index of contractile status.

Figure 5.11 PV loops obtained changing the afterload and maintaining the same preload (left) or changing the preload and afterload (right): the ESPVR does not change (black line). In other words, the contractility does not change appreciably changing preload and/or afterload.

If contractility decreases the ESPVR will shift on the right (blue line) or if contractility increases the ESPVR will move on the left (red line). In these cases of contractility variation, the end-systole of PV-loops (not shown in the figure) would touch the corresponding ESPVR. In other words, the pressure-volume loop cannot cross over the ESPVR because that relationship defines the maximal energy that can be generated at a given inotropic condition (*see also Figure 5.10*).

6

Cardiac Output and the Venous Return to the Heart

6.1 The Cardiac Output

Cardiac output (CO) is the amount of blood that a ventricle pumps in one minute in the large artery. CO is the product of the stroke volume (SV) and heart rate (HR), that is the product of the amount of blood that a ventricle pumps in a beat (SV $\approx$70 mL at rest) multiplied by the number of beats in a minute (HR $\approx$70 b.p.m. at rest):

$$CO = SV \times HR$$

Studying the cardiac cycle we have seen that, with an HR of about 70 b.p.m., about 70–75% of the blood that will form the next SV returns to the ventricles in the rapid ventricular filling phase, while the remaining 25–30% returns during the slow filling phase with a little contribution of the atrial systole. We have also seen that variations in the HR lead to variations in the duration of diastole and in particular to the slow filling phase. Therefore, a variation in frequency is accompanied by variations in the opposite direction of the ventricular filling and SV. It follows that increases in HR do not lead to a parallel increase of CO. The possible relationships between variations in HR and variations in SV have already been described in Chapter 5.5 and will be recalled studying the arterial pressure (Chapter 8).

In a 70 kg resting subject the CO is about 5 L/min. Cardiac output may be reduced by about 20% in the transition from a clinostatic position (lying down person) to an orthostatic position (standing person). This CO reduction, which is accompanied by a decrease in blood pressure, is due to an increase in blood "hosted" in the veins and therefore subtracted from the venous return (*pooling effect*). The pressure reduction is immediately corrected by a reflex increase in sympathetic activity which leads to an increase in cardiac contractility with the use of a part of the systolic residue. Increased sympathetic activity will decrease the capacity of the veins thus improving the venous return and SV as well as the HR with consequent CO enhancement. This will be discussed in Paragraph 6.2. The reflex modalities will be described in Chapter 12 together with the nervous control of the cardiovascular system.

Given the organization of the cardiovascular system, we can say that, in steady-state conditions, for a ventricle the CO is equal to the venous return. Cardiac output and venous return vary physiologically in many conditions. For example, CO can increase slowly by 40–45% during pregnancy and can suddenly increase by 20–40% during excitement or fear. During exercise, CO can increase up to 25 L/min in a sedentary person and 30–35 L/min in Olympic champions. During acute changes, CO and venous return may be transiently different, but in a few seconds, the venous return and CO will be identical and remain so during a steady-state condition. Blood entering the ventricle must be pumped!

6.2 Factors Determining the Venous Return to the Heart

As said before, venous return to a ventricle is the CO of that ventricle: In other words, venous return and CO refer to the quantity of blood that goes in and goes out in one minute, simply observed in veins and arteries, respectively.

Note that, the VDRV is the amount of blood that enters a ventricle during single diastole. As already mentioned in Chapter 5.5, the VDRV is equal to the SV in steady-state conditions. VDRV (mL) is different from the venous return (mL/min), as SV (mL) is different from CO (mL/min).

The principal factor that determines the venous return is the *vis a tergo*. It is the pressure gradient between the blood pressure that enters the venules from the capillaries and the central venous pressure (CVP). Strictly speaking, CVP is the pressure in the big veins and right atrium. The pressure gradient is mainly due to the contractile activity of the heart, which, by pumping blood into the circulation ends up being the main cause of its filling.

Other factors affect this gradient and the venous return to the heart: the *muscular pump* especially in the lower limbs, the *depression of Donders* and, the *gravity* that influences the return to the heart from the districts located above or below the heart of humans in an upright position. The *vis a tergo* is the only factor sufficient to determine a venous return in a resting and laying subject, in the absence of all the other factors that can affect venous return.

The *muscular pump* is represented by the massage action that contracting skeletal muscles determine on the veins within the muscle mass. The veins tend to relax (they are *capacity vessels*) due to the blood pressure inside them. In an upright individual, this pressure is particularly high in the veins of the lower limbs due to a hydrostatic component which can cause vein distension, resulting in an increase in the vein capacity and the blood contained in them. This is described as *venous pooling* and is the amount of blood (about 500 mL) subtracted from the venous return adopting a standing position, in the presence of compensations. The venous expansion by the hydrostatic pressure is limited by the contraction of the large muscles of the lower limbs, which, acting from the outside on the veins, push the blood upwards. For this blood to go upward, small muscle contractions are necessary, as those occurring shifting the weight of the body from a lower limb to the other. This avoids the venous stagnation of the blood and favors its progression towards the heart.

The muscle pump is particularly active during exercise when it increases CO. This will be discussed in Chapter 8.11. The progression to the heart of the blood contained in the lower venous districts is also possible due to the presence of venous valves (*dovetailed valves*) along the veins. The valve placed between the *saphenous* and the *femoral veins* is of particular clinical importance. When this valve is insufficient, the blood tends to stagnate in the *saphenous vein* which in the lower part, corresponding to the lower third of the medial surface of the leg, is distended and presents alterations of the wall. In this way, local pathological dilatations, called *varices* are formed, in the genesis of which two physical principles are important, namely the Laplace's Law and the Bernoulli's theorem (see BOX 6.1 and Figure 6.1).

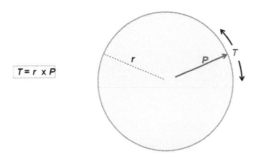

Figure 6.1 Laplace's Law. See the explanation in the text.

Box 6.1 Laplace's Law and the Bernoulli's Theorem Explain the Varices and Aneurysms

According to the *Laplace's Law* (Figure 6.1), in a hollow structure, the circumferential tension (T) depends on the product of the radius (r) for the dP (internal pressure, iP, minus the external pressure, eP). However, we can forget the eP as it is usually low. We consider the only iP and call it just P. The formula reported in Figure 6.1 shows that for certain blood pressure, P, an increase of r causes an increase of T. This increase in T tends to make the wall of the vein to expand thus further increasing r. In this way, a vicious circle is determined so that varices expand progressively more rapidly.

Theoretically, in a steady flow, the sum of all forms of energy in a fluid along a streamline is the same at all points on that streamline. According to the *Bernoulli's principle*, in a conduit, the sum of the kinetic energy ($1/2\ mv^2$) and the lateral pressure (P) is constant: $1/2mv^2 + P = k$.

If along the conduit there is an enlargement (a varice) at that point there will be a decrease in the velocity, *v*, of blood flowing and therefore of the *kinetic energy* with consequent increase of P. This increase in lateral pressure causes an enhancement in the circumferential tension which in turn will further increase the volume of the enlarged duct. This increase in volume will result in a further increase in P with the exacerbation of the situation. It follows that, together with the *Laplace's law*, the *Bernoulli's theorem* also explains the increase in the volume of varices rapidly increasing. What has been said for the volume increase of venous varices can also be applied to *arterial aneurysms* with more dramatic effects because of higher pressure. Even if a rigorous application of *Laplace's law* concerns infinitely thin walls and a perfectly circular shape, with a certain approximation it can still be applied to the hollow organs. In the case of aneurysms and varicose veins, it may explain the progressive and increasingly rapid increase in volume, especially if it is seen together with *Bernoulli's principle*. If there is an expansion in size or aneurysm, though small, along an artery, there is a decrease in kinetic energy at that point. By *Bernoulli's principle,* there will be an increase in P, which, according to *Laplace's law*, will produce an increase in the circumferential tension T which will further increase the radius of the artery. At this point, a *vicious circle* is created with a further increase of the radius, of the lateral pressure, and the circumferential tension. The increase in aneurysm volume can be faster until its rupture.

The presence of varicose veins in the lower limbs, limiting the venous return to the heart, can cause a reduction of CO and pressure drop, especially in the passage from the clinostatic to the orthostatic position. At this time the patient may experience annoying dizziness with transient visual obfuscation. People with varicose veins can avoid these disorders by bandaging their legs with elastic bandages before getting out of bed, which is when the varices are not yet distended by hydrostatic pressure. The bandage must be done before the change of position. If it is performed after the patient has passed into the orthostatic position, the bandage could trap a certain amount of blood in the lower part of the leg, subtracting it from the venous return.

The flow of the venous blood to the heart could also be favored by a decrease in venous capacity due to a constriction of these vessels determined by increased sympathetic activity. Thus, sympathetic activation decreases the compliance of veins, promotes venous return, and increases CVP.

Box 6.2 The Fainting Royal Guard

For *muscular pump massage* to take place, imperceptible muscle contractions are necessary. In a subject in an orthostatic position (*i.e.,* the position of a Royal guard or a speaker) it is sufficient to shift the weight of the body from a lower limb to the other to avoid the venous stagnation of the blood and to favor its progression towards the heart. It is known to all that the British royal guards, standing at attention, must be motionless for hours. To ensure adequate venous return, it may be sufficient to contract the muscles of the lower limbs imperceptibly under the wide trousers. Of course, these soldiers do not need to know physiology, as muscle contraction has become spontaneous for them or suggested by experience. Young recruits sometimes faint because their muscles are inactive and unable to guarantee an adequate venous return and, therefore, an adequate CO, causing a drop in blood pressure, especially in the hot season which also induces the vasodilatation. In summary, gravitational venous pooling and heat-induced vasodilation compromise venous return and CO, leading to low blood pressure and cerebral hypoperfusion and ... hop ... embarrassing fainting of the soldier. In this case, we must not lift the soldier but his legs to favor the venous return, the CO, and the awakening of the soldier.

The *Donders' depression* is the sub-atmospheric pressure present in the pleural cavity. Donders' depression is not constant over time but varies with the respiratory cycle, going down to -8 cmH$_2$O at the end of inspiration and rising to -5 cmH$_2$O at the end of expiration. During inspiration, the greater depression of intrapleural pressure allows an enlargement of intrathoracic veins, while the diaphragm is going down increasing the pressure in the abdomen, where the pressure increases and exerts compression on intra-abdominal veins. These two factors favor venous return. These intermittent changes in intrapleural and intra-abdominal pressures are the cause of those changes in venous return that are responsible for the second-order oscillations in blood pressure (see Chapter 8.9). The effect of Donders' depression on the venous return can decrease if the intrapleural pressure is increased by the Valsalva maneuver, which consists of a forced expiration at closed airways. The reduction of the venous return that is obtained determines a reduction of the SV that can be detected with a corresponding decrease in the arterial pressure and smaller amplitude of the radial pulse (see below).

Gravity: as already mentioned, the force of gravity favors the venous return from circulatory districts located above and obstacle the venous return from those districts located below the heart. Given the upright position, it is of some importance to humans. Since the venous districts located below the heart are greater and are responsible for the *venous pooling* (about 500 mL of blood) in a standing person the venous return and CO tend to decrease. Despite the limitation of the pooling effect by reflex adjustment of venous tone and skeletal muscle massage, CO is 10–20% lower in standing than a lying down position. In a fainting subject, the *Gravity* can be exploited to promote venous return from districts below the heart. Indeed, in the event of a rapid drop in blood pressure due to sudden peripheral vasodilation, the patient may experience dizziness with visual blurring and may faint. In this case, the patient can be placed in a supine position, taking care that the lower part of the body is at a higher level than the heart. This is the *Trendelenburg position* in which the head is at a lower level than the pelvis. Favoring the venous return, we may favor the restoration of an adequate CO.

6.3 The Central Venous Pressure

The term central venous pressure (CVP) refers to the pressure in the right atrium. Since the venous blood, returning from the systemic circulation to the right atrium, passes freely from the cava veins to the atrium, the pressure present in these cavities is similar and depends on the venous return and cardiac activity. When CVP increases due to increased venous return, it is said that CVP increases in an anterograde way. CVP can also increase in a retrograde way if there is a loss of contractility of the ventricles or ventricular failure. If venous return increases, the CVP also increases, resulting in greater filling of the right ventricle, which will pump more blood into the pulmonary circulation. This results in a greater return of blood to the left atrium and ventricle with a subsequent increase of the ejected blood in the systemic circulation.

When the left ventricle fails, it pumps a smaller SV into the aorta and a greater systolic residue is present in the ventricle. As we will see in more detail in the study of the heart-lung preparation of Starling (see Chapter 7.1), an increase in the systolic residue, with unchanged venous return, is accompanied by a greater VEDV and a higher ventricular tele-diastolic pressure, responsible in turn for a blood pressure rise in the left atrium. Through the pulmonary circle, this greater pressure propagates backward to the right atrium where it can cause an increase in CVP.

The left ventricular diastolic pressure is almost equal to the pulmonary vein pressure. Similarly, CVP governs right ventricular diastolic pressure. These four pressures are similar (not identical) and may collectively be called "filling pressure". If the venous return increases towards a ventricle, the corresponding filling pressure increases first, then the systolic flow increases, and the increase is propagated antegrade to the other filling pressures until a new steady-state is gradually reached.

6.4 The Cardiac Output and the Venous Return

Bearing in mind that in steady-state conditions the cardiac output is equal to the venous return, let us now look at Figure 6.2. The CVP (or filling pressure) is on the abscissa axis. On the ordinate axis, the venous return and the cardiac output are reported. To make Figure 6.2 comprehensible, it is necessary to remember: *1)* the CVP can increase in an anterograde way by increasing the venous return or retrograde way for ventricular insufficiency; *2)* the CVP opposes the venous return, for hydraulic reasons, but it favors the increase in cardiac output by the Frank-Starling's mechanism.

According to Figure 6.2, the normal venous return and CO of 5 L/ min correspond to the point indicated with the letter A when CVP is about 2 mmHg.

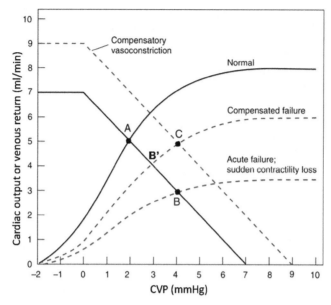

Figure 6.2   This figure displays how cardiac output (CO) varies in response to changes in CVP in normal conditions (black curve) as well as how it can drop during an acute failure of contractility (red dashed curve) and recovery after a while (blue dashed curve). In the meantime, venous return is displayed in normal conditions (solid black line). Venoconstriction and/or liquid retention may shift the venous return to the right and upward (dashed green line). So that CO and CVP cross at A point in a resting healthy person. Since the venous curve is not changed when contractility drops, the working point shift from A to B. After a partial recovery of contractility, it moves to B' and venoconstriction allow a new equilibrium (C point) in which the CO and venous return are equal to the control A, but CVP is higher. This is a compensated heart failure.

The *cardiac output* and *venous return* should be considered together considering that *the reduction in CVP favors venous return* by hydraulic effects (increase the pressure gradients between peripheral venous pressure and CVP), whereas *the increase in CVP favors cardiac output* by Frank-Starling' mechanisms (heterometric regulation, see Chapter 7). For the didactic purpose, we consider first the two curves separately.

Let us try to consider first the *cardiac output curve* at rest (Starling normal curve) and we see how it can vary in response to changes in the anterograde way of CVP. If CVP increases in an anterograde way, the CO increases for each subsequent CVP increments. However, the CO does not increase linearly, since, after the CVP has reached a certain value, CO remains constant at a moderately high level (about 7–8 L/min). When CVP decreases,

CO decreases as the ventricle is less filled. The greater the CVP, and the ventricle stretches (within certain limits), the greater the CO. This is the law of the heart! (see Chapter 7).

If now consider the *venous return*, we see that it can vary in response to variations in retrograde CVP. For high levels of CVP, the *venous return* starts to decrease. For low levels of CVP, venous return increases, and at very low levels of CVP, venous return is maintained at high values that do not change further with the reduction of CVP. This is in part because at negative pressures the veins collapse increasing the resistance that obstacle the venous return. At this point, venous return is even lower that to the maximum of cardiac output (about 8 L/min) that we can observe on the Starling curve of CO when the CVP is increased in an anterograde way at high levels. Fortunately, we know that CO can reach much higher levels in physiologic conditions since the factors that can influence CO are not only dependent on CVP. Indeed, ***the four determinants of cardiac output are heart rate, contractility, preload, and afterload*** (within certain limits an increase in HR, contractility, and preload favor an increase in CO, whereas an increase in afterload hampers the CO). About HR, we must consider that its increase tends to decrease preload and increase afterload (the time to fill ventricle and to lower aortic pressure is shorter) and, therefore to decrease stroke volume (SV). Nevertheless, if the increase in HR is accompanied by an increase in contractility the SV tends to increase and the CO increase is favored.

If the contractility of the left ventricle decreases, the CO curve of the graph moves downwards (dotted red curve), to cross the venous return curve at the point B corresponding to a higher CVP (increased retrogradely). Point B reveals that, in these conditions of decreased contractility, cardiac output and venous return are reduced to the same value. Because if the ventricle becomes insufficient the SV decreases and the systolic residue increases, at the end of each diastole there will be an increase in left VEDV and, in a retrograde way, also an increase in CVP.

If, on the other hand, the contractility of the ventricle increases a little for adrenergic activation, the CO curve moves upwards (dotted blue curve). We can say that the increased use of the systolic residue decreases the CVP to a value that allows an increase in venous return. Indeed, for this new CVP value, the CO and venous return curves intersect at a point B', which indicates that the two variables (CO and venous return) are increased by the same value.

Changes in contractility have shifted the systolic flow curve, while the *venous return curve* has maintained unchanged. On this curve, the point of intersection with the CO curve has simply been moved (from A to B or from

B to B', in the case of drastic reduction and partial recovery in contractility, respectively).

At this point, we can imagine increasing the volume of circulating liquid (Figure 6.2). This can be achieved with a blood transfusion or infusion of a plasma expander, or for water retention. A similar effect can be observed for venoconstriction. If the cardiovascular system contains more fluid, or the venous system is constricted, venous return is higher (increases) for each CVP value. In other words, the venous return curve will move upward (in Figure 6.2 dotted line for compensatory vasoconstriction) and cross the CO curve into a higher C point corresponding to a higher CVP and a higher cardiac output. It can, therefore, be concluded that the venoconstriction and the increase in blood volume brought both venous return and CO to a higher new equilibrium value.

It is interesting to note that, in the presence of an increased volume of liquid or case of venoconstriction, the venous return varies according to a new curve located in a higher position than the control one, while the CO curve position in the graph depends on myocardial contractility. The venous return curve, on the other hand, moves downwards if the volume of circulating fluid is reduced (not shown in Figure 6.2), which can occur in case of bleeding.

Ventricular insufficiency may be accompanied by water retention and venoconstriction with a high displacement of the venous return curve so that the intersection point of the two curves moves to point C (Figure 6.2). In this way, the cardiac output returns to have almost the same value that it had before the insufficiency was established. This situation represents a *compensated insufficiency, i.e.,* in the resting condition CO is not reduced, however, the increase in CVP (in comparison to the value it had in normal condition, A point) reveals the presence of diminished contractility despite the unchanged CO.

Note that the above is only a theoretical simplified description. In Figure 6.2, simple integration of the contractility curve and venous return curve is reported in a situation of acute heart failure, as could occur after myocardial infarction and compensation. In the reality, when the contractility or the filling of the venous system varies, transiently the two values (cardiac output and venous return) do not coincide, and then they move slowly to the new equilibrium point, which represents a new situation of steady-state. We can, for example, say that point B is the transient condition after a myocardial infarct and that several intermediate points between points B and C may occur before that a new equilibrium is reached after certain vasoconstriction, water retention and a partial recovery of contractility occurred. ***This theoretical***

approach reinforces the concept that, at a steady-state, venous return is the CO and vice versa CO is the venous return and that the CVP value is the result of the new equilibrium point. From the above discussion, we also understood that the variation of a single variable (contractility or hydric retention) is largely ineffective in changing CO, which reaches a modest value that cannot increase further. *To have an important increase of CO (e.g., 20 L/min), several cardiac and vascular variables must vary together, for example, contractility, heart rate, venoconstriction, skeletal muscle massage, etc., as we will see in analyzing cardiovascular response during exercise* (see Chapters 8.11 and 12.5).

The reader must understand that *the above curves-analysis does not reveal how the cardiovascular system behaves.* It is only a didactic approach, which aims to reinforce the concept that the cardiovascular system has a capacitive system (mainly the veins), a resistance system (mainly the arterioles), and a pump, but is too simplistic. For example, the model is not adequate to analyze exercise adjustments as during exercise there are several mechanisms such as a decrease in systemic vascular resistance, an increase in cardiac output, a redistribution of cardiac output, nervous reflexes, and humoral readjustments that the simple series model takes in account only partially. The above model may be even more inadequate in heart failure, which is a complex, multifactorial and progressive syndrome, presenting under a large spectrum of dissimilar or overlapping phenotypes and it is often accompanied by one or more co-morbidities.

6.5 The Cardiac Output in the Compensated and Decompensated Heart Failure

Heart failure (HF) can be compensated or decompensated. While in compensated HF the CO at rest is normal, in the decompensated HF the CO is reduced even at rest. What allows recognizing an HF, even in the presence of normal CO at rest, is the presence of an increased CVP (Figure 6.2). Reduced EF can also be present. Moreover, most importantly a failing heart does not adapt the CO to the effort.

We have seen how CVP can increase in an anterograde way by an increase in venous return and a retrograde way for ventricular failure. While in the first case the increase in CVP results in an increase in SV and cardiac output, in the second case it is the consequence of their reduction and represents a limiting factor for venous return.

We have also seen that, in the presence of contractile insufficiency, the contractile curve (Starling curve) that links the CO with the CVP moves downwards (Figure 6.2). In the absence of compensation, the venous return curve does not undergo any displacement, the equilibrium point between venous return and CO will shift from A to B (Figure 6.2), i.e., down and to the right. In other words, the reduction of contractility determines a reduction in CO and an increase in CVP, which indicates that an increase in the systolic residue and the VTDV occurs in the case of reduced SV. Due to the intervention of the Starling mechanism, the increased preload (VTDV or VEDV) may bring back to the normal level the SV and CO, even if the heart contractility has recovered only partially by sympathetic stimulation (C point in Figure 6.2). Due to the increased VTDV, the EF is reduced despite the presence of a normal SV. The condition that is reached is a compensated HF with reduced EF. Another feature of compensated HF is the turgidity of the superficial veins of the neck due to the retrograde effect of the increase in CVP. Decompensation can be revealed during an effort because contractility cannot increase adequately and a further increase in VTDV is not accompanied by an increase in SV (we are close to the plateau of Starling curve).

Patients with normal EF can have HF, and this form of the disease, heart failure with preserved (or normal) EF (HFpEF) is becoming increasingly prevalent especially in diabetes and women (see Box 6.3 in this Chapter).

Box 6.3 Heart Failure with Preserved Ejection Fraction and Heart Failure with Increased Cardiac Output

We have seen how HF depends on a reduction in contractility and can be compensated, with normal CO at rest, or decompensated, with reduced CO even at rest. In both cases, the EF is reduced.

In an apparent paradoxical way, both HF with increased CO and HF with preserved EF have been described.

The *heart failure with increased CO* may be present when, in addition to an increase in HR, there is also an increase in SV, as can occur, for example, in hyperthyroidism. A condition characterized by the presence of "shunts" between small arteries and small veins, which reduces the resistance to flow and avoids the passage of blood through the microcirculation. The persistence of increased CO leads to a chronic increase in heart work, which in the long run, causes a reduction in contractility. The CO can then remain increased compared to a normal situation, while the deficit is revealed by a reduction of the EF and *cardiac reserve* (inadequacy of the CO during an effort).

Patients showing signs of abnormal diastolic function (including abnormal relaxation, filling, or stiffness) also demonstrate EF values ranging more or less around 55%. These patients are often said to have *heart failure with preserved ejection fraction* (HFpEF). This condition is due to the lower distensibility of the heart walls, as can occur in pathologic ventricular hypertrophy. This reduction in distensibility limits ventricular filling and the corresponding VTDV (or VEDV). Although contractility may be unchanged, a reduced VTDV is responsible for reduced SV, and being EF a percentage of the VTDV that is ejected, it may remain unchanged. Pressure-volume loops demonstrate that patients with HFpEF have an abnormality of passive diastolic properties by an upward shift in *end-diastolic pressure-volume relation* (EDPVR, see Figure 5.10). This diastolic dysfunction results from increased LV stiffness because cardiomyocytes in patients with HFpEF are thicker and shorter than normal myocytes, and collagen content is increased. Moreover, reductions in myocardial capillary density with the concentric remodeling with or without hypertrophy have been observed in this condition. The HFpEF is often observed in diabetic patients.

6.5.1 The Valsalva Maneuver

Variations of venous return and CO can be obtained with the *Valsalva maneuver*. This can be done voluntarily with a forced expiration while holding glottis or nose and mouth closed. Thus there is a noticeable increase in intrathoracic pressure with the limitation of venous return. The variations of venous return and CO induced by this maneuver may be used to understand the cardiovascular regulation and the variation of murmurs during the various phases. The hemodynamic effect obtained can be divided at least into four phases (Figure 6.3).

In the first phase, there is a transient increase in arterial pressure produced both by the increase of intrathoracic pressure on the aorta and by blood squeezing from the pulmonary circulation to the left heart.

In the second phase, the arterial pressure decreases, in addition to its average value, also in its pulse pressure value. At this point, the reduced venous return to the right heart, due to the high intrathoracic pressure, has repercussions on the left heart with a reduction of the SV. Because the reduced

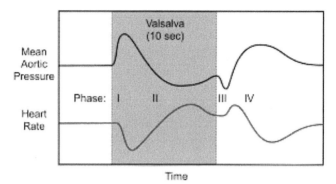

Figure 6.3   A schematic representation of the variation of mean aortic pressure and heart rate during the Valsalva maneuver. The grey part represents the time (10 s) in which forced expiration occurs while holding glottis closed.

blood pressure decreases the stimulation of the baroreceptors, in this phase, we also observe an increase in heart rate.

The third phase characterizes the end of the Valsalva maneuver. It consists of the further shortfall in pressure after the effect of increased intrathoracic pressure on the aorta has ceased and that the sudden re-expansion of the pulmonary circulation retains a certain amount of blood that does not return to the left heart.

In the fourth phase, there is a sudden increase in the mean and pulsating pressure caused by the increased SV after the sudden return to the norm of the intrathoracic pressure promotes the return to the heart of the blood that had accumulated in the venous section of the vascular system. In this phase, we also observe a reduction in heart rate for the increased stimulation of the baroreceptors (see baroreflex in Chapters 8.10.1 and 12.6).

7

Regulation of Cardiac Contraction Force

7.1 Intrinsic and Extrinsic Regulation of the Heart Contractile Force

In Chapter 4, the concepts of contractility and force of contraction of the heart, as well as those of preload and afterload were defined.

In short, and from a didactic point of view, an increase in the *contraction force* may be due to:

(a) an *increase in preload* due to the increase in the VEDV; it is by definition an *heterometric* mechanism;

(b) an *increase in contractility* (or *inotropy*) obtained by increasing the calcium levels in the cytosol of cardiac myocytes during an action potential and/or changing the sensibility of contractile proteins to the calcium. This is done by several factors such as sympathetic or inotropic agents acting on the heart. It is by definition an *homeometric* mechanism;

(c) both an increase in preload and an increase in contractility.

Furthermore, while the *preload* is given by the ventricular telediastolic volume and can also be expressed in terms of ventricular telediastolic pressure, the *afterload* is given mainly by the aortic pressure that is opposed to the ventricular ejection.

Taking in mind these definitions, we can speak of *intrinsic and extrinsic regulation of contractile force*.

The regulation is **extrinsic** when it requires external factors to the heart, such as hormones or nervous control. It can affect contractility and thus the force of contraction.

The regulation is **intrinsic** when it does not require external factors to the heart, but is due to the properties of the myocardium itself. It can be either *heterometric or homeometric*.

The intrinsic regulation is:

* *heterometric* when it depends on initial variations in the length of muscle fibers. This is at the basis of the *Starling's law (Frank–Starling mechanism);*
* *homeometric* when it does not depend on initial length variations of the fibers; such as in the *Anrep and Bowditch effects.*

7.1.1 Intrinsic Heterometric Regulation of Contractile Force

The *"law of the heart"* says that the force of contraction of the heart increases with the increase, within certain limits, of the ventricular telediastolic volume, that is, with the increase in the length of myocardial fibers in diastole. When it was first formulated, at the beginning of the XX century, this *length-tension relation* in the intact heart was against the previous theories that saw in cardiac dilation only a pathological significance.

The heart's law is also famous as the *Frank-Starling's law*, as it is named to two physiologists, *Otto Frank* and *Ernest H. Starling*. In 1895, Frank performed the experiments using the frog heart that was contracting isovolumetrically, that is, without producing any variation of ventricular volume and any SV during contraction. He changed the diastolic volume, but the volume of the heart remained unchanged during systole. This can be obtained inserting a balloon within the heart and filling it with different volumes of liquid. In 1912, Starling and collaborators started to study the effects of various physical parameters on heart performance (Figure 7.1) and they demonstrated that diastolic stretch influence the SV in mammalian hearts. In 1918 Starling published these discoveries as "The law of the heart". It seems, however, that a first formulation of the heart's law is due to the Italian physiologist *Dario Maestrini* who in 1914, began the experiments that led him to formulate the "legge del cuore" or "law of the heart". Thereafter, using the results of Frank, *Sydney W. Patterson* constructed theoretically the curves of Figure 7.2, referred to the dog's heart, which summarizes graphically the heart's law.

The curves obtained by Patterson and the experimental preparation of Starling are described in Boxes 7.1 and 7.2 of these Chapters.

Box 7.1 The Heart-Lung Preparation of Starling

Starling and co-workers studied the relationship between fiber distension and contraction force using the *heart-lung preparation* which can be

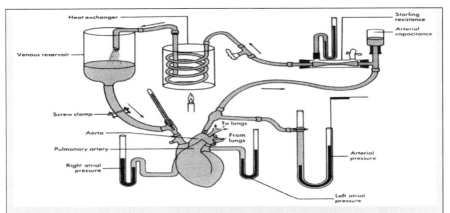

Figure 7.1 The heart-lung preparation of Starling. See the explanation in the text. (After, Knowlton & Starling, J Physiol. 1912 May 6; 44(3): 206–219).

seen in the article published by Knowlton & Starling (1). In Figure 7.1 a simplified version is reported. In this experimental preparation, with intact pulmonary circulation, the anonymous artery was cannulated with a T cannula. The ligature of the subclavian and, more distally, of the aorta, made sure that all the blood pumped from the left ventricle took the path of the T cannula, to which a manometer was connected, thus allowing knowing the arterial pressure.

The T cannula carried blood to a system that included a resistance R (*Starling resistance or resistor*), which consisted of a flexible rubber tube located in a rigid glass tube to which it was glued only at the ends so that air could be introduced allowing a fine regulation of the resistance. Therefore, this resistor simulated *total peripheral resistance* (TPR).

From the resistance, the blood passed into a coil placed in a heat exchanger so that during the extracorporeal circulation it kept the body temperature.

Through a flexible tube, a venous reservoir was connected with the superior vena cava and then with the right atrium. The tube was flexible so that the reservoir level could be varied. The difference in level between the blood meniscus in the reservoir and the right atrium represented the atrial filling pressure, i.e. the CVP, which could be also varied by a screw clamp. The CVP was also measured by a pressure manometer connected to the atrium through the inferior vena cava. A glass bell wrapped the heart and used to measure variation in ventricular

Table 7.1   First Starling maneuver: effects of an increase of resistance (increase in afterload)

Number of beat	1	2	3	4	5
Venous Return	10 mL	**10 mL**	10 mL	10 mL	10 mL
Telediastolic Volume	20 mL	**20 mL**	23 mL	24 mL	24 mL
Stroke Volume	10 mL	**7 mL**	9 mL	10 mL	10 mL
Systolic Residue	10 mL	**13 mL**	14 mL	14 mL	14 mL

volume. Arterial pressure was measured with a manometer inserted in the ascending aorta and total TPR were varied with a Starling resistor positioned around the abdominal aorta. An arterial compliance system is also added to the aorta.

With this preparation Starling performed two different maneuvers:

In *the first maneuver*, Starling increased the TPR and consequently the afterload, while the level of the venous reservoir, and therefore the venous return, remained unchanged.

As reported in Table 7.1, under steady-state conditions (beat 1) SV was 10 mL, in front of a telediastolic volume of 20 mL. Therefore, the residue volume was 10 mL (EF= 50%). After an abrupt increase in resistance (beat 2), SV suffered an immediate reduction to 7 mL. This beat left in the ventricle a systolic residue of 13 mL (20 − 7 mL; EF= 35%).

Since a lower SV was accomplished against an increased afterload, the work for beat (P × V) could be considered unchanged (See Chapter 9). Moreover, since the blood level in the reservoir was always the same, in diastole the same amount of blood returned first to the right heart and then, through the pulmonary circle, to the left heart. In this way the telediastolic volume, that is the preload, passed from 20 to 23 mL in the subsequent beat.

In this third beat, even if the afterload increase persisted, the SV rose to 9 mL (EF = 39%), but the systolic residue rose to 14 mL. In the beat n. 4, the telediastolic volume now rose to 24 mL. Therefore, the SV rose to 10 mL and the telediastolic volume remained equal to 24 mL (EF= 42%). At this point, the SV and venous return are settled at 10 mL, as in the initial control situation and they remained in this way in subsequent beats with a preload of 24 mL. A flow returned to the control value pumped against increased pressure indicated that the heart now performed more work, *i.e.* it developed more contraction energy than control. It should be noted that we are talking about contraction force when the developed pressure is considered, as in Frank's preparation, while we are talking about contraction energy when we consider the work done. Note that in beat 4, EF is lower than EF in beat 1. This shows us that EF is not a good index of contractility when afterload changes.

From the various stages along which the heart restores a normal SV after the increase of the resistance and the afterload, we see that a fundamental role must be attributed to the increase of the preload or the distension or stretching of the fibers. An excessive increase of the preload, and therefore of the telediastolic volume (not seen in the table), ends up determining a decrease, rather than a further increase, of the SV and the cardiac work.

The *second Starling maneuver* consisted of raising the reservoir in the presence of an unchanged resistance. As a consequence of the lifting of the reservoir, the CVP increased and then the ventricular filling, initially of the right and subsequently, through the pulmonary circulation, of the left ventricle. In this case, since the afterload was unchanged, the increase in preload increased the SV which, occurring against a constant afterload, was accompanied by an increase in the work per beat. However, ejection fraction (EF) is unchanged, because contractility is not varied appreciably.

In both cases, the maneuvers of Starling increased work per beat: in the first maneuver was due to a normal SV against an increased afterload; in the second maneuver was due to an increase in the SV against a normal afterload.

What the two maneuvers had in common was an increase in preload: in the first case was due to an increase in the systolic residue, in the second case was due to an increase in venous return. It can also be said that in the first case the increase of the preload occurs in a retrograde way, while in the second case it happens in an anterograde way. An increase in preload, whit the same contractility increases the force of contraction and the SV.

Increased venous return augments the ventricular filling (telediastolic volume) and therefore preload, which is the initial stretching of the cardiac myocytes before contraction. Myocyte stretching increases the sarcomere length, which causes an increase in force generation and enables the heart to eject the additional venous return, thereby increasing SV.

Analyzing the EF, we can see that it is not an excellent contractility index, since in both maneuvers, as for definition, the contractility has not been modified, but the EF vary slightly and slowly with the second maneuver (variation of preload). Instead, with the first maneuver (afterload increase), the EF is reduced and slowly recovers in the following beats. This observation led to some considerations that we will analyze in the Anrep phenomenon (see below).

1. J Physiol. 1912 May 6; 44(3): 206–219)

Box 7.2 Patterson Curves

In Figure 7.2 on the abscissa axis the VTDV or VEDV is shown, while on the ordinate axis the pressure values are shown. Given a diastolic

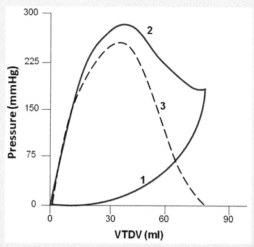

Figure 7.2 Variation of the telediastolic (curve 1), systolic (curve 2), and developed (curve 3) ventricular pressures for each variation of the ventricular telediastolic volume (VTDV). Curve 1 indicates the pressure values of the heart in diastole, curve 2 refers to the pressure values reached in systole and curve 3 to the pressure values developed, i.e. the pressure added by systole to the pre-existing diastolic pressure values. Curve 3 can be obtained by subtracting the values of diastolic pressure from those of systolic pressure, at each volume.

volume on the abscissa axis, the tele-diastolic (curve 1), systolic (curve 2), and developed (curve 3) ventricular pressures can be determined.

As can be seen from curve 1, the increase in VTDV, which is the degree of filling of the heart and stretching of the fibers, leads to an increase in the telediastolic pressure according to the viscous-elastic characteristics of the myocardium.

When the heart contracts, for each VTDV, the pressure rises to a value of curve 2, starting from a value in curve 1.

Curve 3 is obtained by subtracting the values of diastolic pressure from that of systolic pressure, for each VTDV. Each point of curve 3 is the *developed pressure*, *i.e.,* the pressure added by systole to the pre-existing diastolic pressure value.

If we look at the three curves, we see how the ventricular telediastolic pressure starts to be higher than zero only when its volume exceeds about 15–20 mL, in this theoretical case. In this volume range, when the heart contracts we observe, as indicated by curve 2, a progressive increase in systolic pressure. When we subtract the values of curve 1 from those of curve 2, the developed pressure (curve 3) coincides exactly with the systolic pressure (curve 2), until the values of diastolic pressure are equal to zero. When the telediastolic volume exceeds 20 mL, the diastolic pressure values become different from zero, so the difference between systolic pressure and diastolic pressure gives a developed pressure value lower than that of the systolic pressure. The curve 3 (developed pressure) is therefore located below the curve 2.

The developed pressure is the pressure generated actively during contraction. Initially, *albeit* differently, the curves 2 and 3 go up until a maximum. After that, both curves begin to decrease. Indeed, the increase in developed pressure to its maximum is due to an *increase in the contraction force determined by the greater telediastolic volume*, which is the distension of the fibers. The Figure 7.2 clearly shows how, *once certain limits are exceeded, a further stretching of the fibers will no longer determine an increase, but a decrease in the force of contraction*, just as stated by the Frank-Starling law of the heart.

Figure 7.2 also illustrates how increasing VTDV the diastolic pressure increases. Usually, this relationship becomes steeper when developed pressures begin to decrease. The diastolic pressure (curve 1) progressively increases and the systolic pressure (curve 2) decreases so that at a certain point the respective curves meet, and the developed pressure becomes equal to zero, as indicated by curve 3. In these conditions, the stretching of the fibers is exaggerated so that they can no longer contract and therefore do not generate an active force.

In summary, fiber stretching increases the pressure developed up to a certain point after which the developed pressure decreases and finally becomes zero.

7.1.1.1 Mechanisms of the Frank-Starling law

An increase in ventricular volume leads to a more powerful contraction. This is the well-known Frank-Starling mechanism that allows the heart to maintain the SV despite a greater afterload or to increase its output after a rise in preload. Indeed, VEDV can increase either increasing afterload with a subsequent increase in systolic residue or by increasing anterograde venous return.

To explain why the distension of the myocardial fibers increases the contraction force, some hypotheses have been formulated that are not mutually exclusive.

The first hypothesis: takes into account the overlapping in the sarcomere of the thick filaments and the thin filaments, pointing out the importance of the changes in the number of overlapping actin and myosin units within the sarcomere, as in a skeletal muscle. According to this hypothesis, changes in the force of contraction do not result from a change in inotropy. However, other mechanisms consider that the changes in preload are associated with modifications of calcium handling and troponin C affinity for calcium. Therefore, *a sharp mechanistic distinction cannot be made between heterometric (length-dependent) changes and homeometric (length-independent) changes (inotropic mechanisms) in contractile function.* Nevertheless, it is useful to keep them separated because when an heterometric regulation occurs, the Laplace phenomenon must be considered. This influences oxygen consumption and *heart efficiency* (see below Paragraph 7.5). Moreover, heterometric effects are not accompanied by *lusitropic effects,* which are, however, present when some inotropic stimulator is considered (see Paragraph 7.2).

As can be seen in Figure 7.3A, the maximum force development cannot take place if the resting length of the fiber is too short and the thin filaments of one side overlap with the thin filaments of the opposite side thus interfering with the correct sliding of the sarcomere filaments. If the stretching of the fiber causes a sarcomere elongation, all the transversal bridges will be able to

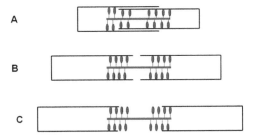

Figure 7.3 Possible positions of myofilaments to explain the mechanism of the Starling law. See the explanation in the text.

insert themselves in the most appropriate way to produce the correct sliding of filaments and then the maximum force development (Figure 7.3B).

The Figure 7.3 also illustrates why the force of contraction increases until a certain point and then it decreases, in the sense that there is a reduction in the force developed when an excessive stretching moves the thin filaments away from the center of the sarcomere so that some transverse bridges will not have any contact (Figure 7.3C).

Another hypothesis that can explain the mechanism of Starling is the increase of the *affinity of troponin C for Ca^{2+}* caused by stretching, so that if a greater amount of Ca^{2+} binds to troponin, the sarcomere can develop a strength increased because the inhibition action exerted by troponin and tropomyosin on the ATPase activity of myosin thick filaments is reduced. This results in an increase in the rate of cross-bridge attachment and detachment, and the amount of tension developed by the cardiac fibers. This effect of increased sarcomere length on the contractile proteins is named *length-dependent activation.*

A third hypothesis consists of the activation of an enzyme called neuronal NOS (nNOS), which synthesizes more nitric oxide (NO) in response to the stretching of the fiber. The activation of the nNOS, located in the myocardial fiber between the transverse tubule and the sarcoplasmic reticulum, would determine the production of NO capable of opening the ryanodine channels for Ca^{2+}. NO may also affect contractile protein affinity for Ca^{2+}.

The Frank-Starling phenomenon can be described in simple terms as a mechanical phenomenon explained by the length-tension and force-velocity relationships for cardiac muscle, as for skeletal muscle. Increasing preload increases the active tension developed by the muscle fiber and increases the velocity of fiber shortening at a given afterload (greater than zero) and fixed inotropic state. However, as said also an homeometric regulation (*length-dependent activation*) takes place simultaneously and the contractile force increases for both mechanisms: an increase in preload and an intrinsic increase in contractility.

Due to the intrinsic properties of the myocardium which are responsible for the Frank-Starling mechanism, the heart can automatically accept an increase in venous return and consequently pump more blood. The mechanism is extremely important from a functional viewpoint because it serves to adapt the flow of one ventricle to that of the other. If this adaptation mechanism did not exist and, for example, the right ventricle pumped more blood than the left ventricle, the blood would accumulate in the pulmonary circulation with catastrophic consequences of congestion of blood in the pulmonary circulation and pulmonary edema (see Chapter 15).

7.1.2 Intrinsic Homeometric Regulation of Contractile Force: Anrep Phenomenon and Bowditch or Scale Phenomenon

Homeometric autoregulation of force of contraction can be defined as the variation of contractile force without changes in preload.

7.1.2.1 The phenomenon of Anrep

The myocardial stretch elicits a rapid increase in developed force, which is mainly caused by myofilament overlapping and an increase in myofilament calcium sensitivity (Frank-Starling mechanism). In addition, a slow gradual increase in force takes place with the first and the second maneuver of Starling (see Box 7.1). This slow force response to stretch constitutes the *Anrep phenomenon*, which is an example of homeometric autoregulation of the heart. It has been initially described as an increase in mechanical heart performance after a sudden increase in aortic pressure. After the initial rise in force of contraction, myocardial performance continues to increase over the 10–15 min following the sudden stretch. As a consequence, the SV is reduced a little despite the increase in pressure, which exceeds the decrease in SV; thus the work per beat (pressure multiplied SV) is increased. Since over this 10–15 min there is not an increase in diastolic ventricular volume, this type of regulation is considered to be homeometric. A progressive decline in VEDV can be observed.

The mechanism underlying the phenomenon of Anrep could be the same as *Gregg's phenomenon*. The latter is an increase in myocardial metabolism which follows an increase in coronary perfusion pressure. The two phenomena can be explained by the passive expansion of the intramyocardial vessels, which, once increased in diameter, would cause distension of the myocardial fibers with an increase in the contraction force. In other words, the distension of myocardial fibers from increased aortic pressure would evoke a sort of Starling mechanism in miniature by acting in the thickness of the cardiac walls rather than in the ventricular cavity. Anrep effect is known to be the result of an increase in the calcium transient amplitude and several mechanisms have been proposed. The chain of events triggered by myocardial stretch includes *a*) release of angiotensin II, *b*) release of endothelin, *c*) activation of the mineralocorticoid receptor, *d*) transactivation of the epidermal growth factor receptor, *e*) increased formation of mitochondria reactive oxygen species, *f*) activation of redox-sensitive kinases upstream myocardial Na^+/H^+ exchanger (NHX), *g*) NHX activation, *h*) increase in intracellular Na^+ concentration and *i*) increase in Ca^{2+} transient amplitude through the Na^+/Ca^{2+} exchanger (*Am J Physiol Heart Circ Physiol. 2013;304(2): H175-82*).

7.1.2.2 The Bowditch effect or staircase phenomenon

The *Bowditch or staircase phenomenon* has already been described in Box 5.2 in Chapter 5. Also known as Treppe's phenomenon, in short, it consists of a progressive increase in contractility and developed pressure, which occur when the heart begins to beat at a higher frequency. The increase in frequency likely enhances the chances of Ca^{2+} entry through the sarcolemma, especially during phase 2 of the action potential.

In Chapter 5.5, it was also explained how in the very first beats after the increase in frequency there is a transitory reduction in contractility. This is likely due to the sudden reduction of the time available to Ca^{2+} to pass from the intake compartment to the release compartment of the sarcoplasmic reticulum, limiting the increase in cytoplasmic concentration of Ca^{2+}. In other words, the phenomenon is related to Ca^{2+} handling and involves proteins that participate in the excitation-contraction coupling, like the SR calcium transport ATPase (SERCA) (see Box 7.3 on SERCA).

Of note, in *heart failure*, the reduction in contractility is not transitory. In this condition, the reduced release of Ca^{2+} from the reticulum likely prevails over the entrance from the sarcolemma so that the increase in frequency corresponds to a protracted and progressive decrease in contractility. In this case, we can speak of *reverse Bowditch phenomenon (negative staircase)*.

Box 7.3 Sarcoendoplasmic Reticulum (SR) Calcium Transport Atpase (SERCA)

The SERCA is a pump that transports Ca^{2+} from the cytoplasm into the SR. The SERCA pump is encoded by a family of three genes, SERCA1, SERCA2, and SERCA3, which are highly conserved but localized on different chromosomes. Alternative splicing of the transcripts, occurring mainly at the COOH-terminal, dramatically enhances the SERCA isoform diversity. There are at least 14 different SERCA isoforms. These isoforms exhibit both developmental and tissue specificity, suggesting that SERCAs contribute to the unique physiological properties of the tissue in which they are expressed. It seems that SERCA1 isoform is mainly expressed in adult and neonatal skeletal muscle, while SERCA2a and SERCA2b in cardiac muscle. In particular, SERCA2a seems to be the dominant isoform [1] and it is reduced in heart failure [2]. SERCA2b seems ubiquitously expressed in all cell types, while SERCA3 has been found co-expressed with SERCA2b in selected cell types such as endothelial cells, lung, pancreatic β-cells, and Purkinje neurons of the cerebellum. In the myocardium, SERCA2 main function is to re-uptake Ca^{2+} from the cytosol in diastole. Several processes may regulate SERCA function. In cardiac and skeletal muscles, the best-characterized regulator is the endogenous molecules *phospholamban* (PLN), which inhibits the function of the SERCA pumps. In the heart, it seems that SERCA2 activity is low when PLN is bound to the pump. Increased *β-adrenergic stimulation* promotes the *phosphorylation of PLN* by activating *protein kinase A* (PKA), thus reducing the association between SERCA2 and phosphorylated PLN (PLN-P), upon dissociation Ca^{2+} re-uptake by SERCA2 pump increases consuming more ATP.

[1.] Periasamy M & Kalyanasundaram A. SERCA pump isoforms: their role in calcium transport and disease. Muscle Nerve. 2007 Apr; 35(4):430–42.
[2.] Hovnanian A. SERCA pumps and human diseases. Subcell Biochem. 2007; 45:337–63.

7.1.3 Extrinsic Nervous and Humoral Regulation

The extrinsic regulation of the force of contraction of the heart can be both nervous and humoral. It acts on the contraction force as it regulates *contractility* without requiring a change in preload.

7.1.3.1 The Nervous Regulation

The heart is innervated by sympathetic and parasympathetic (vagal) fibers. The overall action of the sympathetic and parasympathetic on the cardiovascular system will be studied in Chapter 12.

In brief, in the heart the neurotransmitter of postganglionic sympathetic fibers is noradrenaline. This acts on myocardial adrenergic receptors for the vast majority of type $\beta 1$, but also on the very few $\beta 2$ present and, in some cases, also on the $\beta 3$ receptors overexpressed in heart failure. The presence of α receptors on the myocardium is very modest. They are more present on vessels.

Parasympathetic fibers directed to the heart are preganglionic fibers contained in the two vague nerves, right and left. These fibers originate in the bulb from cells located mainly in the ambiguous nucleus and, to a lesser extent, in the dorsal motor nucleus. The vagal fibers make synaptic contact with postganglionic neurons distributed on the surface of the heart on which they act through very short postganglionic fibers. The right vagus mainly innervates the right atrium and the sino-atrial node, while the left vagus mainly innervates the left atrium and the atrio ventricular node. Few fibers are instead directed to the ventricles on which the vagal innervation has, therefore, a scarce action. The neurotransmitter of the parasympathetic postganglionic fibers is acetylcholine (ACh), which acts on the heart through muscarinic cholinergic receptors.

In this Chapter, the nervous control of the heart is referred exclusively to the force of contraction of the myocardium on which the sympathetic acts increasing, and the vagus decreasing the contractility of inotropy. For anatomic reasons, the action of the vagal nerves concerns mainly the contractility of the atria. Other effects, such as the modulation of HR, are described in Chapter 12.

An increase in sympathetic tone causes an increase in SV with a reduction in the systolic residue or VESV (ventricular end-systolic volume). If the diastolic ventricular filling did not change, the VEDV would decrease exactly like the VESV. Theoretically, the beats following the first after an increase in sympathetic discharge could no longer expel an increased SV. This only happens if the sympathetic activity concerns only the fibers directed to the myocardium. Usually, an increase in sympathetic activity involves all the cardiovascular systems, including the venous system, reducing venous capacity, and, thus increasing the venous return. In this way, as happens in the *fight or flight reaction*, the increase in venous return causes the VEDV to

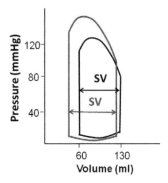

Figure 7.4 Ventricular pressure-volume loop: in the control condition (in black) and after adrenergic stimulation (in red). SV: stroke volume. In control condition, VEDV is 130 mL and VESV is 60 mL, thus SV is 70 mL. After adrenergic stimulation, VEDV is about 120 mL and VESV is about 40 mL, thus SV is 80 mL. See further explanation in the text.

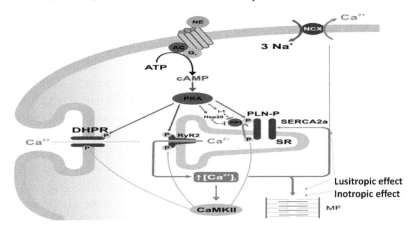

Figure 7.5 Action of norepinephrine on $\beta1$ myocardial receptors. The central role of protein kinase A (PKA). Ryanodine receptors (RyR2) are ion channels that are responsible for the release of Ca^{2+} from the sarco/endoplasmic reticulum (SR). NE: norepinephrine; Gs: stimulator membrane protein-coupled to membrane receptor; PLN-P: phosphorylated phospholamban; Hsp20: heat shock protein 20; I-1: Inhibitor-1. See explanation and other acronyms in the text.

be reduced less than the VESV, *i.e.* the SV is increased. The pressure-volume loop shown in Figure 7.4 illustrates this concept.

The inotropic action of the sympathetic nerves on the myocardium is exerted mainly through the *action of noradrenaline on $\beta1$-receptors*. With the mediation of Gs proteins, these receptors activate adenylyl cyclase (AC) which in turn transforms intracellular ATP into cAMP (Figure 7.5).

The cAMP activates a PKA which has several targets (Figure 7.5). In particular, PKA has three important targets: *1)* phosphorylates and opens L-type channels (also known as DHPR: dihydropyridine receptor) increasing the entry of Ca^{2+} into the fiber, *2)* phosphorylates and opens ryanodine receptors (RyR2) increasing the release of Ca^{2+} from SR, and *3)* phosphorylates, and therefore deactivates, the PLN which in basal conditions (non-phosphorylated) limits the activity of SERCA2a.

This triple-action is potentiated by the Ca^{2+}/calmodulin-dependent protein kinase II (CaMK II). The results of this triple-action are an increase in the intracellular concentration of Ca^{2+} in systole and a more rapid recovery of this by the SERCA2a into the SR in diastole, so that, in addition to a more rapid contraction (*positive inotropic effect*), there is also a more rapid relaxation (*positive lusitropic effect*). Troponin I (TnI) is also phosphorylated by PKA modifying cross-bridge cycle velocity and thus inotropy and lusitropy.

The negative inotropic effects of *parasympathetic vagal stimulation* are limited mainly to the atria. By acting on muscarinic receptors, ACh activates Gi proteins, thus causing opposite effects to those of sympathetic activity. Gi proteins inhibit AC thus lowering cAMP levels, and consequently attenuate the effects described above for PKA. Moreover, ACh reduces the inward Ca^{2+} currents and activates muscarinic potassium K_{ACh} channels, which reduce atrial contractility. On nodal tissue, these effects together with the ACh-mediated inactivation of funny current reduce HR, and velocity of impulse conduction (Chapter 12).

7.1.4 The Humoral Regulation

The *humoral regulation of the heart* is due to circulating substances affecting cardiac contractility. It is mainly due to the action on the heart of catecholamines (noradrenaline and adrenaline) released by the adrenal medulla. Circulating catecholamines have the same effect of sympathetic nerves on cardiac contractility. Their secretion generally increases when sympathetic activity increases. The adrenal medulla receives preganglionic sympathetic fibers contained in the large splanchnic nerves. These fibers, which like all preganglionic fibers, are cholinergic, release acetylcholine which stimulates the activity of the secreting gland cells.

Another substance that acts on cardiac contractility is *angiotensin II*. As will be seen in Chapter 12, following a drop in blood pressure, increased sympathetic discharge may occur on the juxtaglomerular apparatus of the kidney.

A drop in renal pressure and the increase in sympathetic discharge determine an increase in renin secretion by the juxtaglomerular cells. Renin is a proteolytic enzyme that, acting on the plasma angiotensinogen, transforms this precursor into angiotensin I. This, in turn, is transformed into angiotensin II by the *Angiotensin Converting Enzyme* (ACE) which is present on the surface of the endothelium mainly of the pulmonary vessels. Following further enzymatic actions, angiotensin III and IV can be also formed.

Acting on myocardial receptors AT1 *angiotensin II*, and probably also angiotensin III and IV, increase the intracellular concentration of Ca^{2+} which leads to an increase in contractility, even though it has been seen that angiotensin can also reduce the inotropic effect of-adrenergic stimulation on the papillary rat muscle probably through the production of nitric oxide.

Nitric oxide (NO), which will be discussed in detail in Chapter 13.2 exerts a modest positive inotropic action at low concentrations (about 0.05 μM) and a negative inotropic action at high concentrations (> 1 μM). Although the effects of NO on cardiac contraction are moderate, they may be important in the regulation of myocardial function. Either impairment of endogenous NO production by endothelial dysfunction or increased NO production by inducible NO-synthase could result in cardiodepressive effects. Mechanisms underlying the inotropic effects of NO on cardiac muscle include the activation of soluble guanylyl cyclase (sGC) and generation of cyclic guanosine monophosphate (cGMP or GMPc), which can inhibit phosphodiesterase III (PDE III) the enzyme that hydrolyzes cAMP, thus resulting in accumulation of cAMP, which mediate the positive inotropic effects (see above). However, cGMP leads to the activation of protein kinase G (PKG) resulting in reduced inward Ca^{2+}-current and negative inotropic effect.

Other active substances on myocardial contractility will be discussed in Chapter 13.

7.2 Combined Effect of Heterometric and Homeometric Regulation

The combined action of intrinsic heterometric and extrinsic homeometric regulation by sympathetic stimulation was highlighted in an interesting experiment by *Sarnoff and Linden* using an experimental set up similar to the Starling preparation (Figure 7.2).

As can be seen from Figure 7.6, the *ventricular volume* (VV) variations were recorded on smoked paper. The upper edge of the VV trace indicates

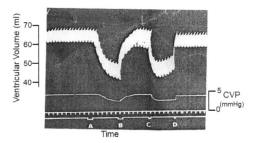

Figure 7.6 Combined effect of sympathetic stimulation and degree of ventricular filling on the stroke volume (SV). Due to the low recording speed, the set of beat-to-beat ventricular volume (VV) variations appears as a single white stripe on a black background (upper trace). **A**: sympathetic stimulation decreases the systolic volume (lower edge of the VV trace) more than the diastolic one (upper edge of the VV trace) with an increase in SV; **B**: the increase of the central venous pressure (CVP; second trace from the top) by lifting the Starling reservoir in the presence of sympathetic stimulation, brings the diastolic volume to control value, but the systolic volume increase less (less systolic residue): this is a further increase in the SV; **C**: stimulation is interrupted and CVP is returned to control; **D**: after a couple of minutes, the systolic flow returned to the control situation. It should be noted that the only sympathetic stimulation between A and B decreases the CVP which is brought back to normal by the lifting of the reservoir. In **C** the return of the reservoir to the normal position and the simultaneous interruption of sympathetic stimulation results in the persistence of the sympathetic effect only, because the norepinephrine has not yet been metabolized (From Linden R. J., 1968).

the VEDV, while the lower margin indicates the *systolic residue* volume, *i.e.* the blood that remains in the ventricle after systole. It is then clear how the vertical distance between the upper and lower margins represents the SV. The continuous line at the bottom is the blood pressure determined by the Starling reservoir level, *i.e.* the *central venous pressure* (CVP).

At the instant *A,* the stimulation of the cardiac sympathetic begins. This causes the reduction of both the diastolic VV and of the systolic VV. As the systolic volume decreases more than the diastolic volume, there is an increase in the distance between the two edges of the VV trace, indicating an increase in SV. The sympathetic stimulation making the contractions more vigorous due to the increase in inotropism, adds a certain amount of blood to the normal SV. This added blood, in basal conditions and with unchanged venous return, is a part of the *systolic residue*.

The increase in contractility slightly decreases the CVP. At instant **B**, while sympathetic stimulation continues, the initial VEDV is restored by increasing venous return by raising the reservoir. Under these conditions also the CVP returns to the control value. If we carefully observe the trace of the ventricular volumes, we can see how the SV is higher than that observed both in the control conditions and in the presence of sympathetic stimulation only. What further increased the SV was the distension of the diastolic VV of the ventricle. At this moment a combined effect of heterometric and homeometric regulation is present.

In **C** the sympathetic stimulation is interrupted and the reservoir is returned to the control position. Under these conditions, while the heterometric adjustment is lacking, the *posthumous* effect of sympathetic stimulation remains due to the slow degradation of norepinephrine. For a short time, the situation returns to be similar that observed immediately before B. Finally, in **D** we can see how, after a certain time (about 2 min) during which the recording is suspended, diastolic and systolic ventricular volume (*i.e.* VEDV and VESV) and SV return to the control level.

In Figure 7.7, we have the graphical representation of the combined effect of the heterometric and homeometric adjustments. On the x-axis, the ventricular telediastolic pressure (VTDP), as preload index, is reported, while on the ordinate axis an index of heart performance, namely the work per beat, is shown. In the control situation, the progressive increase of the VTDP increases the work per beat according to Starling's law. In the presence of maximal cardiac sympathetic stimulation, the increase in work per beat is higher for each VTDP value (red curve of Figure 7.7). Note that the difference

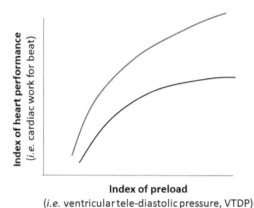

Index of preload
(*i.e.* ventricular tele-diastolic pressure, VTDP)

Figure 7.7 Combined effect of homeometric and heterometric adjustments. Basal situation (black curve) in the presence of sympathetic activation (red curve). The enhancement of preload in basal conditions increases the work per beat (black curve). For each value of VTDP, sympathetic stimulation increases the work per beat by shifting it to a corresponding point on the red curve.

between the two curves is especially pronounced for the higher VTDPs. Suggesting a synergistic effect between the combined action of intrinsic heterometric and extrinsic homeometric regulation.

7.3 The Regulation of Cardiac Contraction Force and Heart Failure

Most often, heart failure (HF) is caused by a myocardial infarction (MI). In Figure 7.8 (similar to Figure 6.2) the processes used to establish a compensated failure of the left ventricle following an acute MI are reported. Acute MI involves an immediate loss of contractility and therefore of the contractile force. As can be seen in Figure 7.8, the CO from point **A**, located on the highest curve, drops to point **B**, located on the lowest curve. Point **B** is also shifted to the right compared to **A**, thus indicating an increase in CVP. It is important to note that since the venous return curve remains in the same position, a new equilibrium can be established between venous return and CO at this lower level.

Now we can imagine that some initial compensatory mechanisms occur:

The first mechanism is the increase in the systolic residue and therefore of preload, which is the cause of the increase in CVP. This mechanism involves the use of Starling's law which, at the same contractility, increases the heart contractile force of stretched fibers and then SV can also partially recovery (the heart can pump more than the loss of contractility would allow in the absence of Starling mechanism).

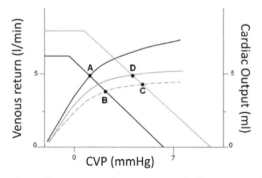

Figure 7.8   Changes in cardiac output and venous return in the compensation of a ventricular failure caused by infarction. Control condition (in black); compensated failure (solid green curve) and non-compensated failure (green dashed curve); increase in venous return due to venoconstriction and water retention (blue line). See the explanation in the text.

A second mechanism is an increased sympathetic tone due to less stimulation of arterial baroreceptors (see Chapter 12.6). The increase in sympathetic tone determines a venoconstriction with a consequent increase in venous return. In a second time also the water retention contributes to increasing venous return. In this way a new venous return curve is found, which is shifted upward and to the right. Venous return curve intersects CO curve at point **C**, located a little upper and to the right of point **B**. This new location of the equilibrium point indicates the presence of a certain recovery in cardiac output without myocardial conditions have improved.

However, the lower stimulation of arterial pressure receptors, with a consequent increase in sympathetic tone and attenuation of vagal tone, determines an increase in contractility and heart rate. These two phenomena determine an increase in the CO for each CVP value, a relationship described by a curve (solid green curve) situated in an intermediate position compared to the first two. On this new curve, the CO value corresponds to point **D**. Compared to point **C,** point **D** is shifted to the left revealing that the recovery of the CO is also due to a partial increase of the inotropism and less than before to the Starling mechanism. This compensation indicates that the heart has some functional reserve to exploit yet. Note that the increase in inotropism, in addition to sympathetic stimulation, may also be due to some recovery of the heart from *myocardial stunning* (a depression of contractility due to ischemia/reperfusion; see Chapter 17). Of course, we described the mechanism step by step, but in real-life they occur all together with their characteristic time course (*i.e.* heart rate increases first and fluid retention will take longer).

Since the increase in CVP tells us that the VEDV has increased, the ejection fraction is reduced, despite the recovery of resting SV to a value close to normal. The situation in **D** is, therefore, a *compensated heart failure*, where the normality of the CO does not exclude compromised contractility. Indeed, the depressed contractility will not allow an adequate CO in case of exercise.

If the conditions of the myocardium are seriously compromised, the mechanisms described above will not be able to restore a normal CO, thus leading to a progressive further reduction of the CO to values not compatible with life. Figure 7.9 illustrates this eventuality: the horizontal line, fixed at 5 L/min, indicates the value of CO that must be reached to have the compensation. If the myocardium is severely damaged there can be no recovery of contractility despite sympathetic stimulation.

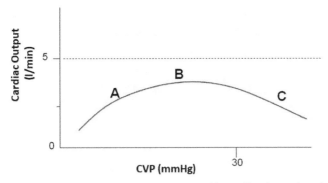

Figure 7.9 Trend of cardiac output in a decompensated heart. See the explanation in the text.

Let us imagine that the initial situation of our observation is that indicated by the point **A** (Figure 7.9). The CO is considerably reduced and the CVP is already quite high: this is a heart failure condition. Since the SV is reduced, there will be a further increase in the systolic residue and CVP (fluid retention). The CO will also increase from **A** to **B** without reaching the compensation value. As the fluid retention and the systolic residue continues to increase if compensation is not reached, the heart progressively encounters an increase in the VEDV, and therefore to dilatation, until the limits of Starling's law are exceeded without the CO returning to normal. From point C onwards the CO progressively decreases with further increase in CVP and aggravation of the failure.

Hydro-saline retention by the kidney contributes significantly to cardiac dilation. This effect, as will be seen in Chapter 12.4, is mainly due to the increased secretion of the proteolytic enzyme *renin* by the juxtaglomerular apparatus. As seen above, *renin* determines the transformation of *angiotensinogen* into *angiotensin I* which will be converted into *angiotensin II* by *ACE*. *Angiotensin II*, in turn, determines **aldosterone** secretion from the adrenal cortex. The action of *aldosterone* on the renal tubules results in greater reabsorption of NaCl followed by greater reabsorption of water favored by the increased release of ***anti-diuretic hormone (ADH)***.

We should also recall that increase in ventricular volume and ischemic cell after a MI lead to a multiphase reparative response in which the damaged tissue is replaced with a fibrotic scar produced by myofibroblasts and fibroblasts. These also induce geometrical, biomechanical, and biochemical changes in the uninjured ventricular wall eliciting a reactive remodeling process that includes interstitial and perivascular fibrosis. Although the initial reparative fibrosis is crucial for preventing the rupture of the ventricular wall, an exaggerated fibrotic response and reactive fibrosis outside the injured area are detrimental and maladaptive as they lead to progressive impairment of cardiac function and eventually to overt heart failure.

Given the maladaptive role triggered by an increase in preload (overload), the fundamental therapy of heart failure should aim to reduce the volume of circulating fluid and improve contractility. To this purpose, a variety of diuretic drugs can be used. The improvement of contractility can be obtained from inotropic drugs. In some clinical conditions, *beta-adrenergic agonists* (dopamine, dobutamine, epinephrine, isoproterenol, norepinephrine, dopexamine), *PDE inhibitors* (amrinone, milrinone, enoximone), and the recently introduced calcium sensitizers (levosimendan) can be used. Old inotropic agents, namely cardioactive glycosides or digitalis are still in use; they act on the myocardial fiber membrane by inhibiting the sodium/potassium pump (this jeopardizes cell function and digitals must be used with caution). The consequent accumulation of sodium inside the fibers slows down the activity of the Na^+/Ca^{2+} exchanger by increasing the intracellular concentration of Ca^{2+} with a subsequent increase in inotropism.

Currently, inotropic drugs are used predominantly in emergency conditions, since although they determine certain improvements, they do not prolong the patient's life. Nowadays in compensated HF and initial decompensation, it is preferred to improve the working conditions of the heart, reducing both sympathetic stimulations using *beta-blockers* and/or afterload, intended as an obstacle to ventricular ejection, with the administration of *ACE inhibitors*. The reduction in the plasma concentration of angiotensin II causes, in fact, a decrease in the arterial pressure and therefore of the afterload. Furthermore, by inducing a lower release of aldosterone, the hydro-saline retention is reduced. Indeed, *diuretics* are also part of the armamentarium to improve the quality of life of HF patients.

7.4 Heart Efficiency: Effort Required by the Heart for Ventricular End-Diastolic Volume

Regardless of the effect on the Starling mechanism, ventricular dilatation modifies the working conditions of the heart, in the sense that *a greater effort is required for a dilated heart to sustain a certain pressure* compared to a normal-sized heart.

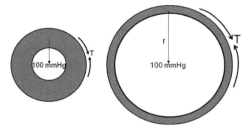

Figure 7.10 The circumferential tension (T) required to support the same intracavitary pressure is lower in a normal heart (left) than in a dilated heart (right). See the explanation in the text.

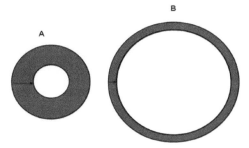

Figure 7.11 Variations of the radius (arrows) for the production of the same stroke volume in a normal heart (A) and dilated heart (B). Note that the red area of A is quite similar to that of B and represents the SV. See the explanation in the text.

If we imagine with some approximation that the ventricle has the shape of a sphere (Figure 7.10), we can apply Laplace's law, according to which the *circumferential tension T* is proportional to the product of the *intracavitary pressure P* for the *radius r* of the sphere:

$$T = P \times r \quad \text{and} \quad P = T/r$$

If the heart is small and therefore its radius r is short, a pressure P can be sustained by a modest tension T. If instead the heart is enlarged, it will have a greater r, so the same pressure P will require a greater tension T to be sustained.

If we apply the Laplace's law to a ventricle and consider the effort that it must make to develop a certain pressure, we see how a larger heart must make a greater effort to develop a greater tension T to obtain the same pressure P. In other words, the larger the heart, the less efficient it is (requires more energy to obtain a certain P; see also Chapter 4 on Laplace's law applies to the heart).

Different cardiac volumes also behave differently to produce a certain SV. If we still compare a ventricle to a sphere, we see how to achieve the same volume reduction, a small sphere needs a radius reduction greater than that required by a large sphere (Figure 7.11). It follows that a dilated heart can improve its SV even with a minimum increase in the diastolic-systolic excursion of its walls. In case of a decrease in afterload, there is a little increase in the diastolic-systolic excursion with an improvement of CO that has nothing to do with either an increase in contractility or the Starling mechanism. It can happen that patients in conditions of severe decompensation have an SV increase with apparent clinical improvements just before dying.

8

Arterial Pressure

8.1 The Device and the Law of Poiseuille

In Physics, the pressure is the force f applied on the surface s, that is, the pressure is force acting on a surface unit:

$$P = f/s$$

In a tube in which a liquid flows, the pressure is the force that the liquid exerts on a surface unit of the wall of the tube itself.

Poiseuille used the device in Figure 8.1 to relate **pressure**, **flow** and **resistance to flow** to each other. As can be seen, the Poiseuille device consists of a vertical tank, from the base of which a horizontal tube with a length l and a radius r depart. Vertical pipes (*manometers* M1-M5) are inserted along this horizontal tube. So the horizontal tube is divided into segments of equal length. In Figure 8.1 each segment is indicated with the letter R and a number from R1 to R6. As can be seen, segment R1 is located between the tank and the first vertical pipe, while the last segment R6 is located between the last vertical pipe and the right-end of the horizontal tube. At this end, there is a tap, Rb2, which can be closed or gradually opened (*i.e.,* the flow resistance can be changed).

When the Rb2 tap is closed and the system is full of water, for the principle of communicating vessels, the water level will be the same in all pipes. In other words, the pressure is the same in the tank and in the vertical pipes, which can be considered as many pressure gauges (manometers), from M1 to M5.

The opening of the Rb2 tap (Figure 8.2) causes the flow and the exit of the liquid from the system. By opening the Rb1 tap, the water enters the tank as it comes out of Rb2. We can see that from the tank to the various pressure gauges, the pressure gradually drops. If we join the various liquid meniscuses (*the meniscus is the upper surface of a liquid*) with a line, we

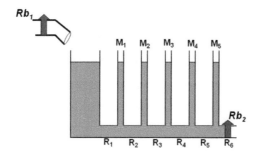

Figure 8.1   *The Poiseuille device.* In the absence of flow (both Rb1 and Rb2 taps are closed) the pressure is the same in the tank and the various pressure manometers (M1-M5) as indicated by the level of water (blue).

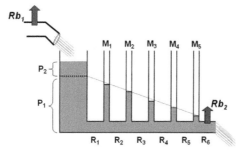

Figure 8.2   *The Poiseuille device in the presence of input and output flow.* The pressure linearly drops in the manometers. The line that joins the meniscuses of the various manometers (from M5 to M1) allows individuating in the tank a P1 and a P2 pressure. The potential pressure, P1, represents the pressure that pushes the water into the horizontal tube. The kinetic component, P2, represents the pressure that imparts speed to the water. The two taps are open and in steady-state, the inflow is equal to the outflow.

obtain an oblique line that descends until it reaches the horizontal tube at the Rb2 tap. The slope of this line between manometers is an index of resistance.

From the side of the tank, the line does not reach the water meniscus but cuts the liquid column into two parts: one of greater height located at the bottom (P1) and one of a lower height located at the top (P2). While P1 corresponds to the pressure necessary to overcome the resistance to water flow along with the system, P2 corresponds to the pressure necessary to give the water the sliding speed (the flow rate of the liquid; indeed, the flow is equal to the velocity multiplied sectional area of the tube). Therefore, with the same total pressure, the lower is the total resistance; the lower will be P1, and the higher P2 with a consequent increase in flow rate.

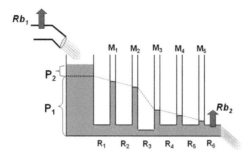

Figure 8.3 *The Poiseuille device after increasing the resistance in section R3*. The pressure drop between M2 and M3 is due to the viscous resistance encountered by a real fluid. The inflow is equal to the outflow, but the mean flow here is lower than the flow in Figure 8.2. The potential component P1 increases at the expense of the kinetic component P2 that decreases. Here the total pressure (P1+P2) is unchanged compared to the previous situation (*i.e.,* in the absence of the additional resistance in R3), because the flow was reduced partially closing Rb1. See the explanation in the text.

If an additional resistance is placed between two pressure gauges (Figure 8.3), consisting of a narrowing of the horizontal pipe, we see that the upstream meniscuses are aligned at a higher level, while those located downstream of the narrowing are aligned at a lower level. The slope of the two lines is reduced, while the difference in height between the meniscus that immediately precedes and the one that immediately follows the resistance increases, indicating that the additional resistance increases the upstream pressure and decreases the downstream pressure. If care is taken to regulate Rb1 so that the level in the tank remains unchanged, the quantity of water that passes through the system in the unit of time is reduced. From the observation of Figure 8.3, we can also see how the line that joins the meniscuses upstream of the resistance between the two pressure gauges meets the liquid column of the tank at a higher point compared to Figure 8.2. In other words, P1 increases and P2 decreases in response to additional resistance. This new intersection point tells us that the fraction of pressure necessary to overcome the overall resistance of the system has increased and that the one necessary to give speed to water has decreased, depending on the fact that the overall resistance has increased while the mean velocity of water flow along the system has been reduced. Note that the velocity is increased only at point R3 because of the radius reduction of this segment of the horizontal pipe.

At the equilibrium situation (*i.e.,* when the mL/min of liquid going in and out from Rb1 and Rb2 is the same), the flow Q in the system is proportional to the driving pressure ΔP, and to the fourth power of the radius r^4 multiplied by π and it is inversely proportional to the length of the pipe l and the liquid viscosity η multiplied by 8:

$$Q = \Delta P \cdot \pi r^4/8\eta l,$$

this is the *Poiseuille's law*. Poiseuille formula has been obtained empirically, changing the length, the radius of the tubes, and/or changing the liquid, and the pressures used.

From this formula the expression of the driving pressure, ΔP, can be derived:

$$\Delta P = Q \cdot 8\eta l/\pi r^4$$

Since the expression $8 \eta l/\pi r^4$ indicates the resistance R, we can say that the ΔP is given by the product of the flow for resistance:

$$\Delta P = Q \cdot R,$$

this is called *Darcy's law*. Therefore, the resistance, R, is given by the relationship between pressure and flow:

$$R = \Delta P/Q.$$

Importantly, since the resistance R is equal to $8 \eta l/\pi r^4$, we can infer that *the most important factor that intervenes in the inverse regulation of the resistance of a single tube is its radius*. In other words, small variations of the radius cause important variations of the resistance of a tube in the opposite direction, by a 16 factor (*i.e.,* when the radius doubles, the resistance decreases 16 times).

In the cardiovascular system, the pressure ΔP is the mean arterial pressure (MAP) minus the CVP (MAP-CVP = ΔP), the flow Q is the cardiac output, and the resistance R is constituted by the *total peripheral resistance* (TPR). Since the cardiac output is given by the product of the stroke volume for the heart rate, the formula of the pressure comes to be

$$\Delta P = SV \times HR \times TPR.$$

Since CVP is close to zero, for simplicity we can write that MAP is equal to ΔP and we get the following expression:

$$MAP = SV \times HR \times TPR.$$

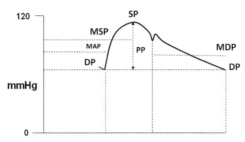

Figure 8.4 Aortic pressure values. DP: diastolic pressure; SP: systolic pressure; PP: pulse pressure; MAP: mean aortic pressure; MSP: mean systolic pressure; MDP: mean diastolic pressure.

Moreover, the fact that the blood flow from the ventricle to the aorta occurs in a pulsatile manner causes the arterial pressure to be pulsatile (not stationary). For this reason, as can be seen in Figure 8.4 in the aortic pressure curve, we can recognize the diastolic pressure (DP), the systolic pressure (SP), and the pulse pressure (PP) values. We can also calculate the MAP, as well as the mean systolic pressure (MSP) and the mean diastolic pressure (MDP).

- *Diastolic aortic pressure*: as already discussed in other Chapters, the DP is the pressure present in the aorta in the instant in which the aortic valve opens, which is the end-diastolic aortic pressure. It is, therefore, the pressure that we find in the passage between the end of the isovolumic systole and the beginning of the ventricular ejection.
- *The systolic aortic pressure:* SP is the maximum pressure of the cardiac cycle. It is usually recorded at the end of the ejection phase with acceleration, *i.e.* at the beginning of the ejection with deceleration.

The SP does not always coincide exactly with the transition between the two ejection phases, but it is very often slightly delayed concerning this moment. The cause of this delay is many, including the reflection of the head of the same pressure wave from the periphery, probably from the resistance vessels. Some authors place the site of reflection in the point of the abdominal aorta from which the renal arteries detach (see also below).

- *Pulse pressure:* PP is the difference between systolic pressure and diastolic pressure (SP-DP).
- *The mean arterial pressure:* MAP is the mean value of all the infinite values that the aortic pressure assumes during the cardiac cycle. This value can be obtained by a graphic integration of pressure curves. Empirically the mean pressure can be obtained with a good approximation by

the formula MAP= (PS+2PD)/3 or MAP = PD+ 1/3PP. These formulas imply that diastole lasts more or less twice the systole at usual HR.

- *The mean systolic pressure:* MSP is the mean value of all the infinite values that the aortic pressure assumes during ventricular ejection. We will see in Chapter 9.1 how the MSP is used to calculate the cardiac work per beat.
- *The mean diastolic pressure:* MDP is the mean value of all the infinite values that the pressure assumes during diastole, which is from the instant in which the incision appears, until the beginning of the following ejection. We will see in Chapter 14.1 how the mean diastolic pressure is used to calculate an index of coronary vascular resistance.

8.2 The Mechanical Factors of Arterial Pressure

Stroke Volume (SV), heart rate (HR), and total peripheral resistance (TPR) are defined as the determinants or the *mechanical factors of arterial pressure.* Variation of these factors varies the values of the pressure, in particular, the values of DP, SP, PP, and MAP. The values of MSP and MDP vary about these pressure values. For didactics convenience, we separately increase SV, HR, and TPR.

8.2.1 Increased Stroke Volume

The first effect is the increase in systolic pressure. Since this is usually reached at the end of the ejection phase with acceleration and since, in the presence of a greater SV, a greater quantity of blood is pushed into the aorta already at this stage, the systolic pressure will be increased.

The diastolic pressure, on the other hand, increases less than the systolic pressure. At the first glance, it could be thought that the increase in DP could be equal to the increase in SP, since, if the increase in systolic flow was a certain number of mL in excess, with the same duration of diastole the same number of mL should be present in the aorta at the end of the diastole. However, a certain part of that excess mL has moved away from the aorta and arterial vessels through microcirculation due to that increase in SP, which represents the driving force capable of pushing blood forward. In Table 8.1 the increase in DP is therefore indicated with only one + signs of normal size. Since the increase in SP is the immediate consequence of the increase in SV, in Table 8.1 the + signs are in bold and larger than those in the same row.

Table 8.1 Changes in systolic (SP), diastolic (DP), pulsatile (PP), and mean arterial pressure (MAP) following an increase in each mechanical determinant (SV, HR, or TPR). For each determinant and pressure, the number of + signs refers to the relative increase of that value compared to the others, while the bold + sign indicates the value of the pressure which increases immediately after the variation of the factor considered

	DP	SP	PP	MAP
SV +	+	**++**	+	++
HR +	**++**	+	-	++
TPR +	**+**	+	=	++
TPR ++	**++**	+++	+	+++

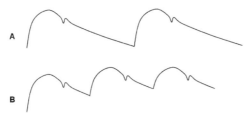

Figure 8.5 Effect of heart rate (HR) increase in arterial pressure values. In A we observe the arterial pressure at basal HR, in B we can see how, after an increase in HR, the early onset of each successive beat encounters a higher diastolic pressure because it has not had time to fall to the basal value. Obviously, the mean aortic pressure results increased.

If the SP increases more than the DP, the difference between the first and the second, *i.e.* the PP, will also be increased. No doubt that the MAP has also increased (MAP = SV × HR × TPR).

8.2.2 Increase in Heart Rate

We can obtain an increase in HR with an artificial pacemaker. To understand the effects due to the single increase in HR, we must consider the variations in contractility that may occur due to the Bowditch phenomenon to be negligible. We must also consider that SV is not reduced due to the reduction of the ventricular filling time so that the cardiac output does not remain constant. In such a condition, as can be seen in Table 8.1, in the event of a controlled increase in HR the pressure increases in its DP, SP, and MAP values. Figure 8.5 illustrates the mechanism by which this increase occurs.

Following a controlled increase in HR, each ventricular ejection begins before the diastolic pressure, following the previous systolic ejection, has the time to be reduced to the usual value. The DP is therefore increased as an immediate effect of the increase in HR. If an unchanged SV is pushed into an aorta in which the DP is increased, the SP should also show the same increase in DP.

In real-life, the increase in systolic pressure is instead less than the increase in systolic pressure since the ejection tends to be reduced due to the shortening of the time available for ventricular filling (*i.e.* reduced preload) and the increased DP (*i.e.,* increased afterload). Of course, an increase in SP that is lower than that in DP leads to a decrease in PP, while the increase in both DP and SP implies an increase in MAP. Since this increase in DP is greater than, and come before that, of the SP, it is indicated with a double bold + signs of increased size in Table 8.1.

What has been said is not valid when an increase in HR exceeds the *upper critical limit of Wenckebach*, which, as we have said, is about 180 b.p.m. in humans. This can occur, for example, in the case of pacing or high-grade supraventricular (or even ventricular) tachycardia in which there is no effect on inotropism due to an ineffective sympathetic activity. In this case, the shortening of the filling time, after having canceled the slow filling phase, also affects the rapid filling phase. With such a reduction in venous return the SV and, consequently, the CO are compromised. Therefore, when the *Wenckebach* upper critical limit is exceeded, there is a fall in CO and a decrease in blood pressure.

8.2.3 Increasing Total Peripheral Resistance

The increase in TPR has the immediate effect of increasing DP. In Table 8.1, this increase is indicated with greater ++ bold signs. The cause of this immediate effect is due to the slower decreasing pressure during diastole (Figure 8.6) caused by the reduced blood flow from the arteries through the precapillary resistance vessels.

Faced with an increased DP, the SV can be normal only following the appropriate adjustments due to the Starling and possibly Anrep mechanisms. SP may, therefore, present the same increase in DP, with unchanged PP and increased MAP.

What has just been described applies in the case of a small increase in diastolic pressure. If, on the other hand, PD is increased to 100 mmHg or more due to an excessive increase in TPR, SP may increase much more than PD with a consequent noticeable increase in PP even in the presence of an unchanged SV (Figure 8.6). The reason for such an increase in SP is to be

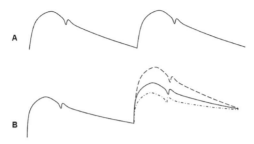

Figure 8.6 Effect of TPR increase in arterial pressure values. A control situation, in B increased resistance. When the TPR increases the DP increases first because the blood flows with greater difficulty from the large arteries to the microcirculation. The SP increases by the same amplitude as the DP (continuous line) as the distensibility of the arteries remained unchanged. Otherwise, SP increases more than the DP if the latter stretches the arteries so much that the modulus of elasticity is increased (upper dashed line). Finally, if a ventricular failure occurs for the **prolonged** *and exaggerated increase in afterload*, the SV is reduced and the SP increases less than DP (lower dotted line).

related to the increase in the *modulus of elasticity* (see Chapter 1.3) caused by excessive distension of the aorta and large arterial vessels. The increase in the modulus of elasticity limits the expansion of these vessels when the SV is ejected. The excessive distension, without changing the structure of their wall, reduces the compliance and the *windkessel effect* (arteries distend when the blood pressure rises during systole and recoil when the blood pressure falls during diastole, see Chapter 1.3). Therefore, a greater part of the energy impressed by the heart to the blood increases the systolic pressure instead to be spent stretching the arterial walls (already stretched).

Finally, a prolongation of the excessive afterload may lead to maladaptive hypertrophy and ventricular failing. In this case, the SP increases to a lesser extent than the DP (Figure 8.6). Often heart failure patients have a high DP and a hypertrophic heart.

8.3 Measurement of Resistance

Since the P/Q ratio gives vascular resistance, we can, therefore, have the total peripheral resistance if we know the MAP and the cardiac output CO. Yet we can obtain the resistance of each vascular district if we know the flow and the pressure in that district.

In the clinic, the pressure is expressed in mmHg and the blood flow is expressed in mL/min. Therefore, the *resistance unit* is *mmHg/mL/min*. For instance, TPR can be 0.02 mmHg/mL/min, if the MAP is 100 mmHg and the CO is 5000 mL/min. Likewise, district resistance can be calculated by the ratio between the average pressure and the blood flow in a given organ or district (P/F).

Normalized resistance: to compare the resistance in the various organs and districts, the resistances are normalized for 100 g of the perfused tissue of the considered organ or district. Normalization can also be done for the total body weight to calculate normalized TPR. Therefore, if in a 70 kg subject the average pressure is 100 mmHg and the cardiac output is 5000 mL/min, the ratio will be 100/5000 or 0.02. If we refer this value to 100 g of body weight, we have to multiply it by 70,000 and divide by 100. We will, therefore, have $0.02 \times 700000/100 = 14$ *mmHg/mL/min for 100 g of tissue* which is the value of the *normalized TPR*. Resistance in the single district is obtained from the ratio between the mean arterial pressure and the blood flow in the affluent artery to the district. If we consider the coronary circulation of a heart of 350 g, the mean perfusion pressure continues to be 100 mmHg, while the flow is only 280 mL/min. The resistance will then be $100/280 = 0.36$. If we want to refer this value to 100 g of heart, we will have $0.36 \times 350/100 = 1.26$ *mmHg/mL/min for 100 g of tissue*. Note that this normalized resistance is less than that of the TPR not because the coronary circulation is smaller, but because it perfuses an organ that is always active and therefore has a low vascular resistance and a high blood flow per gram. Note that the non-normalized district resistance (0,36 *mmHg/mL/min*) is higher than the non-normalized TPR (0.02 *mmHg/mL/min*), (see the explanation in the next paragraph).

8.4 Parallel Arrangement of District Resistance

In Chapter 2 we saw how the individual district resistance, which together form the total peripheral resistance, are arranged in parallel between them. Figure 8.7A shows three resistances arranged in series and Figure 8.7B shows three resistances arranged in parallel.

If in Figure 8.7A one resistance is brought to infinity the sum of the resistances also rises to infinity and a liquid cannot pass from point P_1 to point P_0. If instead one of the resistance of Figure 8.7B is brought to infinity, the liquid can equally pass from P_1 to P_0 through the other two resistances.

From the above, it can be inferred that, while in the case of *resistance in series* (Figure 8.7A) it can be said that the total resistance (TPR) is given

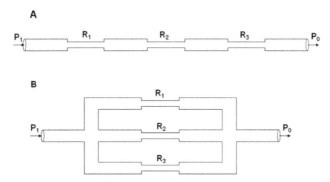

Figure 8.7 Resistances in series (A) and parallel (B).

by the sums of the resistances R1, R2 and R3, in the case of *resistances in parallel* (Figure 8.7B) the reciprocal of TPR (1/TPR, namely the conductance) is equal to the sum of the reciprocals of the district resistances R1, R2 and R3.

Therefore, in the case of the resistances in parallel of the circulatory system the relationship between TPR and district resistances can be represented as:

$$1 / TPR = 1 / R1 + 1 / R2 + 1 / R3 + 1/Rn \ldots$$

This expression indicates that the increase, even indefinitely, of only one of the district resistance, which are very numerous in the circulatory system, has little or no influence on the total resistance and therefore on the pressure. If this were not the case, the amputation of a limb or the occlusion of an artery, raising the resistance of that limb/artery to infinity should increase the pressure considerably! This pressure increase does not occur, for instance, when the brachial artery is occluded during the measurement of the pressure.

To illustrate this relationship, assume that R1 = 5, R2 = 10 and R3 = 20 mmHg/mL/min. In this example, TPR is equal to about 2.86 mmHg/mL/min. This demonstrates two important principles regarding the parallel arrangement of blood vessels:

1. *The total resistance of a network of parallel vessels is lower than the resistance of the vessel within the system with the lowest resistance*. Therefore, a parallel arrangement of vessels greatly reduces resistance to blood flow. *That is why capillaries*, which have the highest resistance when taken as a single vessel because of its small diameter, *for their parallel arrangement constitute only a small portion of the total vascular resistance* of the vascular network.

2. When there are many parallel vessels, ***changing the resistance of a small number of these vessels will have very little effect on total resistance*** for the considered system or district.

Since the reciprocal of resistance is *conductance*, it can be said that in the vascular system, total conductance is equal to the sums of district conductance (with the above numbers conductance is in mL/min/mmHg equal to $0.35 = 0.2 + 0.1 + 0.05$).

8.5 The Regulation of Vascular Resistance

Already in Chapter 1.4, it has been said that vascular resistance is located mainly at the level of arterioles and precapillary sphincters. The resistance vessels, in addition to the arterioles, also include the metarterioles and the proximal tract of the preferential route. These vases are small and not many; they can vary their radius, therefore, for these characteristics, they are considered the site of adjustable vascular resistance.

In basal conditions the smooth muscles of the resistance vessels are in a state of partial contraction, which is the *vasomotor tone at rest*, which guarantees the degree of resistance in basal conditions. The vasomotor tone may increase following an *increase in sympathetic activity* causing vaso-constriction or decrease due to a *reduction in sympathetic activity*, thus inducing vasodilation. Discussing the pressure control mechanisms, we will see how the systemic vasomotor tone is controlled by baroreceptor reflexes (see paragraph 8.10.1.2).

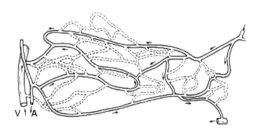

Figure 8.8 *Schematic representation of the microcirculation.* The continuous lines represent the preferential route. The dashed lines represent the capillaries. A: arteriole; V: venule. The arrows indicate the blood flow direction in the preferential route. The preferential route consists of the metarteriole, the proximal and distal tracts, which ultimately end in a venule. The capillaries start from the metarteriola and the proximal tract and drain in the distal tract. At the origin of the capillaries, there are precapillary sphincters, formed by smooth muscle fibers, which ensure or not the opening state of the capillaries. It is believed that in basal conditions only a quarter of the capillaries are open (From Zweifach BW, 1950).

Humoral factors can also take part in the regulation of vasomotor tone and vascular resistance. In addition to adrenal medullary hormones, various prostaglandins and endothelial factors intervene in this regulation, among which particular importance is given to nitric oxide (NO) with vasodilator action and endothelins with vasoconstrictor action.

In many tissues, the vasomotor tone depends on the *metabolic state* of the tissues themselves. "Active metabolites", *i.e.* compounds that cause vasodilation, tend to accumulate in active tissues. These metabolites include H^+, K^+, CO_2, H_2O_2, and adenosine. The latter also inhibits the release of noradrenaline from nerve endings. The increase in lactate and a decrease in pH, as well as a decrease in O_2 tension, contribute to vasodilation and *active hyperemia* due to increased organ metabolism.

The O_2 tension decreases and the metabolites accumulate after an artery occlusion. This determines noticeable vasodilation which, upon reopening of the vessel, causes the flow to increase considerably (*reactive hyperemia*), to remove the metabolites. The higher is the tissue metabolism, the greater is the reactive hyperemia. Both a myogenic response and the release of NO by the endothelium contribute to this flow increase. The first is due to the drop in pressure (during occlusion) and the latter is due to the increase in shear stress (at the reopening), at the arteriolar level (see Chapters 11 and 13).

The *metabolic auto-regulation* theory considers that when the pressure decreases, the consequent decrease in the flow determines the accumulation of the metabolites, which, by vasodilating, cause the flow to return to the starting point. Yet, when the pressure increases, the flow at the beginning increases, and the metabolites tend to be removed. Consequently, the vasomotor tone increases, and the flow returns to the control situation. In other words, the flow to an organ is dictated by the metabolic state of the tissues within a large range of perfusion pressure.

Like other vessels, resistance vessels can also be stretched by increasing pressure. Indeed, the distension is experimentally evident when the pressure rises from 0 to about 60 mmHg. In this case, due to the increase in the vessel radius, a decrease in resistance occurs. It is easy to understand that the flow increases more than it would increase if the radius did not change. However, in some district circles, when the perfusion pressure exceeds 60 mmHg, the *myogenic auto-regulation mechanism* intervenes.

Thanks to this mechanism, an increase in pressure from 60 to about 120 mmHg causes vasoconstriction and *vice versa* the decrease in pressure determines vasodilation. These mechanisms are called myogenic because they are directly due to variations in a stretch of the smooth muscle of the vessels due to changes in *transmural pressure* (the difference in pressure between the inside and the outside of the vessels). In district circles in which myogenic auto-regulation occurs, the flow remains constant even in the presence of pressure changes. Therefore, myogenic and metabolic factors (described above) tend to keep the flow constant in a wide range of perfusion pressure.

The pressure inside the blood vessels at which they collapse and close completely is called *critical closing pressure*. It varies in the different districts as it depends mainly on *transmural pressure* (the difference in pressure between inside and outside the vessel) and vasomotor tone.

As said, sympathetic activity usually induces vasoconstriction. This is usually due to the action of noradrenaline on α-adrenergic receptors located on smooth muscle. In addition to the sympathetic vasoconstriction, sympathetic vasodilation can be observed in some districts. In a district where on vessel smooth muscle $\beta2$-adrenergic receptors prevail over α-adrenergic receptors, thus sympathetic vasodilation can be also observed regardless of the effects on organ metabolism. In the coronary circulation, the sympathetic stimulation can determine an increase of the flow both for a direct $\beta2$-adrenergic receptor-mediated vasodilation and, in particular, for the simultaneous increase of the myocardial metabolism determined by the increase of the HR and the contractility following the $\beta1$-adrenergic receptor-stimulation.

The so-called *sympathetic vasodilator* acting through the release of *acetylcholine* (ACh) may be observed in the resistance vessels of the skeletal muscles. This sympathetic vasodilator system is considered responsible for the first phase of vasodilation that is observed in the muscles at the beginning of a physical exercise in some animal species. Controversy exists regarding the prevalence of sympathetic $\beta2$- and ACh-mediated vasodilation in the various animal species. In humans, both $\beta2$- and ACh-mediated vasodilation are included in the so-called sympathetic vasodilator mechanism of the skeletal muscle which might intervene before the start of the exercise. Nevertheless, during exercise, a *functional sympatholytic* (attenuation of α-mediated sympathetic vasoconstriction) by metabolic and endothelial factors (NO) is considered crucial for a full mediated vasodilatation in exercise-induced active hyperemia.

8.6 The Arterial Windkessel and the Arterial Pressure

At various points in this book, and in particular, in Chapter 1.3, we have mentioned the **windkessel effect** (*bellows effect*) of the aorta and elastic

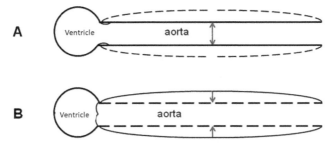

Figure 8.9 Arterial *windkessel*. During the ejection (A) the wall of the artery expands accepting a part of the SV; during the diastole (B) the wall returns to the starting position. Compliance is the change in volume due to a unitary increase in pressure.

Figure 8.10 Loss of *windkessel* effect with increased systolic pressure and reduction of diastolic pressure. See the explanation in the text.

arteries. ***This effect must not be included among the mechanical factors of arterial pressure.*** Although it intervenes on determining DP, SP, and PP, it does not *"in itself"* influence the MAP. At the base of the *windkessel effect* (Figure 8.9) there is the distensibility or *compliance* (C) of the elastic arteries, given by the relationship between the variation in volume dV and the change in pressure dP:

$$C = dV/dP.$$

As we have already seen, the elastic arteries like the aorta, stretching out at the time of the ventricular systolic injection, increase in volume and host in the expansion a part of the ejected blood while absorbing in their wall a part of the energy imprinted on the blood by ventricular contraction.

The energy that is absorbed by the rapid distension (Figure 8.9A) is subtracted from the systolic pressure, which is therefore attenuated compared to the SP value we would reach in the absence of wall distension. Subsequently, the aorta wall slowly returns to the rest position (Figure 8.9B) giving back to the blood the pressure previously absorbed during the expansion and thus preventing an excessive fall of the aortic pressure during diastole.

When the arteries lose their elasticity, as happens for example in the elderly (*arteriosclerosis*), the *windkessel effect* is lost and an increase in the systolic pressure and a decrease in the diastolic pressure occur (Figure 8.10). Furthermore, the fall from the systolic to the diastolic value occurs more quickly. Finally, while there is an increase in PP, the MAP remains unchanged.

The situation that has been reported is typical of the loss of *windkessel effect* only. When the loss of the *windkessel effect* is accompanied by a limited or no decrease in the diastolic pressure, with an increase in MAP, we must conclude that in determining the blood pressure increase there are also other factors such as an increase in peripheral resistance and/or in excess volume (see Chapter 1.2). The loss of the *windkessel effect* with increased peripheral resistance can take place in *atherosclerosis*. Of note, arteriosclerosis is a physiologically aging effect on the vascular tissue which partially loses its natural elasticity. Atherosclerosis, on the other hand, is a pathological condition characterized by the presence of atheromatous plaques which aggravates the arteriosclerosis.

As said above, even an increase in mean pressure without increasing the stiffness of arterial vessels can cause an increase in pulse pressure. If the mean pressure increases, the aorta, and elastic arteries find themselves in a state of greater circumferential tension so that their wall becomes less extensible. In this way, even if the structure of their walls is not altered, the arterial vessel *modulus of elasticity* is increased (the compliance is reduced).

8.7 The Impedance to Ventricular Ejection

Although it helps to explain the genesis of arterial pressure with a good approximation, the concept of resistance is not adequate to explain the real obstacle that the heart encounters in injecting blood into the large arteries during ventricular ejection. Since the ventricles perform their pumping effect in a pulsatile manner, the concept of resistance must be replaced by the concept of aortic impedance to ventricular ejection.

The explanation can start from an electrical analog. Figure 8.11 illustrates an alternating current circuit.

In A, an alternator is represented which generates this type of current. At the exit of A, we find the rectifying diode D which transforms the alternating current into the pulsating current. The circuit also has a capacitor with capacitance C, an inductance I, and a resistor R located in parallel with each other. The impedance Z_l in which *l* represents the amplitude of the pulsation,

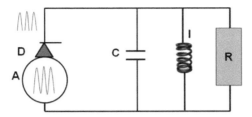

Figure 8.11 Alternating current circuit. A: alternator; D: rectifying diode; C: condenser; I: inductance; R: resistance. The rectifying diode transforms alternating current into the pulsating current. See the explanation in the text.

is given by:
$$Z_l = R^2 + (I - 1/C)^2.$$

If we pass from the electric analog to the ventricle-aorta system, l represents the amplitude of the pulse pressure, R the total peripheral resistance, I the inertial mass of the systolic flow and C the compliance or distensibility of the aorta and the large elastic arteries. In fact, as during a pulsatile increase in the current intensity, a capacitor takes on electrical charges to return them to the circuit when the intensity of the pulsatile current decreases, in analogy the elastic arteries accept blood expanding when the pressure passes from the diastolic to the systolic value for returning it to the downstream part of the circulatory system when the pressure returns to the diastolic value.

8.8 The Velocity of Propagation and the Length of Pressure Wave

Like all elastic waves, even the *pressure wave* or *sphygmic wave* propagates with a certain velocity and has a certain length.

If the vessels were rigid tubes filled with blood in an open system, the speed of propagation of the sphygmic wave would be equal to that of sound in water, *i.e.,* about 1000 m/s. In a closed system with rigid walls pressure would rise instantaneously in all points of the system (*Pascal principle*). This would avoid any pressure gradient and fluid movement. Fortunately, the elastic structure of the arterial walls considerably reduces the velocity of the transmission of the pressure wave. In the young subject, in which the arteries possess a good elasticity, it is about 5 m/s. The velocity increases with ag ing due to arteriosclerosis and the loss of elasticity, so that in the elderly subject it can reach 10 m/s. Importantly, as will be seen later, the velocity of the

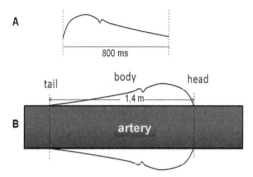

Figure 8.12 Pressure curve recorded in the time domain (A) and pressure wave in the space domain (B).

pressure wave must not be confused with the blood flow velocity which is on average 0.1–0.2 m/s.

The length of the sphygmic wave is about 1.4 m (spatial period of a periodic wave or the distance over which the wave's shape repeats). Of course, wave length must not be confused with wave duration, which is about 800 ms. In common descriptions wave duration is reported on the abscissa axis, while on the axis of the ordinates the pressure values are reported. In Figure 8.12, we compare a pressure curve recorded in the time domain (A) with its representation in the space domain (B). In the space domain, we can distinguish in the sphygmic wave a head, a body, and a tail. On the x-axis, the length l of the wave is included from the beginning of the head (on the right of Figure 8.12) to the end of the tail (on the left of Figure 8.12). Generally, in humans, the sphygmic wave has a length of 1.4 m. Since in humans no artery reaches this length, it follows that during propagation we have a time interval in which the artery is under the action of a pressure wave along its entire length.

If the pulsation of an artery generated by a certain beat is not due to the passage of blood, or a fraction of the blood pumped by that beat, this does not mean that at that moment no blood passes through the explored point. Rather, it means that the blood passing at that moment was an instant earlier in the artery immediately before the point where we perceived/sensed the pulse.

The effects of propagation velocity, length, and reflection of a pressure wave, are illustrated in Figure 8.13, where the spherical part on the left represents the left ventricle. From this starts the arterial system represented as a distensible tube facing to the right. The tube is divided into 5 compartments (C1-C5) and compartment C5 is the place were resistances are located. Above

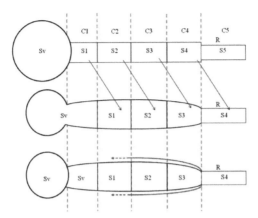

Figure 8.13 Propagation and reflection of the pressure wave in the arterial tree. Once the systole occurred, some of the blood contained in the heart moved to compartment C1, moving the blood contained in the subsequent compartments forward. When blood has been ejected into C1, the sphygmic wave has already reached the end of C4 from where it is reflected in a retrograde way (red curve). C1-C5: vascular compartments; Sv: Stroke volume; S1-S5: the blood of vascular Sections; R: resistance. Therefore, in each beat in our circulation, there are pressure waves traveling forward and backward. Here the Stroke volume is written as Sv, instead of SV, to remember that contrary to what is represented, SV is a fraction of the ventricular volume.

(diastole) the blood contained in the ventricle (Sv) is different from that contained in the various compartments of the arterial tree (S1-S5). In B, after the systole has ended, part of the blood contained in the ventricle (Sv) has passed into C1, while the blood previously contained in C1 has passed into C2 and so on.

In the intermediate position of Figure 8.13 we also see that the diameter of the arterial system is increased and that this increase extends to the end of section S4. The increase in diameter is due to the propagation of the pressure wave, while the length of the expanded vascular tract indicates the part of wave length that has already invaded the arteries. Since the head of the wave has reached section S4 when the blood ejected by the ventricle Sv is still in section S1, it is easy to understand how the sphygmic wave proceeds at a velocity much higher than that of the blood.

In the lower part of Figure 8.13, we see the effect of the wave reflection from the resistance located in the S5 section. Along the tube, in addition to the increase in diameter due to the presence of the sphygmic wave in anterograde movement, we also observe the arrival from the right of a further expansion that proceeds retrograde, as it is reflected by the resistance vessels (S5)

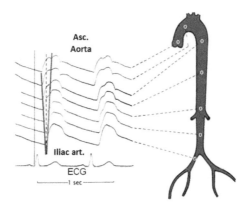

Figure 8.14 *Pressure wave along the arterial tree.* The red line indicates the diastolic pressure which coincides with the beginning of the ejection. The black dashed line corresponds to the end of the ejection phase with acceleration. The green line indicates the overlap of the reflected pulse wave on the main pressure wave. We can see that the reflected pulse wave is more temporally delayed as we get closer to the heart.

towards the heart. The reflected component of the sphygmic wave overlaps to the anterograde component which has not yet been exhausted. It is obvious that the reflected component will be registered, near the heart, in the rather late part of the pressure wave as it arrives at that point after reaching the resistance vessels and then re-traveling for a large part of the arterial tree in a retrograde direction.

Figure 8.14 illustrates how pressure curves vary as you move away from the heart. The pulse pressure, far from dumping out as one might imagine, actually grows taller and steeper for some distance.

If only the *windkessel effect* of the aorta and large elastic arteries and the viscous properties were active, we would simply have damping of the oscillations towards the periphery. On the contrary, despite the progressive reduction of the mean pressure, both an increase in the systolic pressure and a decrease in the diastolic pressure can be observed. These modifications are due to the wave component reflected from the periphery. In particular, this component, proceeding in a retrograde sense, is recorded belatedly at the origin of the aorta (ascending aorta), and earlier in the distal arterial tracts (*e.g.* iliac artery) due to the different length of the path between the points of pressure recording and wave reflection point. For this different occurrence of a reflected wave on the anterograde wave, the pulse pressure continues to increase as far as the third generation of arteries (*e.g.* femoral artery) but beyond this, it is progressively dampened by viscous properties of the arterial wall and blood. The pulse oscillations continue to diminish traveling in the arterioles and then disappear when the wave reaches the systemic capillaries.

8.9 The Oscillations of I, II and III Order of Arterial Pressure

The diastolic-systolic variations of the arterial pressure we have described are considered first-order waves. Second-order waves are formed by sets of synchronous oscillations with respiratory acts. Third-order waves are slow sets of waves (about 0.1 Hz) sometimes superimposed to the II order waves, which might be generated by chemoreceptor and baroreceptor reflex control systems.

Second-order oscillations, also called *Traube and Hering waves*, are characterized by an increase in pressure that begins in the second phase of inspiration and continues in the first phase of expiration and by a decrease that begins in the second phase of expiration and continues in the first phase of inspiration (Figure 8.15).

With the onset of inspiration, the pressure in the pleural cavity decreases as the thorax and lungs expand, retaining a certain amount of blood in the pulmonary circulation to reduce the venous return to the left heart for a few moments. As a result of this reduced venous return, there is a reduction in stroke volume and, consequently, in arterial pressure.

The decrease in pressure in the pleural cavity also causes an increase in the venous return to the right heart which, starting from the second phase of inspiration, will affect the venous return of the left heart through the pulmonary circulation. It then follows that starting from the second phase of inspiration; the SV and arterial pressure begin to increase. The increase in pressure continues in the first phase of expiration, when the lungs, after having expanded, begin to decrease in volume, reducing the bed of the pulmonary circulation that generates a greater return to the left heart which in turn favors an increase in SV.

Figure 8.15 Second-order waves of arterial pressure. I: inspiration; E: expiration. See explanation in the text.

However, the expiration also increases the pressure in the pleural cavity with a reduction in venous return to the right heart. As we have seen for the increase, also the decrease in venous return has repercussions after a few moments on the left heart, thus leading to that decrease in arterial pressure which extends from the second phase of expiration to the first phase of the next inspiration.

The III order oscillations, or *Mayer waves*, are very slow waves. Sometimes present in physiological conditions, they often appear in pathological conditions such as during severe anemia or after copious bleeding. It is thought that in these oscillations the pressure increase phase is due to the arrival of poorly oxygenated blood to the aortic and carotid chemoreceptors (remember the amount of oxygen in the blood depends mainly on the presence of hemoglobin). The activation of these receptors would stimulate the rostral ventrolateral medulla (RVLM) through afferent pathways contained in the vagal and glossopharyngeal nerves. Stimulation of RVLM would increase efferent sympathetic activity and therefore in an increase in HR, contractility, and vasomotor tone. The consequent increase in pressure would improve the arrival of blood and oxygen to the chemoreceptors located within the aortic and carotid bodies or glomus (the glomus arteriole does not constrict), which would therefore no longer be stimulated and would allow the reduction of systemic pressure to the initial values. At this point, the cycle is repeated. Similar intermittent stimulation of baroreflex has been also suggested as a cause and effect of pressure wave oscillations.

8.10 The Mechanisms of Control of Arterial Pressure

The heart activity is controlled by sympathetic and parasympathetic nerves and humoral factors. The tone of blood vessels is also controlled by sympathetic nerves and by many local and humoral factors. Of course, being MAP = HR $\times$ SV $\times$ TPR at any level of filling of the cardiovascular system, only changing these mechanical factors (HR $\times$ SV $\times$ TPR) and/or the level of filling of the system, MAP can be varied. The stability of arterial pressure and its values in the normal subject is linked to the functioning and integration of a series of nervous and humoral control mechanisms. Often the control mechanism activity is directed at stabilizing blood pressure. When a change in pressure occurs this triggers a response that returns mean blood pressure towards its previous value (control value or set point), this is a *negative feedback mechanism*. To understand the functioning and importance of each of the controlling mechanisms it is necessary to consider three of their characteristics: *the gain, the range, and the time course.*

The gain indicates the efficiency of the system. If, after a perturbation, the mechanism under examination can bring the arterial pressure back to control value, the gain is infinite, while if it does not correct the effect of the perturbation at all, the gain is zero. The gain can be determined with the following formula:

$$G = (Ei - Er)/Er,$$

where G indicates the gain while Ei and Er indicate the induced error and the residual error, respectively.

If a perturbation brought the average pressure from 100 to 180 mmHg, the Ei corresponds to 80 mmHg. If after a certain time the pressure returns to 100 mmHg, Er will be zero, and G will be infinite, while if the pressure remains equal to 180 mmHg, Er will be equal to 80, *i.e.,* equal to Ei, and G will be equal to zero. If the pressure drops from 180 to 110 thanks to the control mechanism, Er will be equal to 10, so the difference Ei - Er will be equal to 80 - 10 = 70, and G will assume the value of 7 (G = 70/10).

The range indicates the pressure range within which the mechanism is effective (Figure 8.16). For example, a mechanism may be operative when the pressure is between 60 and 180 mmHg, not below or above this range. These are the threshold and saturation value for some baroreceptors, which

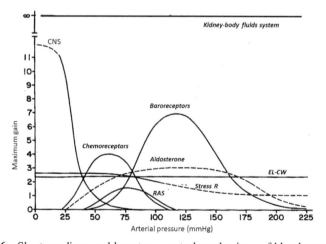

Figure 8.16 Short, medium, and long term control mechanisms of blood pressure reported in the pressure domain (from Pinotti O., 1963). RAS (renin-angiotensin system); Stress R (stress relaxation); CNS (central nervous system); Exchanges of liquid through the capillary wall (EL-CW).

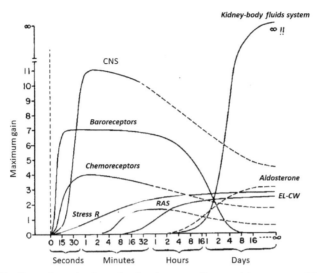

Figure 8.17 Control mechanisms in the short, medium and long term of blood pressure reported in the time domain after a sudden variation of pressure. (from Pinotti O., 1963). RAS (renin-angiotensin system); Stress R (stress relaxation); CNS (central nervous system); Exchanges of liquid through the capillary wall (EL-CW).

start to discharge when arterial pressure is above 60 and stop when it is around 180 mmHg.

The time course refers to the time taken by each mechanism to reach the maximum gain (Figure 8.17). If we classify the various control mechanisms according to the time course, we obtain *short, medium, and long term mechanisms*, depending on the time taken to reach the maximum gain. The time course also includes the adaptation time, which however is not taken into account in the definition of the control mechanisms. For instance, after hours of pressure increase baroreceptors adapt and discharge as before the pressure increase.

8.10.1 Short-term Control Mechanisms

There are three short-term control mechanisms and they are all neural regulation. They are the ischemic response of the central nervous system, the baroreceptor reflex, and the chemoreceptor reflex. These mechanisms will be considered also in Chapter 12. As can be seen in Figure 8.17, the *maximum gain* of these systems is reached in a few seconds.

8.10.1.1 Ischemic response of the central nervous system (CNS)

The *ischemic response of the CNS* is considered the last resort in the protection against serious falls in pressure, such as those caused by severe blood loss that brings the so-called *mean circulator filling pressure* (MCFP, see Chapter 1.2 and Figure 1.3) to values close to zero (*i.e.* when a 30% of blood is lost by hemorrhage). In situations of serious fall in pressure, ischemia of the RVLM area leads to their neuron activation. There is, therefore, a generalized sympathetic activation that results in arteriolar vasoconstriction responsible for an increase in TPR, and venular constriction with reduction of venous capacity and improvement of venous return.

Since the sympathetic also increases HR and contractility, the SV and CO can tend to increase and help to limit the fall in pressure. As shown in Figure 8.16 and Figure 8.17, the *maximum gain* is located between 35 and 0 mmHg, while the range concerns a reduced pressure (shock range). It is therefore very difficult for this mechanism to restore a normal pressure value if no other actions are taken such as appropriate blood transfusions.

Even when the intracranial pressure increases for a *space-occupying mass* (*e.g.* tumors or extradural hematoma), there is hypoperfusion and hypoxia of RVLM with a consequent increase in discharge from these neurons. This leads to an increase in sympathetic discharge and blood pressure which in turn, by stimulating the baroreceptors, will result in bradycardia (*Cushing reflex*) due to a prevalence of parasympathetic effects on the sinus node (see also next paragraph).

8.10.1.2 Baroreceptor system

Arterial baroreceptors are always operative in physiological conditions and function "to inform" the autonomic nervous system of beat to beat changes in arterial pressure. As can be seen in Figure 8.16, the range of baroreceptor control mechanism is very wide, ranging from 40 to 180 mmHg in MAP with the maximum gain when pressure variations occur around a mean pressure of 90–110 mmHg. The pressure level at which the maximum gain is found indicates that this system constantly tends to maintain pressure around normal values.

The extension of the pressure range suggests that this mechanism protects both from increases and decreases in pressure through greater or less stimulation of the baroreceptors, respectively.

The cause of the extension of the range of activity of the baroreceptors lies in the fact that the afferent fibers that depart from them are at least of

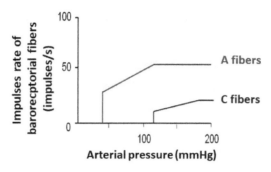

Figure 8.18   Trend of discharge rate of baroreceptors in fibers A and C.

two types (Figure 8.18): low threshold and low saturation A-type myelinated fibers, and high threshold and high saturation unmyelinated C-type fibers. The A-type fibers have a stimulation threshold of 40 mmHg and are therefore the first to be stimulated. Their discharge frequency increases linearly with the pressure increase up to 120 mmHg. For higher pressure value, their discharge frequency does not increase further, *i.e.,* are saturated. The C-type fibers, on the other hand, have their threshold at about 110 mmHg and their saturation at 180-200 mmHg. Fibers with a threshold and a saturation level around 60 and 180, respectively, also exist (not reported in Figure 8.18).

The stimulation/activation of the aortic and carotid baroreceptors by the increase in pressure causes inhibition of the sympathetic and stimulation of the vagal tone. The former allows the reduction of TPR and cardiac contractility, the latter of HR. These effects contribute to bringing the arterial pressure back to the control level; reducing it (this is called the *buffering effect* of baroreflex). *In the case of a drop in pressure, the lesser stimulation/deactivation of the baroreceptors will increase the pressure through an increase in sympathetic and withdrawal of vagal tone*. The former allows the increase of TPR and cardiac contractility, the latter of HR. The activation or deactivation of the baroreflex is usually so rapid that there are no appreciable changes in pressure.

Nevertheless, the time course of the baroreceptor activity in the pressure control indicates that the complete *adaptation* (baroreflex reset) takes place after a few days, even if adaptation starts a few hours after the disturbance (Figure 8.17). In other words, if the pressure is increased and for some reasons is not corrected, the baroreceptors adapt to the new pressure value and start to discharge at the same rate observed during control conditions. This observation has led for many years to the misleading concept that baroreceptors are not the long-term controller of arterial pressure. However, recent evidence suggests that adaptation is incomplete, and sustained baroreceptor responses lead to long-term alterations in sympathetic activity, particularly of sympathetic nerves directed towards the kidneys, thus mediating an increase in renal excretory function and a consequent limitation of the increase in blood pressure.

Sometimes, in anxious or in elderly subjects the baroreflex could be compromised. In these subjects, an emotional stimulus would lead to a sudden and conspicuous increase in pressure. This alteration can be at the basis of the so-called *white coat hypertension*, a phenomenon in which patients exhibit a blood pressure and heart rate level above the normal range in a clinical scenario, but not in a familiar environment.

While the ischemic response of the central nervous system very often fails to return to the normal values the pressure of a subject after the loss of a considerable quantity of blood (*e.g.,* blood pressure around 35–40 mmHg), the mechanism of the baroreceptors almost completely restores the pressure after a severe but not excessive bleeding (*e.g.* blood pressure around 55–60 mmHg). Due to the reduced stimulation of the baroreceptors, it is possible that a few hours after bleeding, some subjects may have a very high HR (*e.g.,* 150 b.p.m.) and a little reduced pressure (*e.g.* pressure is restored to about 90 mmHg).

8.10.1.3 Chemoreceptor system

The chemoreceptors, located in the aortic and *carotid bodies* (glomus or glomi) are stimulated when the O_2 tension (pO_2) decreases in the arterial blood or, to a lesser extent, when the CO_2 tension (pCO_2) increases. Figure 8.16 illustrates how this system is characterized by a range between 25 and 110 mmHg, with maximum gain at a pressure value of 60– 70 mmHg. It is easy to understand how the chemoreceptor mechanism protects against pressure drops only. If the average pressure falls below 70–75 mmHg, a smaller amount of blood perfuses the capillaries of the aortic and carotid bodies where these receptors are located. A lower blood supply leads to a lower oxygen supply.

In the presence of hypoxia, the chemoreceptors, while acting mainly on the respiratory system where they increase lung ventilation, also act on the cardiovascular system where, through inhibition of the vagus and activation of the sympathetic nerves, induce tachycardia and increase peripheral resistance, respectively, and favor then the recovery of the pressure.

While the increase in TPR is a *"pure"*, *"non-mediated"* chemoreflex, the increase in HR is mediated by the increase in pulmonary ventilation. I t has been observed that, if variations in pulmonary ventilation are prevented in an anesthetized dog, the stimulation of chemoreceptors by reduced pO_2 leads to bradycardia rather than tachycardia.

Remember that the fibers coming from the chemoreceptors are contained in the same nerves where that afferent by the baroreceptors are contained: the fibers coming from the aortic bodies are contained in the *vagus* while those originating in the carotid bodies are contained in the glossopharyngeal nerve.

Compared to the baroreceptors that protect against both the increase and the decrease in pressure, the chemoreceptors intervene in emergencies in the face of serious falls in pressure. This function has been tested in experimental preparations placed under general anesthesia. If in these experimental preparations the conduction along the vagal and/or glossopharyngeal nerves is prevented by cooling these nerves, there is a dramatically different result depending on whether the pressure is normal or seriously reduced:

1. *If the pressure is normal*, only the buffer fibers coming from the baroreceptors are activated, while those coming from the chemoreceptors are inactive, so that the cooling of the vague and glossopharyngeal cells causes an increase in pressure (the buffering effect of baroreflex is removed by the cooling).
2. *If instead the pressure is very low*, such as after copious bleeding, the fibers coming from the chemoreceptors are highly active, while those coming from the baroreceptors are inactive. In this case, the response to nerve cooling will be a further irreversible fall in pressure with the death of the animal (the activating effect of chemoreflex is removed by the cooling).

8.10.2 Medium-term Control Mechanisms

The medium-term control mechanisms are the exchange of liquid through the capillary wall, the renin-angiotensin and aldosterone system (RAAS) and the stress-relaxation phenomenon. In these mechanisms the maximum gain, although modest, is reached after several minutes or hours from the disturbance.

8.10.2.1 Exchanges of liquid through the capillary wall

It intervenes after a few minutes from the disturbance and reaches the maximum gain a few hours later. It has an infinite range and has no adaptation (Figures 8.16 and 8.17).

It is known that the balance between blood pressure in the various sections of a capillary, the colloidosmotic pressure of plasma and interstitial proteins, and the pressure of interstitial fluids regulates the net filtration of fluid in the interstitium from the capillaries. Indeed, already in normal conditions, a more or less large quantity is drained by the lymphatic system and subsequently returned to the bloodstream through the thoracic duct (see Chapter 18).

In the presence of an increase in blood pressure, due to an increase in excess volume and average systemic pressure, it is possible that filtration prevails over reabsorption along the entire length of the capillary and exceeds the amount that the lymphatic circulation returns to the bloodstream. In this case, the excess volume is reduced and the pressure tends to return to normal. If, on the other hand, the pressure is reduced due to the reduction of the excess volume, the reabsorption prevails over filtration and the pressure tends to rise again. Theoretically, the fluid exchange mechanism through the capillary wall has the same gain for all pressure values and does not undergo adaptation.

8.10.2.2 Renin-angiotensin-aldosterone system

Although the renin-angiotensin mechanism is often indicated separately from that of aldosterone which is referred to as a long-term mechanism (Figure 8.17), it is considered useful to treat them together for the integrated way in which they occur.

Through mechanisms that will be explained later (see Chapter 13.4) a fall in pressure induces renin production by the juxtaglomerular apparatus of the kidney. The RAAS is a critical regulator of blood volume and systemic vascular resistance.

In addition to determining mid-term vasoconstriction, angiotensin II also induces aldosterone production by the granular cells of the adrenal cortex. Aldosterone increases the renal absorption of sodium and therefore also of water by increasing the excess volume and thus favoring the rise in pressure. *Vice versa*, the increase in pressure reduces the basal renin release, and finally less angiotensin II and less aldosterone. In other words, the RAAS may be activated or inhibited to protect against both decreases and increases in blood pressure, respectively.

Even if the maximum activity of aldosterone (Figure 8.17) is reached a few days after the fall of the pressure and therefore it is a long-term mechanism, for the reasons explained above it has been treated among the medium-term mechanisms. It should also be kept in mind that the RAAS is counterbalanced by atrial and brain natriuretic peptides (see Chapter 13.3).

8.10.2.3 Stress-relaxation phenomenon

If a spring is stretched, its tension increases. After a certain time, however, while the stretching continues, the tension undergoes a certain reduction. This is the stress-relaxation phenomenon that can be observed with elastic bodies.

The blood vessels have elasticity. When the pressure increases, the radius of vessels also increases, causing distension of their wall which presents an increase in circumferential wall tension. After a while, the circumferential tension decreases. The decrease in wall tension ends up attenuating the increase in pressure. On the contrary, when the pressure decreases, the radius decreases and the wall see its tension decrease. After a certain time, the circumferential wall tension increases slightly. As a result of the increase in tension there is a certain recovery of pressure inside the vessel. This mechanism has little importance in blood pressure regulation.

All the medium-term control mechanisms we have examined so far have a limited gain and, with the sole exception of the exchange of liquids through the capillary wall, they undergo some adaptation (Figure 8.17).

8.10.3 The Long-term Control Mechanisms

8.10.3.1 Kidney-body fluids system

A variation of blood pressure corresponds to consensual diuresis and natriuresis. This is the only true long-term control system and is linked to the RAAS and other hormones and nerves regulating renal function. The *kidney-body fluids system* represents the most efficient mechanism in controlling blood pressure. It starts a few hours after a pressure variation and reaches an infinite gain after a few days (Figure 8.17). This gain remains the same for all pressure values. The system also does not undergo adaptation.

The kidney-body fluid system certainly uses the RAAS that we have instead placed among the medium-term mechanisms, even if the intervention of aldosterone is rather late and, as can be seen in Figure 8.17, it has a partial chronological coincidence with the mechanism we are discussing. It also uses natriuretic hormones and ADH. Without hormones and nerves, the mechanism is less efficient.

The characteristics that distinguish the kidney-body fluid system from the RAAS are infinite gain and lack of adaptation of the former mechanism. These characteristics suggest that the kidney must also use a mechanism of its own. This mechanism is believed to use the pressure gradient between collecting ducts and peritubular capillaries and *vasa recta*. If the blood pressure in peritubular capillaries and *vasa recta* is increased, the amount of water that is reabsorbed by the collecting ducts decreases. Consequently, diuresis increases, and excess volume is reduced. Conversely, if the pressure in the peritubular capillaries and *vasa recta* is low, the absorption of water is favored and diuresis decreased, so that excess volume, MCFP, and arterial pressure end up increasing.

8.11 Hemodynamic and Arterial Pressure Modifications in the Physical Exercise

In an individual during physical exercise, an increase in CO is normally observed. In dynamic isotonic exercise, CO can be quadrupled in the sedentary subject and can have a sevenfold increase in the well-trained athlete in the case of maximum exercise.

The increase in the CO is mediated by a greater venous return to the heart, as well as by an increase in heart rate and contractility. The increase in venous return to the heart is due to the massage of the skeletal muscles that contract on the veins that cross these working muscles and by the constriction of these and other veins by the sympathetic activation that accompanies the exercise.

In addition to the veins, the sympathetic also acts on the heart where it determines an increase in HR and contractility. Since the greater venous return accelerates the ventricular diastolic filling, the SV increases even if, due to the increase in HR, the filling time is reduced, while the better contractility determines a greater use of the systolic residue. In summary, in physical exercise, we are faced with a situation, in which an increase in HR is associated with a conspicuous increase in CO, because of both HR and SV increase.

Despite the increase in CO, in healthy subjects, physical exercise generally does not cause a significant increase in pressure. A possible increase in systolic pressure is generally accompanied by a decrease in diastolic pressure. Despite this, in fact, the generalized activation of the sympathetic also affects the precapillary resistance, reducing the blood flow especially in the mesenteric territory, the conspicuous metabolic vasodilation that takes place in the larger vascular territory, *i.e.,* in the skeletal musculature, ends up determining a reduction in TPR. It is noteworthy that what was said applies to the healthy individual, while in hypertensive subjects the increase in pressure during exercise is a possible event.

It is interesting to examine how muscular vasodilation occurs. At rest, in the pale-fiber muscles, the blood flow is about 2–4 mL/100 g/min. In stress conditions this flow can rise to 70–80 mL/100 g/min. This increase occurs in two stages. At the beginning of the exercise, or even in the preparatory phase of the exercise, the flow can increase up to 30 mL/100 g/min. This initial increase is due to the vasodilation that affects the precapillary arterioles and the proximal tract of the preferential path that surrounds the capillaries but does not affect the sphincters of the capillaries that arise from the proximal tract and end in the distal venous tract of the same route (Figure 8.9). This vasodilation increases the flow in the preferential path but not in the capillaries, therefore it does not represent a greater supply of nutrients to the muscles that contract. This first phase of muscular vasodilation is due to the so-called sympathetic vasodilator with cholinergic mediation or sympathetic beta2 mediation.

So what is the functional meaning of the initial vasodilation? The meaning is essentially hemodynamic in that it allows a reduction in TPR thus avoiding the increase in pressure that would occur as a consequence of the conspicuous increase in cardiac output.

The second phase of the increase in blood flow to the muscles occurs when the catabolites produced by the contraction determine the release of the smooth muscular fibro-cells that form the precapillary sphincters. In this way the flow increases in the capillaries, ensuring greater oxygen and nutrients supply to the muscles. In the Chapter on district circulation, the mechanisms responsible for vasodilation due to exercise in skeletal muscles will be discussed in detail (see Chapter 14.4).

The values of muscle blood flow that have been considered above concern the pale-fiber phasic muscles, which make up 75% of the organism's striated muscle mass. In the red fiber tonic muscles, the resting flow is about 20–30 mL/100 g /min and can rise to 150 mL/100 g /min under intense effort. These values are made possible by the greater vascularization of these muscles, for which a greater supply of blood and oxygen allows the persistent tonic contractions necessary for posture without the occurrence of *muscle fatigue* due to the exhaustion of high-energy phosphate compounds. The exercise described above is the so-called dynamic isotonic exercise in which there is a rapid sequence of short isotonic muscle contractions.

In the static exercise sustained and prolonged muscle contractions occur. These are directed at overcoming high resistance, as happens, for example, in the case of weight lifting or physical bodybuilding exercises. From a hemodynamic point of view, static exercise is characterized by a modest increase in heart rate and a significant increase in blood pressure. The increase in pressure, mainly due to the increase in sympathetic activity on the resistance vessels, favors the perfusion of vessels compressed from the outside by muscles that contract in a prolonged, mainly isometric, way.

8.12 The Measurement of Arterial Pressure in Humans

In humans, arterial pressure is measured using the Riva-Rocci sphygmo-manometer, both in its original mercury version and in the Bourdon version in which the mercury manometer is replaced by a metallic tubular spiral gauge with an elliptical section that can be stretched by the increase in pressure inside it, moving an arrow placed in the center of a dial.

Through special tubes, the pressure gauge is connected both to a rubber cuff wrapped in a cloth and to a pump that allows blowing air into the system. Given the connections, the pressure is the same in the pressure gauge and the cuff.

After the cuff has been placed around an arm, the air is pumped into the system to increase the pressure in the cuff and compress the brachial artery until the blood flow is completely blocked. To make sure that the artery is adequately compressed, the operator must check the disappearance of the radial pulse.

Once the occlusion of the artery is ascertained, the operator places the membrane of a *phonendoscope* downstream of the cuff along the medial margin of the distal biceps muscle tendon where the pulsation of the artery could be felt, if the artery is not completely occluded.

When the artery is closed, the operator does not hear any sound at the phonendoscope. By letting a certain amount of air to flow from a valve placed near the pump, he perceives the appearance of a sound (first Korotkoff's sound) in the phonendoscope as soon as the pressure in the cuff has dropped a little below the systolic pressure (Figure 8.19). The genesis of Korotkoff's sound is because the pressure succeeds in making the blood pass into the section of the artery compressed by the cuff only in systole, that is when the vessel is transiently and partially opened. Because of this incomplete opening, vortices are formed in systole in the artery that can generate sounds. The pressure value indicated at this moment on the pressure gauge is practically that of the systolic pressure.

The further decompression of the cuff allows the artery to remain partially patent for a progressively longer time during the cardiac cycle. The sound becomes longer and better audible (phases II, III, and IV of Korotkoff' sounds, see Figure 8.19). Then when the pressure in the cuff becomes lower than the diastolic pressure, the blood flows throughout the cardiac cycle through a completely patent artery. The Korotkoff' sounds at this point disappear, and the operator can read on the manometer the diastolic pressure.

Currently, for clinical use other types of pressure gauges are used, mostly with electronic transduction: some of these pressure gauges are easily used

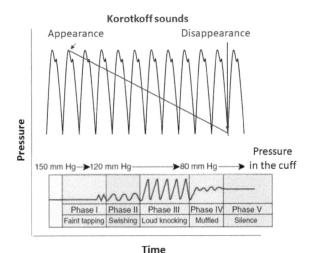

Figure 8.19 Measurement of arterial pressure using the Riva-Rocci method. The oscillating black curve is the arterial pressure. The oblique red line represents the decrease in pressure due to the deflation of the sphygmomanometer cuff. The violet arrows indicate the appearance (systolic pressure) and the disappearance (diastolic pressure) of Korotkoff sounds. The various phases of the Korotkoff sounds are reported. It is said that the systolic pressure must be read at the appearance of the first sound and diastolic pressure can be read on the manometer at the passage between phase IV and phase V.

by the same patient, while others, which allow the recording of the pressure curve, are used only in specialized centers also for their cost.

A correct determination of arterial pressure in humans must take into account heart rate and venous return to the heart. While there are no problems in assessing heart rate, changes in venous return can be indirectly assessed by repeating the measurement, first in the lying patient, and immediately after in an upright position.

It should also be remembered that oscillations of pressure synchronous with the respiratory rhythm can explain the small variations that can be obtained in subsequent measurements performed under unchanged patient conditions. The preferential use of the left arm for the determination of arterial pressure has no other reason than to standardize the procedure.

8.13 The Arterial Pulse

The arterial pulse is that increase in tension of an artery that can be felt manually in conjunction with the passage of a pressure wave at the point that is explored with palpation. The arterial pulse begins with aortic valve opening and ejection of blood from the left ventricle into the aorta. The nature of arterial pulse depends on left ventricular SV, ejection velocity, compliance, distensibility, and capacity of the arterial system.

Being the propagation of a pressure wave, or sphygmic wave, the arterial pulse does not reflect the passage of ejected blood in the explored point. The blood flow rate is much lower than the speed of propagation of the sphygmic wave. It follows that, when a heartbeat is revealed by the pulsation of a peripheral artery, a pressure wave arrives in the explored portion of the artery but not the blood ejected into the aorta from that beat. The pulse is transmitted by a wave of arterial wall distension at about 5 m/s, while the blood advances only 0.2 m in 1 s.

Pulse registration can be done using various techniques. Since a pulse wave has the same morphology as the pressure wave that generated it, we can observe an ascending or *anacrote phase* and a descending or *dicrote phase* (Figure 8.20). The dicrote phase shows the incisura due to the closure of the aortic valve and the dicrote wave due to the recoil phenomenon.

The apex of the systemic pressure wave when recorded not too far from the heart is rather rounded (*plateau*). Therefore, an initial part and a late part can be distinguished. The initial part coincides with the maximum ventricular ejection speed, while the late one is due to a reflection of the wave from the distal arteries and resistance vessels (see above). In some recordings, the apex of the curve may be even split into two waves. The one that corresponds to the maximum ejection speed is called the *percussion wave*, while the one that corresponds to the reflected wave is called the *tidal wave*.

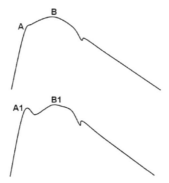

Figure 8.20   Pulse wave (sphygmic wave) recorded by an artery not far from the heart. A and A1: percussion waves; B and B1: tidal waves. In the bottom trace, the 2 waves A1 and B1 are more dissociated compared to the waves A and B of the above tracing. The dissociation can be due to a delay in the propagation backward of the reflected wave.

With aging, arteriosclerosis, and hypertension, there is decreased compliance, increased vascular resistance, and vasoconstriction of the arterial tree. The noncompliant arterial tree contributes to increased pulse wave velocity and the tidal wave becomes more sustained. Of note, the reflected component of the pressure wave regards the same wave which generated it. To understand this phenomenon, it is necessary to keep in mind the velocity and the site of reflection of a traveling wave (see Paragraph 8.8.).

In the clinic, using two fingertips several aspects of the pulse could be assessed, such as upstroke, systolic peak, the diastolic slope of the pulse and stiffness of the arterial wall, the rate, the rhythm, the amplitude, and the volume, as well as the pulse contour analysis with minimally invasive devices.

8.14 The Venous Pulse

The venous pulse can easily be registered by one of the external jugular veins. It is due to the retrograde propagation of pressure changes in the right atrium or CVP and is therefore characterized by three positive waves *a*, *c*, and *v* and two negative waves (depression) *x* and *y,* as in the atrial pressure. In the clinic, the venous pulse can be visible, rather than palpable, observing the neck of the patients.

Like the waves of CVP and left atrial pressure (Figure 5.1), the *a*-wave is due to the contraction of the atrium while the tricuspid is still opened. The positive *c* wave appears instead wider and longer-lasting than the one that can be recorded in the atrium. Indeed, not only the closure of the tricuspid but also to the transmission, through the tissues, of the internal carotid pulsation contributes to this wave. The *v* wave is due to the filling of the atria while the tricuspid is still closed.

Of the two negative waves the *x* is due to the displacement towards the apex of the atrial floor during ventricular ejection and the *y* to the rapid fall of blood from the atrium to the ventricle at the time of opening of the atrio ventricular valve. Cannon *a*-waves, can be observed in the jugular vein of humans when the atria contract against a closed tricuspid valve, as can occur in some arrhythmias (see also atrial pressure in Chapter 5.3).

Box 8.1 What is Normal Blood Pressure?

Normal blood pressure values and hypertensive classification are epidemiologic concepts that may vary in different populations and historic periods.

Tsimane Indios, inhabitants of Bolivia Amazon forest, have systolic/diastolic blood pressure values of about 116/73 mmHg throughout life. Also, the Kuna Indios who live on small islands outside mainland Panama in the Atlantic has blood pressure values of about 100/70 mmHg throughout their life. When they move to Panama capital, Panama City, they do show the age-dependent increase in blood pressure typical for the Western population. The Yanomami Indios who live across the border in the Brazilian Amazon region, also has similar blood pressure values throughout their life span.

If naturally living people have lifelong blood pressures of 100–120 in systolic and 70–80 mmHg in diastolic pressure, are we all hypertensive? Why occidental countries people have so much consistent increase in systolic blood pressure with age? Several factors might contribute: genetic changes, sedentary lifestyle, the composition of our diet, and the obesity epidemic are likely involved. This suggests that exercise, lifestyle, and diet modification can help in lowering our blood pressure values.

Is the true target value, therefore, <120/80 mmHg? For sure with blood pressure there are limits; at least there must be enough pressure while standing to perfuse the brain. This value is probably around a MAP of 80–90 mmHg in young and healthy individuals. From several clinical studies, it appears that to achieve the clinical benefit, the level of baseline blood pressure, and the degree of blood pressure-lowering are essential factors. The most recent US Clinical Practice Guidelines for High Blood Pressure (November 2017) changed the recommended target levels of blood pressure significantly. Indeed, they now consider <**120/80 mmHg or less as normal**, 120–129/80 mmHg as elevated, 130–139/80–89 mmHg as stage 1, and >140/90 mmHg as stage 2 hypertension. The 2018 ESC/ESH Guidelines for the management of arterial hypertension use a little different classification and they consider <**120/80 optimal,** systolic **120–129 and/or** diastolic **80–84 Normal**; 130–139 and/or 85–89 High normal. Hypertension is classified as: Grade 1 hypertension 140–159 and/or 90–99; Grade 2 hypertension 160–179 and/or 100–109; Grade 3 hypertension $\geq$ 180 and/or $\geq$ 110, considering systolic and diastolic values respectively. Isolated systolic hypertension is considered when systolic $\geq$ 140 and diastolic < 90 mmHg.

Perhaps, in hypertensive management, a more personalized medicine should be applied to overcome the transatlantic debate on single numbers that is not very fruitful for optimal management. For sure, when optimal management of blood pressure therapy is reached a reduction of cardiovascular risk is achieved.

Box 8.2 A Bit of History: From the Discovery of Blood Pressure to the Discovery of Hypertension

Although in his fundamental work, Anatomical Exercise *by Motu Cordis et Sanguinis in Animalibus* published in 1628 William Harvey (1578–1657) had suggested a driving force for circulation, it took more than a century before blood pressure was measured. In 1733, Steven Hales (1677–1761) inserted a glass cannula into the carotid artery of a horse with help from his assistant, and for the first time determined the level of blood pressure in a living awake animal and demonstrated its changes during the cardiac cycle. Unfortunately, his fundamental experiment had little influence in the clinical practice of the physicians of that time as this approach is a little invasive and difficult if not impossible to use in "relaxed patients". Furthermore, any understanding of the importance of hypertension for the development of cardiovascular disease was missing at that time. This began to change in 1896 when an innovative Italian doctor, Scipione Riva-Rocci (1863–1937) built a device that with some modifications is still in use today, namely the *sphygmomanometer*. At first, Riva-Rocci could only assess systolic blood pressure with his fingertips on the radial artery; however, immediately after a Russian military doctor, Nikolaj Sergeeviè Korotkov (also known as Nikolai Sergejewitsch Korotkow, or as Nikolai Korotkoff 1874–1920), described the vascular noises that are named after their discoverer. In November 1905, during a conference of the Imperial Military Medical Academy, Korotkoff reported his discovery. In 1939, the Joint Committee of the American Heart Association and the Cardiac Society of Great Britain and Ireland recognized officially the *Korotkoff and Riva-Rocci's method* for blood pressure determining. Therefore, blood pressure measurement slowly entered clinical practice at the beginning of the last century. However, as long as the seminal Framingham Study was initiated in a suburb of Boston in the post-II war years, many physicians were convinced that having high blood pressure was *"essential"*, hence the name of *"essential hypertension"*, even today used for the hypertensive conditions of who we do not exactly know the cause, which, represents over 95-98% of the cases of hypertension that a physician finds during his professional activity. Essential, hypertension is also called idiopathic or primary hypertension and can be defined as high blood pressure in which secondary causes such as aldosteronism, pheochromocytoma, renal failure, renovascular disease; aorta coarctation or other causes of *secondary hypertension* or Mendelian forms (monogenic) are not present. Nowadays, essential hypertension is considered a heterogeneous disorder, with several putative causative factors leading to hypertension in the various patients who can be placed in different syndromes. Very often in hypertension, there is *endothelial dysfunction,* which may be the cause and/or the consequence of hypertension itself.

9

Work and Heart Metabolism

9.1 The Work of the Heart

At each beat, each ventricle performs a work consisting of two components: the potential energy and the kinetic energy. Of these two components, the largest is the *potential energy* which is represented by the pressure (P) given to the volume (V) of blood that is moved from the ventricle into the great artery that starts from it. The *kinetic energy* is represented by the velocity (v) given to the mass (m) of blood *during the ejection phase.*

The total work of a ventricle (W_t) will therefore be:

$$W_t = P \times V + 1/2 mv^2$$

where $P \times V$ is the potential component and $1/2 mv^2$ is the kinetic component.

Indeed, the potential energy corresponds to the area enclosed by the pressure-volume loop (Figure 5.7) and gives to the blood the energy necessary to move along the vessels during ejection and the following diastole (when the ventricles are resting, *i.e.* are not working).

Of course, $P \times V$ and $1/2 mv^2$ have both the physical dimensions of a work. We can show that $P \times V$ is equal to a work in which there is a displacement (length, l) in the direction of the applied force, f.

Indeed, the pressure, P, is a force f applied on a surface, s, and is given by the ratio

$$P = f/s.$$

Therefore, the force, f, can be indicated as a product:

$$f = P \times s.$$

Since the displacement has the dimension of a length, l, we can write:

$$\text{Work} = P \times s \times l.$$

Since the product of a surface s for a length, l, is a volume V, it will be

$$\text{Work} = P \times s \times l = P \times V,$$

where V represents the *stroke volume*.

Therefore we can write it also as:

$$\text{Work} = \text{MSP} \times \text{SV},$$

where MSP is mean arterial pressure in systole and SV is stroke volume.

Similar reasoning can be done for the *kinetic energy*. If we consider the work of the left ventricle, we see that under resting conditions the potential component represents about 98% of the total work, while the kinetic component amounts to only 2%. It has been calculated that in the case of physical exercise the increase in ventricular ejection velocity can increase the kinetic component up to 25%.

In absolute terms, the *potential component of the work* of the right ventricle is about one-sixth of the potential component of the left ventricle, due to the low pulmonary pressure. Instead, in absolute terms, *the kinetic component* is similar because the velocity of ejection and the blood mass ejected by the two ventricles are similar (Figure 9.1).

The sum of the work performed by the two ventricles in the same beat, *i.e.,* the total cardiac work (CW_t) will, therefore, be left ventricular work (LvW) plus right ventricular work (RvW). If the considered pressure, P, is the *mean*

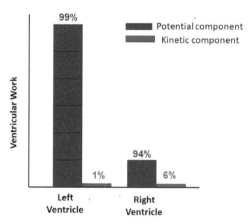

Figure 9.1 Potential and kinetic components of the work of the left and right ventricle. The column heights are proportional to the absolute values of the two work components in the two ventricles.

aortic pressure in systole, we can write:

$$LvW = P \times V + 1/2mv^2$$
$$RvW = 1/6P \times V + 1/2mv^2 =$$

$$CW_t = 7/6 \, P \times V + mv^2.$$

We have seen that for the left ventricle in resting conditions, the kinetic component represents about 1–2% of the total work performed by the left ventricle. Since in absolute terms, the kinetic component is almost the same for the two ventricles, it derives that the kinetic component represents about 6–12% of the total work performed by the right ventricle (Figure 9.1).

As above mentioned, the pressure that must be considered in the calculation of the potential component of the work of the heart is the *mean systolic pressure*, *i.e.* the mean value of aortic or pulmonary pressure during ventricular ejection. Indeed, this is the only phase of the cardiac cycle in which there is a movement of blood from the ventricle to the big artery and, thus, the only phase in which an external work is performed by the ventricles.

The mean systolic pressure is the average of the infinite pressure values that the blood assumes in the aorta (or in the pulmonary artery) during the ejection phase. Also, the velocity, v, of blood ejection is the average of the infinite values that the velocity assumes during ejection. Therefore, the external work can be represented as changing volume, pressure, and velocity of blood in the arterial system. To better underline these continuous variations of these parameters it may be appropriate to report the two components of cardiac work as an integral:

$$W = \sum_{E1}^{E2} P \times dV + \sum_{E1}^{E2} 1/2dmv^2,$$

where the sign $\sum_{E1}^{E2}$ indicates the integral from the beginning E1 to the end E2 of the phase of ejection.

Nevertheless, it is possible to calculate the real work of the heart using the general formula, above reported ($W = P \times V + 1/2 \, mv^2$).

Since $P \times V$ corresponds to the area enclosed by the pressure-volume loop (Figure 5.7) it can be inferred that this area, and thus the work, can be modified by varying either the filling (preload) or the pressure in the aorta (afterload). However, even if the areas obtained can be the same, for example reducing afterload an increasing preload, the cardiac efficiency will be different, depending on the changed parameter (see below).

Box 9.1 Calculation of Cardiac Work

We can estimate the work performed by the ventricle considering that about 70 mL of blood (0.7×10^{-4} m^3) is ejected into the aorta producing an average systolic pressure of 100 mmHg (1.33×10^4 N/m^2) at the average outflow velocity of 0.5 m/s.

Calculating the *potential component of the work* of the left ventricle we see that it performs less than 1 joule of mechanical work. In other words, the arterial system gains *1 joule* of potential energy which serves to move the blood in the vascular system along the entire cardiac cycle. To get an idea of how much a joule is, we should think about the work required to lift a mass of 102 g (a small apple) of one meter, opposing the Earth's gravity.

The *potential work of the left ventricle* can be calculated multiplying MSP for SV (pW = MSP $\times$ SV = P$\times$V). To evaluate the *potential work of the right ventricle*, in the absence of an exact measurement of the pressure in the pulmonary artery, one can divide by 6 the P$\times$V work of the left ventricle:

To calculate the *kinetic component of work* as $1/2mv^2$, we should consider that 70 mL of blood have a mass of about 80 g and is ejected at a velocity of about 0.5 m/s. Therefore, the kinetic energy is about 0.01–0.02 kg m^2 s^{-2} or *0.01-0.02 joule*. This is 1–2% of the external work of the left ventricle and can be neglected at rest. The kinetic energy is about is 6–12% of right ventricular work.

Given the negligible value of kinetic energy, the sum of the work of the two ventricles per beat results from the sum of the potential components of the two ventricles (*1.167 J*). Referred to 24 hours the work of the heart is about *117634 joules*. You can get an idea of how much work is done in one day by our heart thinking that it corresponds to throwing an apple at a height of one meter from our hand for about 110,000 times per day.

9.2 Heart Performance

In physics, the efficiency (E) is given by the relationship between the external work (W) produced and the energy (En) consumed to produce it:

$$E = W/En.$$

This formula also applies to the heart, where, it is used the *myocardial oxygen consumption* (MVO_2) as an index of consumed energy. MVO_2 (mL O_2/min) measured according to the Fick-principle by multiplying coronary sinus blood flow (mL/min) by the arteriovenous oxygen content difference.

We have seen that the main component of the work of the heart is that given by the $P \times V$ product, *i.e.*, the product of the pressure P during the ventricular ejection (mean systolic pressure) for the volume of blood ejected V (stroke volume).

The efficiency of the heart is rather low (only 10–25% of the energy invested is converted to external power) mainly because in the *isovolumic systole* the ventricles consume energy without producing work since at this stage there is no movement of blood. It is likely pushing a really heavy object that does not move at all. The residual energy mainly dissipates as heat.

The $P \times V$ product can have the same value either for a low value of P and a high value of V, or, *vice versa*, for a high value of P and a low value of V. However, a heart beat against a high aortic pressure or afterload (a beat "predominant in pressure") is accomplished with lower efficiency (low stroke volume) compared to a beat performed against a low aortic pressure which yields a high stroke volume (a beat "predominant in volume"). In other words, the higher the value of the pressure that the blood must overcome to eject the blood into the circulation, the more is the energy cost of a phase (the isovolumetric phase) in which the oxygen consumption does not correspond to any work (it is said that during the isovolumetric phase an internal work is performed without displacement). In general, beats "predominant in pressure" are those beats that take place in hypertensive subjects and those that occur at a high frequency that does not allow diastolic pressure to fall. These are both conditions of low heart efficiency. On the contrary, beats "predominant in volume" are those beats occurring in hypotensive subjects, and those occurring at a low frequency that, due to the long diastole, allow diastolic pressure to reach rather low levels. These are both conditions of higher heart efficiency. Among the poor performant beats are also included in the very early extrasystoles, which may even have an efficiency equal to zero, when, due to their extreme precocity, they fail to produce any systolic ejection (*i.e.* the systole is only isovolumetric).

During dynamic exercise, the heart efficiency can improve, being even the double of the resting conditions (about 20–30% vs about 10–15% at the resting condition). This is because during this type of exercise the stroke volume can even double; while the diastolic pressure is low and the mean aortic pressure does not change appreciably. On the contrary, high blood pressure increases oxygen demand and reduces efficiency. Finally, it must be remembered that, due to the lower pressure of the pulmonary circle, usually the right ventricle works with higher efficiency compared to the left ventricle.

Another, important aspect to consider regarding heart efficiency is the level of ventricular volume. Indeed, considering the Laplace's law (see Chapter 4) we have learned that to develop a certain pressure P a heart of enlarged radius, r, has to exert a greater wall tension, T. Indeed, the Laplace's law for a hollow sphere is $P = 2T/r$. Since T is equal to wall stress, S, multiplied for wall thickness, w, the formula can be also written as $P = 2Sw/r$. This implies that to develop a certain P during contraction, an increase in S and T and therefore in MVO_2 is necessary, with a consequent decrease in efficiency, in the presence of cardiac dilation such as that occurring in heart failure. That is why it is advisable to reduce cardiac dilation by diuretic therapy in heart failure. Diuretics also reduce blood pressure thus improving efficiency. Of course, the ventricular volume cannot be reduced drastically; otherwise, the heterometric adjustment could be compromised. Even the blood pressure cannot be reduced too much otherwise the perfusion of the organs could be endangered. A balanced effect should be pursued to improve cardiac efficacy without compromising the CO and perfusion of the vital organs. The expression $P = 2Sw/r$ tells us that $Pr/2w$ is equal to S. The latter decreases as the thickness of the ventricular wall increases, for example, in athletes. Therefore, despite an increase in r, paradoxically, an increase in w mitigates the increase in S and MVO_2. Indeed, systolic wall stress, S, represents a primary determinant of MVO_2 in pressure and volume overload. However, the MVO_2 further increases in these diseases when hypertrophy becomes inappropriate about the pressure and volume demands imposed on the left ventricle.

9.3 The Metabolism of the Myocardium

As we have already mentioned in Chapter 4.4, the energy that is used in the myofilament sliding is produced by the hydrolysis of ATP. In this hydrolysis, the release of 7.5 kcal/mole is obtained.

$$ATP \Leftrightarrow ADP + P_i + 7,5 kcal/mole.$$

The splitting of ATP is also necessary for the cell maintenance and the transport of ions by Na^+/K^+ and Ca^{2+} cell membrane pumps as well as by the Ca^{2+} pump of the sarcoplasmic reticulum, SERCA2.

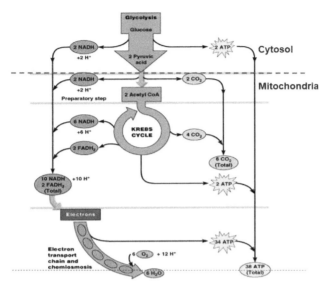

Figure 9.2 ATP resynthesis from glycolysis, Krebs cycle, and oxidative phosphorylation. The scheme shows that there is a theoretical yield of 38 ATP molecules per glucose during cellular respiration. Modified from Pearson Education Inc.

Indeed, most of the ATP is used for the myofilaments sliding which has been described in Chapter 4.1. The majority of ATP is synthesized by oxidative phosphorylation in the abundant mitochondria. About 40% of the cardiomyocytes volume is made by the mitochondria. For this operation of resynthesis, it is necessary that energy sources, namely glucose, fatty acids, and oxygen, are supplied continuously. As can be seen in Figure 9.2, the development of energy by the glucose begins with anaerobic glycolysis which gives rise to the formation of pyruvic acid in the cytosol. In this first metabolic phase for each glucose consumed, 2 moles of ATP are resynthesized, a rather modest amount for the needs of the heart.

The pyruvate decarboxylase enzyme transforms pyruvic acid into oxalacetic acid, which, in the presence of acetyl-coenzyme A (acetyl-CoA), enters the *Krebs cycle* (tricarboxylic acids cycle), leading to the re-synthesis of another 2 moles of ATP. Acetyl-CoA is mainly produced by the *β-oxidation of fatty acids*, but it can also derive from pyruvic acid due to the action of pyruvate-dehydrogenase which catalyzes the process of oxidative decarboxylation. At the end of the Krebs cycle, in addition to the formation of ATP and CO_2, oxalacetic acid is still present, while acetyl-CoA has been removed. The Krebs cycle is followed by the *oxidative phosphorylation* responsible for the resynthesis of 32–34 moles of ATP. Therefore, for each mole of glucose consumed, from glycolysis to the end of oxidative phosphorylation, 34–36 moles of ATP can be reformed within mitochondria. In eukaryotic cells, the theoretical maximum yield of ATP generated per glucose is 36 or 38, depending on how the 2 NADH generated in the cytoplasm during glycolysis enter the mitochondria and whether the resulting yield is 2 or 3 ATP per NADH.

It is said that the heart is something an opportunist, increasing its utilization of whatever substrate is currently available in the bloodstream. However, usually, the heart draws energy mainly from the lipids, which is the β-oxidation of fatty acids. Therefore, at rest from lipid metabolism derives about 65–70% of the energy required, while from the metabolism of glucose and lactate derives only 30–35% of necessary energy. During endurance exercise, the quote of fatty acid used may increase. The heart has a scarce capacity to use anaerobic glycolysis. Therefore, if oxygen is not delivered, the heart stops beating in less than 3 or 4 minutes, although in hypoxia it can use anaerobic glycolysis to a greater extent than usual. Also, after a meal rich in carbohydrates it can, for some time, use prevalently glucose. In diabetic acidosis, the heart can also draw energy from the oxidation of ketone bodies, while during some type of exercise it can oxidize mainly the lactic acid which is formed by the reduction of pyruvic acid. The oxidation of lactate involves a specific lactic dehydrogenase (LDH-H4), which is typical of the cardiac muscle. This enzyme is checked in the blood of patients with symptoms of myocardial infarction, to ascertain the diagnosis because dying cardiomyocytes releases it in the blood together with other enzymes, including creatine phosphokinase. This latter enzyme is important in the use of the small quantity of energy stored as creatine phosphate (phosphocreatine). Also, a small quantity of oxygen is stored in the cardiomyocytes as oxymyoglobin (an oxygenated form of myoglobin). These stores allow some cells to survive for a few minutes when the coronary flow is interrupted (see Chapter 17 on ischemia/reperfusion injury for more information on this topic).

10

Electrocardiogram

10.1 The Definition of Electrocardiogram and Dipole Theory

The electrical activity of the heart produces *electric fields* transmitted to the surface of the body. ***The electrocardiogram (ECG) can be defined as the graphic recording of these electric fields by electrodes placed on the skin.***

The electrical activity of the heart is due to the processes of depolarization and repolarization during a heartbeat. Both processes propagate from a region to the other of the heart. Indeed, the ECG is the recording of origin and diffusion of both heart excitement (depolarization) and its recovery to resting conditions (repolarization). Therefore, ECG is not the registration of action potentials as such but the consequence on the surface of the body of the processes of the traveling of depolarization and repolarization within the heart, which result in several variations of the electric field that determine a modular repetition of similar waves.

While the electrical activity is located in several parts of the heart (sinusatrial node, sinoatrial, interatrial and atrioventricular conduction systems, myocardium), *the ECG is the result of the processes of depolarization and repolarization **traveling** within the various parts of the heart that produces cyclic variations of the electric field on the body surface. These cyclic electrical fields can be registered by electrodes on the surface of the body.* From these considerations on the definition of ECG, it emerges that ECG can give information on the *diffusion of the impulse* (the traveling depolarization/ripolarization) *within the heart mass* but cannot give direct indications on the mechanical activity of the heart.

Since we record a phenomenon occurring outside of the cells, for the ECG we can consider only the electrical phenomena on outer sides of the fibers. In Chapter 3.1 we have seen that a resting myocardial fiber is polarized, that is, it has positive charges outside and negative charges inside the cell membrane

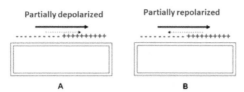

Figure 10.1 Partial depolarization (A) and repolarization (B) of a myocardial tissue. The dashed arrows represent the sense of movement of the vectors. The continuous arrows are vectors having the tip pointing towards the part of the polarized fibers (positive resting or repolarized part). Vector length is proportional to the difference in potential between the two outer sides of the fibers. When the fiber is completely depolarized (all negative outside) or repolarized (all positive outside) there is no vector.

(Figure 3.1A). When it is in a state of excitement, the cell membrane is depolarized, that is, it has negative charges outside and positive inside the cell membrane (Figure 3.1B).

While the excitement is propagating along the tissue, we see that some of the fibers are depolarized on the left, but still polarized (resting) on the right (Figure 10.1A). The traveling of the excitation is therefore direct from left to right and can be represented by an arrow (dotted arrow in Figure 10.1A), which moves from left to right and has the tip pointing to the right. In Figure 10.1B, the traveling of the repolarization is direct from right to left and can be represented by an arrow, which moves from right to left and has the tip pointing to the left (dotted arrow in Figure 10.1B).

The electrical phenomena generate electric fields which have vector characteristics: *size* (length), *direction* (position within the coordinate system), and *polarity*.

The tip of the vector (continuous arrow in Figure 10.1) pointing to the right does not indicate the direction of movement, but the fact that it is facing the still polarized fibers. Indeed, this vector is a representation of the electrical field and it is always represented with the tip pointing towards the positive field, regardless of the direction of its movement. It follows that, after complete depolarization, if the fibers undergo repolarization starting from the last depolarized tract, *i.e.* from the right end, which is again positive outside, the vector will still be represented with the tip pointing to the positive field (to the right of the reader). However, now the repolarization moves backward (i.e. in the opposite direction to the previous sense of depolarization) as indicated by the dotted arrow (Figure 10.1B). For simplicity, we can say that the tip of the vector is like the headlights of a car that first goes forward and then reverse. In this example, an electrode on the left side of the patient "will see" the *vector tip (or the lights* if we refer to the example of the car) during both the depolarization and the repolarization.

By convention, an electrode connected to the positive pole of the electrocardiograph (the entry of galvanometer) is named *positive electrode*. By convention, a positive electrode determines an upward (positive) deflection in the ECG if it is in a positive electrode field. Therefore, in our example, if the electrode on the left side of the patient is a positive one, the ECG wave will be upward for both depolarization and repolarization (see also below).

The length of the vector is an indication of the *size (voltage)* of the electrical field and depends on the difference between the positive and negative electrical charges present on the outside of the membrane, *i.e. the difference in potential between two parts both located outside the fiber*. The magnitude of this potential difference is one of the main factors (not the only factor; see below) determining the amplitude of the wave recorded by the ECG. Of course, when the fibers are completely polarized or completely depolarized as in Figure 3.1, no vector is represented; in this case, the ECG will not record a difference of potential and will not produce any wave.

The way in which the depolarization and repolarization of the myocardium give rise to the electrocardiographic waves that can be recorded at the surface of the body can be studied with the *dipole theory*. A dipole consists of two points of opposite electrical charge separated by a very short distance.

Box 10.1 The Dipole Theory

An *electrical dipole* can be created with a container filled with a saline solution, *i.e.,* having the ability to conduct electricity (Figure 10.2). Since it is three-dimensional, the solution is a conducting volume. The uncovered tips of two electrodes represent the two opposite electrical charges, generated by a battery connected to the electrodes. In this way, electric fields will form in the solution. The positive field around the anode and the negative one around the cathode.

Since they are immersed in a conducting volume, the two electric fields will be represented by eccentric isopotential (equipotential) spheroidal surfaces. For simplicity, here instead of a spheroid, just isopotential eccentric oval circles on a single plane are considered. The isopotential surfaces of each field have a voltage which decreases with the square of the distance from the respective pole. The potential differences can be measured by two electrodes aligned across the two poles and connected to a voltmeter.

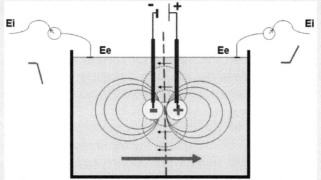

Figure 10.2   Artificial dipole scheme. The red arrow indicates the orientation of the dipole. The black arrows indicate the direction of the current represented by the black dotted circles. The red vertical dashed line represents the potential-free plane. Eccentric oval circles (solid circles) indicate isopotential surfaces. The passage of current causes opposed deviation of the galvanometer needle depending on the position of the electrodes. For convention, when the Ee is in the positive field the red trace of a recorder (*e.g.* an electrocardiograph) goes upward, *viceversa* the trace goes downward when the Ee is in the negative field. Ei: the indifferent electrode connected to exit of galvanometer; Ee: the elecrode connected to entry of galvanometer; See explanation in the text.

The vertical plane passing through the center of the dipole is at zero potential at each point. Two electrodes aligned on this plane will not record a potential difference when connected to a voltmeter, whatever is the potential difference generated by the dipole.

If the two electrodes are connected to entry and exit of a voltmeter or galvanometer and we place one exploring electrode (Ee; the one connected to entry of galvanometer) on the surface of the conductive volume close to the anode side and the other indifferent electron (Ei; the one connected to exit of galvanometer) at zero potential. The galvanometer needle will be deviated from the side indicating the passage of a current from Ee to Ei (Figure 10.2). If the electrode Ee is instead placed on the cathode side, the deviation of the needle will take place in the opposite direction.

The more Ee will be close to the anode or cathode, the greater will be the voltage of the isopotential surface in contact with the electrode, so that the greater will be the intensity of the current and the deviation of the galvanometer needle.

In addition to deflecting the needle of a galvanometer, the passage of a current can also be represented by a graphical recording (red traces in Figure 10.2 and 10.3). The recorders (*e.g.* the electrocardiograph), by convention, are calibrated so that, when the Ee is in the positive electrical field of the dipole, the deflection is represented by an upward movement of the trace from the baseline (a positive wave), while the opposite occurs (a negative wave or downward movement of the trace) when the Ee is in contact with the negative field (Figure 10.3).

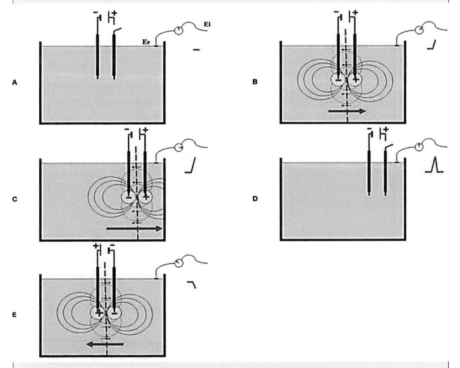

Figure 10.3 Genesis and displacement of an electrical dipole in a conducting volume. A: no dipole; B: dipole appearance; C: displacement of the dipole towards the exploring electrode Ee; D: disappearance of the dipole; E: dipole with inverted polarity. The deflection of the galvanometer needle and of the graphic trace (in red), according to the positivity or negativity of an electric field in a conducting volume, is reported. Ee: exploring electrode; Ei: indifferent electrode. See explanation in the text.

If we move the dipole into the container and we follow the graphical deflection from the baseline, we will see the trace to go up and down. For instance, if in the initial situation the circuit is open, there is no dipole, and therefore, the trace remains at the baseline levels, i.e., it remains at the *isoelectric level* (a putative zero level, Figure 10.3A).

If we close the circuit (Figure 10.3B), the positive and negative electric fields appear. The Ee is now reached by an isopotential surface rather distant from the positive pole and therefore with a rather low voltage. Under these conditions, there is a modest upward deflection of the graphic recording trace.

If we move the dipole to the right (Figure 10.3 C), the Ee comes progressively in contact with progressively higher voltage isopotential surfaces; therefore the deviation of the trace becomes progressively more marked. If now the circuit is opened (Figure 10.3 D), the two electric fields disappear and the trace returns to the isoelectric position.

The next step is to bring the dipole back to the center of the container (Figure 10.3E), taking care to reverse its polarity. As long as the circuit is open, the galvanometer does not signal any current passage and the trace remains isoelectric. When the circuit is closed, the electric fields are formed again, obviously with inverted polarity with respect to the previous situation.

As a synthesis of the maneuvers that have been described, we can say that the trace has a positive (upward) deflection each time the Ee sees the positive pole approaching and a negative (downward) deflection if it sees the negative pole approaching.

Up to this point, we have examined the approaching of the dipole to the Ee. Now let's consider what happens when the poles move away from Ee. Let's return to the situation in which we have opened the circuit and the trace is on the isoelectric line (Figure 10.3 D). As soon as the circuit is closed, a maximum upward deflection occurs due to the contact of Ee with a high voltage isopotential surface (Figure 10.3 C). If the dipole is then moved faraway and towards the left side of the container, the deflection progressively decreases (Figure 10.3B). Of course, the deflection disappears if the battery is disconnected again (Figure 10.3A). With this maneuver, we see that a positive deflection is present not only if the pole with positive charges approaches Ee, but also if it moves away from it. If we then repeat this last maneuver after having inverted the position of the dipole we can observe that Ee allows recording a negative deflection not only if it "sees" the negative field of the dipole approaching, but also if it "sees" the negative field of the dipole moving away from it.

Reading the Box 10.1: ***"The dipole theory",*** we have understood that *an ECG deflection is positive (upward) when a "positive" **Ee** is in the positive field of a dipole, regardless of whether the positive field approaches or moves away from the electrode, and the deflection is negative (downward) if the Ee is in the negative field of the dipole, regardless of whether the negative field approaches or moves away from the electrode.*

Since in electrocardiography the various dipoles that are formed in the processes of depolarization and repolarization of the various parts of the myocardium can be represented by vectors with the tip pointing towards the positive and the tail facing the negative pole, respectively, we can also say that *the deflection is positive whenever **Ee** "sees" the tip and negative every time it "sees" the tail of the vector*. Remember: **Ee** is the electrode connected with the entry of the voltmeter and we can call it *the positive electrode*.

Moreover, since the difference in potential between the positive and the negative field of the dipole can be more or less wide, its amplitude is represented by the length of the vector. Since the vertical plane passing through the center of the dipole is at zero potential at each point, the two electrodes aligned along this plane will not record a difference of potential when connected to a voltmeter, whatever is the potential difference generated by the dipole. Therefore, *the difference in voltage actually detected by the galvanometer (dV) in addition to the electric charge amplitude of the dipole (V) also depends on the spatial orientation of the electrodes with respect to the orientation of the vector that represents the dipole. In other words, dV detected by the galvanometer of the ECG depends on the cosine of the angle (α) formed between the vector and the axis that connects the two electrodes, namely the lead axis* (Figure 10.4). ***We can say, dV is equal to the voltage magnitude of the dipole (V, represented by the length of the vector) multiplied for the cosine of the α angle between the axis of the vector and the axis of the lead:*** dV= V $\times$ α cosine.

If at a given instant the vector and the axes of the electrodes are perfectly aligned, *i.e.* the angle is zero degrees and the *cosine* is 1, the voltage difference gives rise to the maximum deflection possible. If the vector and the axes of the electrodes forms an angle of 90°, the *cosine* is zero and voltage difference (dV) recorded is zero, regardless of the magnitude of the dipole. Of course, also the distance of the Ee from the center of the dipole will influence

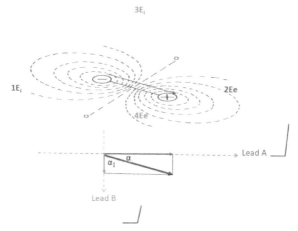

Figure 10.4 Properties of a dipole and graphical resolution of a cardiac dipole vector. Dashed lines are isopotential or equipotential circles generated by an electrical dipole, whose potential is inversely proportional to the square of the distance from the center of the charges. The red arrow represents the dipole potential and its spatial orientation. It can be translated and used for graphical resolution with axes of electrode leads. The dashed blue arrow represents the spatial orientation of the Lead A obtained with the 1Ei and 2Ee as recording electrodes. The dashed green arrow represents the spatial orientation of the Lead B obtained with the 3Ei and 4Ee as recording electrodes. The voltage difference (dV, represented by blue arrow) detected on Lead A depends on dipole voltage magnitude and α angle cosine between dipole vector and Lead A axis. The voltage difference (dV, represented by green arrow) on Lead B depends on dipole voltage magnitude and α_1 angle cosine. The amplitude of the dark red graphical traces recorded on Lead A and Lead B is proportional to dV detected on the considered Lead (dV =V $\times$ α cosine).

the voltage detected. Also the characteristic of the conducting volume will be important.

If instead of an artificial dipole, like the one described above, we consider myocardial fibers, we see that these can be a dipole only if there are positive and negative charges on their external surface, which is the fibers are partly polarized and partly depolarized (Figure 10.5). If the fibers are uniformly polarized or uniformly depolarized (Figure 10.5A and D, respectively), there is no dipole and in the recording, the trace is in an *isoelectric position* (the theoretical zero line of the ECG).

Let us now imagine that the myocardium is divided into two parts: the right one is completely polarized and the left one completely depolarized (Figure 10.5B). The positive field of the dipole is the sum of all the positive charges of the polarized part, while the negative field is given by the sum of

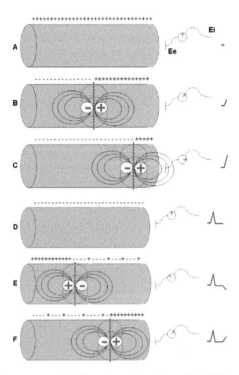

Figure 10.5 Dipole in a myocardial fiber. A: resting fibers (the ECG trace is in the isoelectric position); B: partially depolarized fiber; C: almost totally depolarized fibers; D: totally depolarized fibers (the ECG trace is again in the isoelectric position); E: partially repolarized fibers starting from the depolarized side first (left); F: partially repolarized fibers starting from the last depolarized side (right). Ee: exploring electrode; Ei: indifferent electrode. Some + signs in the negative fields indicate that repolarization is a disorganized phenomenon (does not follow a conduction system). See explanation in the text.

all the negative charges of the depolarized portion. The two poles are located close to each other, separated from each other by an imaginary median plane (the potential-free part). As in the case of the artificial dipole, the relative electric fields are present around the two poles with the respective isopotential spheroidal surfaces.

Let us imagine recording the transition from the situation in which the tissue is entirely polarized to that in which the tissue is depolarized to the left and polarized to the right, the trace of the recorder undergoes a deflection upwards whenever the positive Ee electrode is in the positive field. For simplicity, the *lead's axis* (represented by the straight line connecting the two electrodes used for that ECG lead) is perfectly aligned (the α angle is zero degrees, *cosine* 1) with the vectors created by the depolarization and repolarization. The initially modest deflection is accentuated when the depolarization front advances to the right. The advancement, in fact, leads to contact Ee with increasingly positive isopotential surfaces (Figure 10.5B and C). When all the fibers are depolarized, we find only negative charges on their surface, *i.e.* the dipole no longer exists, and the trace returns on the isoelectric line (Figure 10.5D).

If at some point the repolarization starts from the left side, *i.e.* from the fibers that were the first to be depolarized, the trace will undergo a downward deflection (The Ee is in the negative field, Figure 10.5E) to return to the isoelectric line as soon all fibers will be repolarized, *i.e.,* with the disappearance of the dipole. On the other hand, if the repolarization takes place from the right, which is from the last part that has been depolarized, the deflection will be positive, *i.e.,* the Ee is in the positive field (Figure 10.5F).

To sum up,

1. a wave of **depolarization** traveling

    (a) toward an Ee results in a *positive deflection*;
    (b) away from an Ee results in a *negative deflection*.

2. a wave of **repolarization** traveling

    (a) toward an Ee results in a *negative deflection*;
    (b) away from an Ee results in a *positive deflection*.

3. **Voltage amplitude** of deflection (positive or negative) is mainly related to

    (a) the mass of tissue involved;
    (b) the cosine of the angle beween vector and derivation orientation (a wave of depolarization or repolarization oriented perpendicularly to derivation axis determines no net deflection),
    (c) the distance of the electrods from the dipole center, and
    (d) conductive caracteristics of the tissues around the electrical tissue.

Now let's see what really happens in the heart. If we consider a portion of the non-septal wall of the left ventricle, we observe how the depolarization, which proceeds from the subendocardial surface of the myocardium to the subepicardial one, is directed from the right to the left of the studied subject (Figure 10.6A).

In the heart ventricle, repolarization starts from the last section of the depolarized myocardium (Figure 10.6B), that is, from the left side of the patient, which is to the right of the observer. For what we have said before,

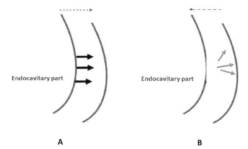

Figure 10.6   **A:** Depolarization (continuous black arrows) of the non-septal wall of the left ventricle from the subendocardial surface to the subepicardial one (as indicated by dotted red arrow); this depolarization will generate the R wave of Figure 10.7. **B:** Repolarization (continuous blue arrows) of the non-septal wall of the left ventricle from the subepicardial surface to the subendocardial one (as indicated by dotted red arrow); this repolarization will generate the T wave of Figure 10.7.

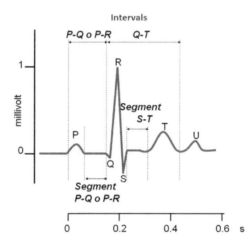

Figure 10.7   Waves, intervals, and segments of the ECG. The case shown in this figure is a typical trace recorded when the Ee is on the left side of the body of the studied subject (*e.g.,* lead II) and the QRS complex can be referred to as qRs. *See explanation in the text.*

since both during depolarization and repolarization Ee sees the tip of the vector, (*i.e.,* the positive field), in both cases are recorded positive waves. This is what happens in the electrocardiogram (Figure 10.7) where the depolarization of the free wall of the ventricle (R wave) has the same orientation (upward) of repolarization wave (T wave) of the left ventricle.

From the observation of Figure 10.7, it can also be seen how the T wave, referred to the repolarization, has a smaller amplitude of the R wave referred to the depolarization. The lower amplitude of the T wave depends on the fact that in the repolarization the difference in potential between the polarized and the depolarized myocardium is less than the difference we observed in depolarization. This is likely due to a partial repolarization that affects all the still depolarized tissue (Figure 10.5F). In other words, the depolarization is a well-organized phenomenon that proceeds along the conduction system, while the repolarization is a disorganized phenomenon that creates less detectable differences in voltage (minor intensity and different angles, as represented in Figure 10.6).

10.2 Morphology and the Meaning of Electrocardiographic Waves

As we know the electrical activity of the heart is generated in the SAN located in the upper part of the right atrium. The impulse generated by these cells diffuses radially through the myocardium of both atria. The conduction of the impulse from the atrial to the ventricular myocardium is allowed by a specific conduction system consisting of the AVN, the His bundle, the right and left branches, and the Purkinje fibers network. The atria and the ventricles are chambers electrically separated from the atrio-ventricular fibrous plane and, under normal conditions; the only conducting structure that allows the impulse to progress from the atria to the ventricles is the AVN-His system. The AVN, consisting of slow-conducting fibers, located on the right side of the lower part of the interatrial septum, allows the slow transmission of the impulse from the atrial to the ventricular side. After passing through the AVN, the impulse reaches the bundle of His and, going through the right and left branches, and the Purkinje fibers network, finally reaches the ventricular myocardium, determining its activation from the endocardium towards the epicardium. Therefore, the impulse activates first the interventricular septum from left to right, then the free wall of the ventricles (from endocardium to epicardium) and finally the base of the ventricles from right to left. Subsequently the phase of ventricular repolarization proceeds in reverse, from the epicardium to the endocardium. This electrical activation sequence (depolarization and repolarization) is recorded from an electrocardiographic point of view as a regular succession of waves and intervals that are called as follows:

P wave, corresponding to the depolarization of the atria;

QRS complex, resulting from the depolarization of the ventricles;

T wave, representing the ventricular repolarization

Sometimes a **U wave** is appreciated, due to the repolarization of the papillary muscles or the Purkinje network. It may be present in normal heart or may be particularly prominent in hypokalemia or hyperkalemia, as well as during treatment with cardioactive-glucosides or with some anti-arrhythmic drugs.

In describing the registration of the processes of depolarization and repolarization of a myocardial fiber, we imagined placing the Ee on the right side of Figure 10.5 (at the left side of the studied subject). The electrocardiographic complex, *recorded from the left side of the patient's body (e.g., lead II)*, presents the characteristic waves that we can see in Figure 10.7, in which we observe the sequence of waves of prevalently positive (upward) waves, namely P, qRs, and T waves. By convention, while large voltage deflections of the rapid ventricular complex are indicated in capital letters, low voltage deflections are indicated by lowercase letters. By definition a *Q wave* is the first negative deflection followed by a positive *R wave*. A *S wave* is a negative deflection which follows a R wave.

Indeed, the morphology of ECG waves is influenced by several factors:

1. The position of Ee on the body surface: the closer the Ee is to the heart the greater the amplitude of the ECG wave.
2. The polarity of the field generated by the dipole: an Ee in the positive filed will determine an upward wave and *vice versa*.
3. The mass of the tissue involved in the depolarization/repolarization: the greater the mass the greater the voltage/amplitude of the wave.
4. The angle formed between the vectors representing the dipole and axes of the lead (Figure 10.4).
5. The velocity of impulse propagation.
6. The conductor characteristics of the tissues surrounding the heart.

As said, in Figure 10.7 it is represented a typical ECG that we can obtain when the Ee is on the left side of the studied subject (*left lead*). In the ECG the abscissas axis represents the time expressed in seconds and, generally, 25 mm is a second, while on the ordinate axis the voltage of the various waves is referred to the calibration of 1 mV/cm. *The durations of waves are usually measured on **lead II** (a left lead, see below). In a left lead ECG, the orientation of the waves is as described below:*

The P Wave is an upward wave that has duration of about 0.08 s and a voltage of 0.1–0.2 mV.

The QRS complex, also called *rapid ventricular complex,* has a duration of about 0.08–0,10 s with a positive R voltage of +0.6 – +1.0 mV. The ventricular depolarization is an organized event which follows the conduction system. Therefore, in a left lead ECG we record a *qRs complex*, in which the *q wave* represents the depolarization of the interventricular septum, the *R wave* is the depolarization of the free wall of the ventricles and the *s wave* represents the depolarization of the base of the ventricles (Figure 10.7). In some cases, the negative waves (Q and S) may not be visible, depending on the low intensity of the electrical fields and *cosine* of the angle formed between vectors and axes of the leads. ***The duration and amplitude of QRS*** depend on the velocity, the mass of the ventricle and the orientation of impulse conduction and propagation. Indeed, the QRS duration and amplitude increase in cases of changes in intraventricular mass and impulse propagations, as can occur in ventricular hypertrophy, in branch blocks or in ventricular extra-systoles (see below).

The **T wave** indicates the repolarization of the ventricles. The gradual beginning of T wave makes difficult to establish its duration, which is sometimes reported to be 0.27 s. The T wave voltage is +0.3 – +0.4 mV. The spatially disorganized repolarization is the cause of the gradual beginning and low voltage of the T wave.

It is said that the repolarization of the atria, which should be indicated by a *Ta wave*, is invisible because it is hidden in the ventricular QRS complex. It must be said, however, that this wave is not visible even when, in the presence of atrioventricular blocks (see below), the PQ segment is elongated and remains flat on the putative zero line. Again, it is likely that the disorganized repolarization determines a slow intensity of the electrical fields and a not optimal angle between vectors and axes of electrodes, with a *cosine* close to zero.

Of note, the left and right ventricles are simultaneously activated by the two branches of the His bundle and the vectors generated by the left and right ventricle have on opposite orientation. However, since the left myocardial wall has a greater thickness than the right one, in the algebraic sum between the two vectors the left vector will prevail, therefore, for simplicity, in the majority of the case; we can consider only the left ventricle and forget for a while the presence of the right ventricle.

According to the sequence of ventricular activation, the interventricular septum is the first to be activated and an electrode placed on the left side of the body (a left lead) sees the tail of the activation vector of the left component of the septum which moves from the endocardium to the right. Therefore, a negative Q wave represents the activation of the interventricular septum.

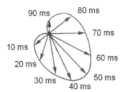

Figure 10.8 The arrows represent the orientation of the resulting vectors which, moment by moment, describe the activation of the ventricles. Only the loop is displayed on a vector-cardiograph monitor. It is as drawn by the moving tip of the vector and is called the *vector loop*.

The positivity and the amplitude of the R wave are also to be related to the activation modalities of the ventricular free walls. Since the free wall of the left ventricle is activated from right to left (Figure 10.6), the resulting vector will be turned to the left with consequent positivity and considerable amplitude of the R wave in the left leads (see Box 10.2: Why the R is so high?).

The negativity and the amplitude of the S wave, in the left leads, are due to the fact that the last region to depolarize is the base of the ventricle, close to the *anulus fibrosus cordis*, therefore the last excitation dipole generates a small vector directed upwards and to the right.

Figure 10.8 shows the representation of the resulting vectors that, moment by moment, describe the activation of the ventricles. On the monitor, the vector appears, travels, growes, wanes and finally disappears. It can be said that the depolarization vectors move counterclockwise and disappear in about 90 ms.

Box 10.2 Why the "R" is so High?

It seems that the diffusion of the impulse to the innermost layers and the myocardium does not give rise to electrocardiographic waves, because of the modalities in which this portion of the heart is activated. In fact, the diffusion of the impulse from the filaments coming from the Purkinje network to the myocardium takes place with the formation of depolarization islets surrounded by the still polarized myocardium. Between the islets and the surrounding myocardium, vectors are formed with the tip pointing towards the polarized myocardium. Since for each islet, there are simultaneous vectors facing in all directions, their algebraic sum is equal to zero (Figure 10.9A). The expansion of these islets (Figure 10.9B) then leads to their fusion from the subendocardial side

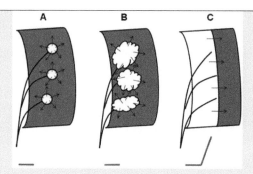

Figure 10.9　Activation of the free wall of the left ventricle. In white the depolarized myocardial portions. A: initial depolarization; B: expansion of depolarized areas; C: a fusion of depolarized areas with depolarization front moving towards the epicardium. See explanation in the text.

of the myocardium (Figure 10.9 C): the various vectors will no longer be oriented in all directions but will be parallel to each other in the endocardium with the tip pointing towards the external myocardium. These will advance all together to invade the remaining part of the ventricular wall. In this way, it will be possible to positively sum together generating the high positive R wave.

10.2.1 Electrocardiographic Intervals and Segments

From a terminological viewpoint, while the intervals comprise the wave, the segments are between waves.

From the beginning of the P wave to the beginning of the QRS complex there is the so-called ***PR interval*** (or PQ in case the Q wave is present). The normal duration of PR interval is between 0.12 and 0, 20 s with an average value of 0.16 s. The PR interval corresponds to the atrio-ventricular conduction time. An elongation over 0.20 s is a sign of First-degree atrio-ventricular block (see Paragraph 10.4).

The ***QT interval*** comprises the QRS complex and the T wave. It is indicated by the term *electrical systole*. Its duration is about 0.43–0.44 s. However, its duration is greatly influenced by the heart rate. Therefore, in a clinical context, we use the so-called *corrected QT* (QTc). It can be shorter or longer than usual in some familiar pathologies (see Box 10.3: Corrected QT and Long and short QT syndromes). QT interval is also elongated in more common pathologies such as coronary insufficiency and may be shortened in an overdosage of cardioactive-glucosides.

Unlike intervals, the duration of the segments is of little clinical importance. Their level is more interesting. The ***PR or PQ segment*** is at the end of the P wave and before the QRS complex. Its duration depends on that of the P wave. The ***ST segment*** is between the end of the QRS complex (*J point*) and the beginning of the T wave. Given the progressive beginning of T wave and the sinuous trend of the ST segment, its duration cannot be defined. In pathological conditions, the ST segment can undergo over-leveling (*ST elevation*) or under-leveling (*ST depression*). Normally, it should be at the same level as the *PR segment* and ***TP segment*** (the segment between the end of T wave and the beginning of P wave). See below the ST elevation myocardial infarction (STEMI) in the BOX 10.10: Myocardial infarction.

Normally, the segments represent the ***isoelectric condition***, which is when the ventricles are completely depolarized (ST segment) or completely repolarized (TP and PR segments). Actually, during PR segment, besides atrial repolarization, a small number of cells of the AVN-Hys system have an electrical activity that cannot be recorded by a traditional ECG. This activity can be recorded, for example, by intracavitary electrodes that are an *electrogram*, which is a tracing of the electrical potentials of myocardial tissue with electrodes placed directly within the heart cavity with catheters, instead of on the surface of the body.

Box 10.3 Corrected QT and Long and short QT syndromes

Various formulas have been proposed for correcting QT intervals as a function of HR, each with limitations (*e.g.*Bazett, Fridericia, Hodges, and Framingham formulas). The *Bazett's formula* ($QTc = QT/\sqrt{RR}$) is one of the most used formulas to calculate the *corrected QT* (QTc) interval. This calculation will "correct" the QT length variation due to HR variations. When the QT-interval is expressed in milliseconds and the RR-interval in seconds, according to Bazet Formula the normal values of Qtc are: between 350–360 and 440–450 ms in males and 360–370 and 460–470 ms in females, respectively (Table 10.1).

Although Bazett's formula is the most widely used correction method in clinical practice, Fridericia's formula ($QTc = QT/\sqrt[3]{RR}$) is

Table 10.1

QTc value (msec)	1–15 years	Males	Females
Normal	<440	<430	<450
Borderline	441–460	431–450	451–470
Prolonged	>460	>450	>470

Table 10.2   LQTS classification

Long QT Syndrome	Defective Channel	Altered Current
LQT1 syndrome	KCNQ1/KvLQT1	Potassium current, IKs
LQT2 syndrome	HERG	Potassium current, IKr
LQT3 syndrome	SCN5A	Sodium current, Ina

recommended by the Food and Drug Administration for clinical trials on drug safety.

Borderline QTc values (Table 10.1) may be a sign of dangerous long QT syndrome (LQTS), which deserves to be carefully evaluated. A patient with LQTS is at risk for *torsade de pointes* (see also Box 10.9: Mechanisms of arrhythmias), ventricular fibrillation and, thus, sudden death. Several forms of LQTS have been described (Table 10.2) which is often due to mutation of one or several genes, which very often lead to channelopathies (alterations of membrane channels). These mutations lead to prolongation of the duration of the ventricular action potential, thus lengthening the QT interval. Abnormalities in KCNQ1 (formerly called KvLQT1) and HERG potassium channels cause LQTS, named LQT1 and LQT2 syndrome, respectively. While homozygous mutations in KCNQ1 cause LQT1 and deafness (Jervell-Lange-Nielsen syndrome), heterozygous mutations cause LQT1 only (Romano-Ward syndrome). Also, sodium channel abnormalities, involving the gene SCN5A, are responsible for an LQTS, named LQT3 (see also late sodium current, in the Box 10.9: Mechanisms of arrhythmias).

A rare short QT syndrome (SQTS) has been also described: it predisposes to dangerous arrhythmias. Also for this syndrome different forms of inherited cardiac channelopathies have been described: gain of function in potassium channels and loss of function in calcium channels have been described. Besides a short QTc interval (around 340 ms or less), a short or absente ST segment and tall, narrow T waves in precordial leads have been described in this syndrome.

In both LQTS and SQTS, there is not a single QTc value to differentiate most cases of syndromes from healthy individuals. QTc intervals close to lower and upper limits (360–450 ms range in males and 370-460 ms range in females), must alert the cardiologist and should only be considered diagnostic of SQTS or LQTS when supported by symptoms or family history of sudden death and syncope, because their QTc values may overlap with those of healthy people.

10.3 Electrocardiographic Leads on the Front Plane and the Electrical Axis of the Heart

The electrocardiographic leads (derivations) of the so-called *12-lead ECG* can be classified as leads on the frontal plane (six leads) and leads on the horizontal plane (six leads). The leads on the frontal plane are further divided into *bipolar limb leads* (*standard leads*) and in *unipolar limb leads*, while the leads on the horizontal plane are unipolar and are called *unipolar precordial or thoracic leads*.

10.3.1 Leads on the Frontal Plane

When considering these leads, the vectors that are formed in the processes of depolarization and repolarization of the myocardium, resulting from the average of the various ventricular vectors, are considered in their projection on the anterior or frontal surface of the thorax.

The *bipolar limb leads* record the difference between the potential changes on a limb towards which the tip of the vector is directed and the potential variations on a limb that sees the tail of the same vector. If we consider that the mean vector of ventricular activation travels from right to left and from top to bottom, we understand how the left arm is placed on the positive side with respect to the right arm, while the left leg is placed on the positive side with respect to both the right and left arm. Unlike bipolar leads, the *unipolar limb leads* do not record the difference between the potential variations of two limbs, but the difference between the potential variations of a limb relative to a potential-free point (*i.e.* a point that has a virtually zero potential).

10.3.1.1 Bipolar limb leads

The *bipolar limb leads* are also called standard leads or leads of Einthoven. They were the first to be used. Depending on the way the electrodes placed on limbs are connected to the galvanometer poles, the standard leads are distinguished in I, II and III lead or derivations, also called D1, D2 and D3.

The *First lead (Lead I or D1)* is obtained by connecting the left arm with the entry or *positive pole of the galvanometer* and the right arm with the exit or *negative pole of the galvanometer*. The *Second derivation (Lead II or D2)* is obtained by connecting the left leg with the positive pole and the right arm with the negative pole. The *Third lead (Lead III or D3)*, finally, is obtained by connecting the left leg with a positive pole and the left arm with the negative

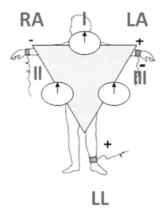

Figure 10.10   Bipolar limb leads. RA: right arm; LA: left arm; LL: left leg. Ellipses with arrows indicate galvanometers. The three limb electrodes define the standard frontal plane limb leads that were originally defined by Einthoven. With the right leg electrode acting only as an electronic reference (not represented in the fig), it serves to improve common mode recording (avoiding unwanted noise). The other three electrodes form pairs of electrodes in each ECG lead. Within each pair, 1 electrode is established as the positive end of the lead, in the sense that current flowing toward that electrode will determine an upward (positive) deflection on the ECG. The other electrode of the pair would be called negative. **Lead I** is defined as the potential difference between the left arm (positive) and the right arm negative (LA-RA), **lead II** is defined as the potential difference between the left leg (positive) and the right arm (negative) (LL-RA), and **lead III** is defined as the potential difference between the left leg (positive) and the left arm (negative) (LL-LA). Whenever net current flow toward the first electrode of the pair which is defined as a positive voltage deflection is recorded on the ECG. *In other words, when the positive electrode is in the positive field an upward deflection is recorded. When the positive electrode is in the negative field a downward deflection is recorded.* According to Kirchhoff's law, the sum of the voltage gains and voltage drops in a closed circuit is equal to zero. Therefore, lead II = lead I + lead III at any instant in the cardiac cycle. This relationship is known as *Einthoven's law*. A fourth electrode placed on right leg (not represented in this figure) is used as the grounding for parasitic currents.

pole. As shown in Figure 10.10, special electrodes are placed on the limbs and connected with the poles of the galvanometer through cables along each of which resistances of 5000 Ω are inserted.

The points on which we placed the electrodes represent the angles of an equilateral triangle that is called the *Einthoven triangle*. As seen in Figure 10.11, each side of the triangle can be considered the axis of a lead, *i.e.* the axis that connects the two electrodes considered.

Identification of the electrical heart axis or "mean vector" of depolarization of the ventricles. We have seen how the vector resulting from the

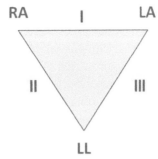

Figure 10.11 Einthoven triangle. RA: right arm; LA: left arm; LL: left leg.

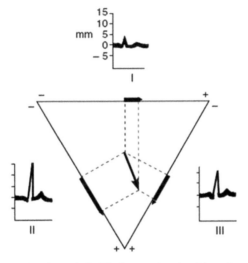

Figure 10.12 Identification through the Einthoven triangle of the orientation of *the electrical heart axis*, which is an average of all depolarization vectors for the ventricles.

activation of the ventricular myocardial mass is prevalently oriented from top to bottom and from right to left. We can draw a *mean vector* at the center of the Einthoven triangle (Figure 10.12). By orthogonally joining the tip and the tail of the vector with the sides of the triangle, we obtain the projections of the vector on the derivation axes. The length of these projections will be maximum in the derivation whose orientation is more parallel to the vector (in our case the II lead). In contrast, the smallest projection will be in the lead that is most nearly perpendicular to the orientation of the vector.

From this observation we can understand how using the Einthoven triangle it is possible to identify with a good approximation the orientation of the *mean vector* on the horizontal plane of the thorax. In this case, we do an inverse process, we start from the ECG waves to calculate the vector orientation in the Einthoven triangle, which is on the frontal plane.

As we have already seen, activation of the ventricular myocardial mass is represented in the ECG by the QRS complex (Figure 10.7). In most of the left-lateral or -inferior leads of a normal ECG, this complex consists of a small downward (negative) deflection called *q* wave, a large upward wave (positive) called *R* wave and a second small negative wave said *s* wave. If we calculate the algebraic sum of the qRs complexes for each lead it is possible to obtain each "resultant deflection", which is indicative of the amplitude of the mean vector, and can be reported on the axis of the respective lead as an arrow with the tail on the central part of the side of the triangle and the tip pointing toward "the positive limb" if the qRs sum is positive or toward "the negative limb" if the QrS sum is negative. Since each of these arrows represents the projection of the *mean vector* on the axis of a derivation, we understand how the orientation of the *mean vector* can be identified by the orthogonal lines starting from the extremity of the arrows in each lead that will cross in points that identify the tip and the tail of the *mean vector.*

10.3.1.2 Unipolar limb leads

The *Unipolar limb leads* were initially constituted by the so-called *Wilson derivations or leads*. The positive pole of the galvanometer (the entry) was connected to an Ee that was placed from time to time on the right arm (VR derivation), on the left arm (VL derivation) and on the left leg (VF derivation), while at the exit (negative pole) of the galvanometer was connected a point with a theoretical zero potential called *central terminal* (CT, Figure 10.13). The acronyms VR, VL, and VF meant respectively voltage right, voltage left and voltage foot. The CT was obtained by applying the *Kirchhoff laws,* according to which "the sum of all currents leaving a node in an electrical network is always equal to zero". The node, or CT, was obtained by joining the cables connected to the three limbs at one point. From this CT a cable was connected to the negative pole of the galvanometer (Figure 10.13A).

Wilson's leads had the drawback of giving rather modest amplitude deflections. The inconvenience was overcome in 1942 with *Goldberger leads*, which allowed a 33-50% *circa* increase in the amplitude of the deflections. This result was obtained by disconnecting the "explored limb" by the CT from time to time (Figure 10.13B). The Goldberger leads are aVR, aVL, and aVF, where the "a" means *augmented*. Wilson's VR, VL, and VF leads had relatively low amplitudes because the potential at the exploring site was also

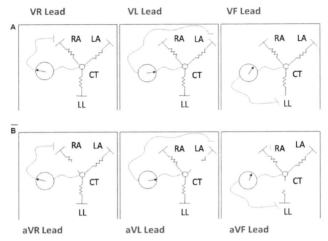

Figure 10.13   Unipolar limb leads according to Wilson (A) and according to Goldberger (B). RA: right arm; LA: left arm; LL: left leg or foot. CT: central terminal. See explanation in the text.

included in the CT. By removing the single exploring potential from the CT, Goldberger produced the "augmented unipolar" limb leads.

From a mathematical point of view, the standard derivation can be obtained by the difference between the voltage variations of the positive left arm and the voltage variations of the negative right arm; therefore:

$$\text{Lead I} = \text{VL} - \text{VF}.$$

This difference is obtained by subtracting at each instant the voltage recorded in VR from the voltage recorded in VL.

We can say that the first-lead corresponds with a good approximation to the difference between aVL and aVR only if two-thirds of the difference is considered, for which the relation is given by:

$$\text{Lead I} = 2/3(\text{aVL} - \text{aVR}),$$

therefore:

$$\text{Lead II} = 2/3(\text{aVF} - \text{aVR}), \text{and}$$
$$\text{Lead III} = 2/3(\text{aVF} - \text{aVL}).$$

The electrical heart axis and the hexaxial reference system. We have already seen how the *electrical heart axis orientation* on the frontal plane can be, with a good approximation, identified with the Einthoven triangle that

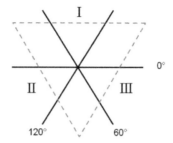

Figure 10.14   Construction of the *triaxial reference system*. The axes of the bipolar leads are translated to the center of the Einthoven triangle. Lead I is by convention at 0°; therefore, lead II is at +60° and lead III at +90°.

refers only to the standard leads. If, in addition to the standard derivations, we also use the unipolar limb leads, we can obtain the *hexaxial reference system* (*Cabrera system*) that allows a more accurate and rapid evaluation of the orientation of the *electrical heart axis* expressed in angular degrees.

Figure 10.14 illustrates the construction of the *triaxial* first and then the *hexaxial reference system*. Since the vertices of the angles of the Einthoven triangle correspond theoretically to the points of the limbs on which the electrodes are placed, they correspond, to the registration points of the bipolar leads, as well as to those of registration of the unipolar leads. If the sides of the triangle are translated to the center of the triangle, we initially obtain the *triaxial reference system*.

In the *triaxial reference system* we consider conventionally situated at 0° the positive point of the First-lead and, proceeding clockwise, we find the positive point of the II derivation at +60° and the positive point of the III derivation at +120°. If we now combine the center of the triaxial reference system with the registration points of the aVR, aVL and aVF leads, we obtain the *hexaxial reference system* (Figure 10.15). In this system we can see that, compared to the I derivation, aVR is at −150°, aVL at −30° and aVF at +90°. By convention, all leads above lead I assume a negative value.

Box 10.4 Identification of Electrical Heart Axis: The Isodiphasic Rule and the Parallelogram Procedure

To identify the *electrical heart axis* by means of the hexaxial reference system with *the isodiphasic rule*, the following rule must be kept in mind:

- the QRS complex has the maximum positive deflection in the electrocardiographic lead parallel to the dipole axis. In other words,

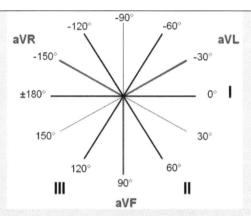

Figure 10.15 Construction of the hexaxial reference system starting from the triaxial system. The unipolar leads (in red) and the bipolar derivations (in black) are reported with the orientation in degrees. The leads above 0 level (I lead) must be marked negative and those below should be marked positive.

the vector dipole moves with its tip towards the *exploring electrode* (Ee) along the lead axis.

- The QRS is isodiphasic in the electrocardiographic lead with lead axis located at 90° with respect to the previous one. A deflection is isodiphasic when the algebraic sum of the positive and negative components is equal to zero. An isodiphasic QRS is recorded when the axis of the electrocardiographic lead is perpendicular to the main orientation of the traveling dipole, *i.e.* the dipole and the axis of the lead form a right angle (90°, cosine 0). While, a perpendicular instant and fixed dipole would not make the electrocardiograph pen move, when it "travels" makes the pen to go equally up and down.

In Figure 10.16A, the deflection has the maximum positive amplitude in the II lead and is isodiphasic in aVL. Therefore, in this case, the electric axis is oriented toward +60°.

In Figure 10.16B, the deflection has the maximum positive amplitude in the I lead and is isodiphasic in aVF. In this case, the electric axis is oriented to 0°, *i.e.* it is horizontal.

In Figure 10.16C, the deflection has the maximum positive amplitude in aVF lead and is isodiphasic in I. The electric axis is then oriented at +90°, *i.e.* it is vertical. It is rotated clockwise on the frontal plane.

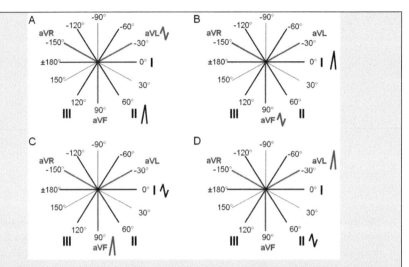

Figure 10.16   Identification of the orientation of the electric axis on the frontal plane by means of the hexaxial reference system. *See explanation in the text.*

In Figure 10.16D, the deflection has the maximum amplitude in aVL and is isodiphasic in the II lead. In this case, the electric axis is oriented at $-30°$, *i.e.* it is rotated counterclockwise on the frontal plane.

Often, the electric axis is not perfectly oriented towards a derivation on the frontal plane. Let's imagine for example that it is oriented to $+75°$ where there is no reference to any lead axis (Figure 10.17). Since the point located at $+75°$ is halfway between the II lead and aVF lead, positive deflections of the same amplitude will be recorded in these two leads. Given the orientation of the electric axis, the isodiphasic deflection should be at $-15°$ where there is also no axis. Since, with respect to the orientation of the vector, the I lead is located at an angle a lower and the aVL at an angle greater than $90°$, in the I lead the deflection will be mainly positive and in aVL prevalently negative. With the concepts presented here, the reader will be able to recognize the orientation of the heart electric axis in the majority of the clinical situations.

If we would calculate as precisely as possible the orientation of the *electrical heart axis*, we could report on each of the axes of the various leads the algebraic sum of the QRS and apply to these arrows the vector sum (the *parallelogram procedure*). In this way, we will obtain a resultant vector oriented on the frontal plane to a certain gradation,

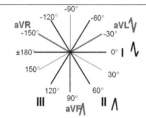

Figure 10.17 Electrical axis oriented at 75° or 77°. See explanation in the text.

which in the example shown in Figure 10.17 could be at +77 degrees rather than at the +75° identified with the previous procedure.

Usually, the orientation of the *electrical heart axis* in most cases is around +60° but it can be often at an angle between 0 and +90°. It is said that *the normal range* could be between −30° and +110°. It is clear how the orientation of the electric axis under normal conditions reflects the position of the heart in the thorax. The orientation of the electric axis allows therefore recognizing possible rotations on the frontal plane: counterclockwise at a degree lower than 0, *i.e.,* towards aVL, or clockwise, at a degree greater than 90°, *i.e.,* toward III leads. The orientation of the electric axis, besides the propagation of the depolarization in the ventricles, is also influenced by anatomic features of the heart and by the thorax and body conformations, as well as by heart movements dictated by respiratory movements of diaphragm muscle.

When the heart electrical axis is outside the normal range, *i.e.,* it goes beyond −30° or +110°, it is strongly indicative of unhealthy cardiac conditions. However, if it falls in the normal range this is not necessarily indicative of healthy cardiac conditions. In general, we can say that we must observe very carefully a long-limbed subject (longitype) with an electrical axis around zero degrees, since this subject, due to its physical characteristics, should have a vertical axis. On the other hand, we will very carefully evaluate a brachitype that has an electrical axis around 90 degrees, as this subject should have a more horizontal electrical axis. For instance, left and right ventricular hypertrophy deviate the cardiac axis to the left and right, respectively.

10.3.2 Leads on the Horizontal Plane

Precordial leads, also known as thoracic derivations on the horizontal plane, are unipolar leads in which *Wilson's central terminal* is compared with an Ee

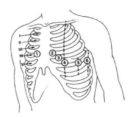

Figure 10.18   Position of the electrodes in the precordial unipolar derivations.

placed at various points of the thorax. They are classified in V1, V2, V3, V4, V5, and V6 (Figure 10.18).

To obtain these leads, six electrodes (Ee electrods) are placed on the chest in the following locations:

V1, fourth intercostal space at the right sternal border;
V2, fourth intercostal space at the left sternal border;
V3, midway between V2 and V4;
V4, fifth intercostal space in the midclavicular line;
V5, in the horizontal plane of V4 at the anterior axillary line or, if the anterior axillary line is ambiguous, midway between V4 and V6;
V6, in the horizontal plane of V4 at the midaxillary line

The left ventricle is so much thicker than the right ventricle that the potential generated by left ventricle depolarization overwhelm and may obscure the potential generated during right ventricle depolarization. Therefore, in the precordial leads, the activation of the left ventricular free wall is seen both from the right (V1–V3) and from the left (V5 and V6). At the V4 there is a passage from the right to the left so that the point of the chest on which this lead is recorded is called the *transition zone*.

If we look at Figure 10.19 in lead V1 we can see how the activation of the free walls of the ventricles appears from the right as a deep negative deflection S, therefore we can have rSr' ventricular complex. Moreover, in the right precordial leads, a small positive wave r represents mainly the depolarization of the septum that in V2 and V3 may also include the depolarization of the most anterior part of the ventricles (right and left).

When the R wave and the S wave have the same amplitude, this lead marks the *transition zone*. This normally occurs in lead V4 or in V3. If the transition zone is between V3 and V4, neither of the two will be isodiphasic, but in one lead will prevail the S wave and in the other one the R wave. Yet, in V5 and V6, the QRS morphology is typical of the left leads and can be somehow similar to those recorded in leads of the frontal plane such as aVL,

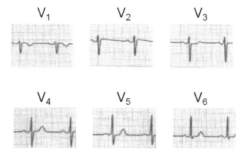

Figure 10.19 ECG in the 6 precordial leads. Note the transition zone in V4. See explanation in the text.

I and II leads. On the other hand, ECG morphology in V1 is somehow similar to that recorded in aVR.

The transition zone is shifted to the right (*e.g.* in V2) when the left ventricle is partially rotated forward and to the right. In this case, this rotation may be indicated as a counterclockwise rotation along the longitudinal axis. The transition zone is shifted to the left (*e.g.* in V5 or V6) when the left ventricle is partially rotated posteriorly and to the left. In this case, may be referred to as clockwise rotation. The rotations are indicated in this way because they are described from the viewpoint of an operator at the foot of the bed of the studied subject. In a recent analysis counterclockwise rotation, the most prevalent QRS transition zone pattern was associated to a low risk of cardiovascular diseases (CVD) and mortality, whereas clockwise rotation was associated with the highest risk of heart failure and non-CVD mortality.

In summary, while the leads on the frontal plane inform us of possible rotations of the heart along the anterior-posterior axis, the leads on the horizontal plane inform us of possible rotations along the longitudinal axis. In addition to the physiological variations of position, rotations of the electrical axis of the heart may also be due to pathological factors, such as hypertrophy and blocks of conduction, that will be only briefly considered in this book.

10.4 Electrocardiographic Aspects of Conduction Disorders and of the Main Arrhythmias

10.4.1 Conduction Disorders

To understand clearly the concepts reported here, the reader has to keep in mind the concepts reported in the Chapter 4 and in particular those reported in Paragraph 4.1.3 on impulse conduction.

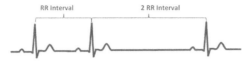

Figure 10.20   Sine-atrial block (II degree). See description in the text.

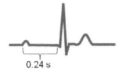

Figure 10.21   First degree atrio-ventricular block: prolonged PR interval. See description in the text.

The sino-atrial, atrio-ventricular, and branch blocks are the conduction disturbance mainly described.

10.4.1.1 Sino-atrial block

In this condition the impulse is generated in the sinus node (the pacemaker) but cannot exit from the nodal tissue to activate atria and ventricles (Figure 10.20). Sino-atrial blocks of first, second and third degree are described on the basis of alteration in the sino-atrial conduction that are recognized only by an intracavitary electrogram and not by an ECG with electrodes on body surface. Similar to the AV block (see below), the I degree sino-atrial block is only a prolongation of sino-atrial conduction, while the II degree is the absence of some P wave so that a sudden doubling or multiple of the usual RR interval is observed on the ECG. The III degree sino-atrial block (absence of P wave) cannot be distinguished from *sinusal arrest (the SAN of the heart ceases to generate the electrical impulses)* as that occurring during the sinus syndrome or the prolonged massage of the carotid bodies.

10.4.1.2 Atrio-ventricular blocks

The AV blocs can be classified in I, II and III degree.

First degree AV block: in this condition an extension of the PR interval over 0.20 s is observed. Although *impulse conduction is slowed in the AV conduction system*, there are no missed beats and every P wave is followed by a QRS complex (Figure 10.21).

The II degree AV block can be defined as a condition in which sometimes the atrial impulses fail to conduct to the ventricles due to impaired conduction in

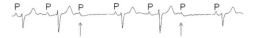

Figure 10.22 Second-degree atrio-ventricular block of the Wenckebach type (Mobitz I type). Note, the presence of Wenckebach periods denotes the elongation of the PR interval. In this case we see two missed ventricular beats (P waves indicated by red arrows, not followed by QRS).

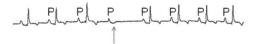

Figure 10.23 Second-degree atrio-ventricular block of the Mobitz II type. The PR interval remains constant until a P wave (red arrow) fails to activate the ventricles.

the AV conduction system. It can be of the Wenckebach or Mobitz type. The *Wenckebach block*, also named *Luciani-Wenckebach block* and classified as *Mobitz I type*, is characterized by the so-called *Wenckebach periods*, that is, a sequence of beats in which the PR interval is progressively longer till a P is not followed by a QRS (Figure 10.22). After the failure to conduct the impulse and to generate a QRS complex, another Wenckebach period begins that also ends with a missed ventricular beat.

The so-called *Mobitz II type* is characterized by one or more P waves that intermittently and unexpectedly are not conducted and therefore are not followed by a QRS complex (Figure 10.23) and no ventricular beat occurs. Of course, when one P is conducted suddenly a subsequent complete ECG pattern (P-QRS-T) and ventricular beat appear. In the Mobitz II block, the PR interval, when present, can have a normal duration, but more frequently it is greater than 0.20 s (*i.e.*, a first-degree block is already present).

Also, an *AV block with minimum increment* has been described, which is a kind of intermediate condition between the Mobitz I and Mobitz II types of block. At a normal electrocardiogram recording speed (25 mm/s), this block does not show any progressive elongation of the PR interval, so it resembles the Mobitz II block. The progressive elongation is instead appreciable if the electrocardiography paper flows at higher speed (*i.e.*, 50 mm/s).

Usually the Mobitz II is due to a subnodal pathology (His-Purkinje System alteration), whereas Mobitz I and first degree AV block are due to nodal pathology. However, only an intracavitary electrogram can distinguish between nodal and Hiss/subnodal diseases, measuring the Atrial-Hiss conduction time (A-H interval, usually lower than 120 ms) and Hiss-Ventricular time (H-V interval, usually lower than 50 ms).

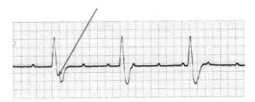

Figure 10.24   Third degree atrioventricular block. The P waves follow one another, at the sinoatrial frequency, without being conducted to the ventricles, which pulsate with a low-frequency called idioventricular rhythm. Some P waves can be masked/buried in the QRST complex (red arrow).

The ***III degree AV block*** is also called *complete atrio-ventricular block* (Figure 10.24). In this block, no impulse is conducted from the atria to the ventricles. It is often due to a Hiss/subnodal disease. In the III degree AV block the atria continue to pulsate at the frequency of sinoatrial rhythm, the ventricles pulse at an *idioventricular rhythm* characterized by a frequency lower than 40 beats per minute (b.pm., generally 25–30 b.p.m.). In the ECG this type of block is represented by the a relatively normal sequence of P waves not followed by normal QRS complexes. *The irregular QRS ventricular complexes have an irregular morphology and no temporal relation with the P waves.* Since at a ventricular rate of 25-30 b.p.m. the heart is not able to produce a sufficient cardiac output, it is necessary to control the heart rate with an artificial pacemaker.

The ***advanced heart block*** or *high-grade AV block* is an intermediate form between second- and third-degree AV block. This occurs when intermittently some sinus node impulse (P waves) are conducted to the ventricles. The P wave can have a regular frequency, but they are conducted with a variable ratio (*e.g.* sometimes with a 4:1 ratio and sometimes with a 2:1 ratio, between P and QRS).

Box 10.5 The Override Suppression and the Preautomatic Pause

A complete AV block (III degree AV block) can be established on the basis of pre-existing I or II degree blocks. In any case, when a III degree AV block occurs, it causes the stopping of the ventricular activity with the consequent fall of the arterial pressure. The period in which there is no ventricular systole is called a *preautomatic pause* and can sometimes last several minutes. This pause is due to the phenomenon of *override suppression*. Overdrive suppression is a phenomenon whereby the dominant SAN removes the pacemaker activity from other *latent pacemaker centers*, having the highest impulse generation frequency. If the cells of the latent centers cease to be "guided" by the SAN (*e.g.,* the III degree AV block), the latent pacemaker is not able to start immediately but it will begin to generate the idioventricular rhythm after a while, *i.e.* after a preautomatic pause. This is due to the fact that the rapid rhythm of the SAN induces in the latent pacemakers *an additional hyperpolarizing current* that requires time to disappear. Thus in the latent pacemaker, a spontaneous depolarization and the generation of action potentials may require several seconds before to impose its own rhythm. The frequency of an idioventricular rhythm increases progressively until it reaches a relatively stable idioventricular rhythm. The progressive increase, albeit limited, of the ventricular rate constitutes the *warming phase:* the additional hyperpolarizing current slowly disappears. This additional hyperpolarizing current is mainly due to a higher velocity of Na^+/K^+ pump during the pacing by the faster pacemaker, which, after the III degree AV block, slowly decelerates.

Therefore, at the end of the preautomatic pause, some idioventricular beats start with QRS complexes not preceded by P waves and having a very different morphology from that of sinus beats. In particular, due to the mode of ventricular activation, these complexes generally have a longer duration and a greater amplitude than those of sinoatrial origin. The different morphology and longer duration are due to the anomalous traveling of the impulse within the ventricle mass which do not follow the conduction system but the working myocardium. The greater amplitude is due to the fact that the impulse originates in a ventricle and travels only towards the other ventricle and, therefore, there are no vectors that travel simultaneously in the two ventricles and in the opposite direction that "blunt" each other (actually, a normal impulse originated in the atrium reaches the interventricular septum first and then the two free ventricular walls almost simultaneously, so the R we can record, for example in D2, results "blunted" by the small vector in the right ventricle).

If the period of *ventricular asystole* (absence of ventricular contractions), that represents the preautomatic pause, lasts more than three minutes, the patient may experience irreversible lesions of the central nervous system. On the other hand, if the duration of the preautomatic pause is not excessive, the patient can recover after a transitory loss of consciousness. Naturally, as already mentioned, an *idioventricular rhythm* does not allow a complete recovery of normal cardiac output and a pacemaker device is necessary.

Box 10.6 Atrio-Ventricular Dissociation

The *idioventricular rhythm* from an III degree AV block should not be confused with *atrio-ventricular dissociation* (Figure 10.25).

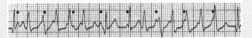

Figure 10.25 Atrioventricular dissociation from increased ventricular frequency to sinus frequency. The P waves, indicated with the black dots, cannot be conducted to the ventricles because these pulsate at a higher frequency due to an ectopic focus.

In this case, the AV conduction may be normal, but the ventricles pulsate with a frequency higher than or equal to that of the atria from which they can no longer be controlled. Dissociation occurs in the presence of *ventricular tachycardia* or, less frequently, *accelerated ventricular rhythm*. The latter is a pathological rhythm of ventricular origin at a constant frequency which, although it is not high, it is superior to that of the sinus node. In both ventricular tachycardia and accelerated idioventricular rhythm, the atrioventricular conduction system can be fully functional.

10.4.1.3 The bundle-branch block

The bundle-branch blocks concern the impossibility of the impulse to propagate along one of the two branches of Hiss bundle. There are therefore blocks of the right bundle and the left bundle. However, the ventricle affected by the block is excited by the impulse propagation through the working myocardium. Because longer time is taken by this abnormal impulse conduction, the QRS complex has an abnormal morphology and a longer duration than normal.

Generally, in the case of *left bundle branch block* (Figure 10.26A), the rapid ventricular complex is widened in all the electrocardiographic leads and a double R wave is present in the left electrocardiographic leads (V6), whereas a wide S wave in V1. A late vector oriented to the left is present, which explain the second R in left leads (Rr's in V6) and wide S in right leads (rS in V1).

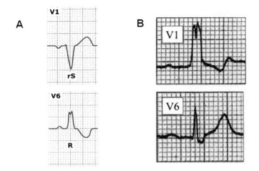

Figure 10.26 *Left branch block (A)* and *right branch block (B)*. See explanation in the text.

Generally, in the case of *right bundle branch block* (Figure 10.26B), the doubled R wave is present in the right electrocardiographic leads (V1), while the S wave lasts longer than the R wave in the left leads (V6). A late vector oriented to the right is present, which explain the second R in right leads (RsR' in V1) and wide S in left leads (qRS in V6).

Several sub-classifications are possible. The block can be complete or incomplete or can be of only a sub-branch (*fascicular block*).

10.4.2 Extrasystoles and Tachyarrhythmias

The most frequent arrhythmias are the supraventricular (mainly atrial) and ventricular extrasystoles and tachyarrhythmias.

10.4.2.1 Extrasystoles

The extrasystoles, also called *premature beats*, are beats that arise outside the sinus node before than the sinus rhythm discharge.

Depending on the site of onset, the extrasystoles are classified into *supraventricular* and *ventricular*. The supraventricular extrasystoles owe their name to the fact that, in addition to the atria, they can arise at any point located above the bifurcation of the His bundle, while the ventricular ones arise below the bifurcation.

Before that the next sinus beat occurs, the extrasystole is followed by an interval longer than that of a normal cardiac cycle, which can be measured from the distance between two R waves (RR interval). This interval is called "*postextrasystolic pause*", which can be compensatory or non-compensatory.

In the *compensatory postextrasystolic pause* the interval between the R wave preceding the extrasystole and the one that follows it, has the same duration of two regular RR intervals. Typically, a ventricular extrasystoles has a *compensatory postextrasystolic pause.*

In the *non-compensatory postextrasystolic pause* the interval between the R wave preceding the extrasystole and the one that follows it has duration shorter than two normal cardiac cycles (RR interval). Typically, supraventricular extrasystoles have a *non-compensatory postextrasystolic pause.*

Depending on the mode of appearance, the extrasystoles can be sporadic, in the form of a bigeminal rhythm or trigeminal rhythm. While *sporadic extrasystoles* occasionally arise during a sinus rhythm, a *bigeminal rhythm* occurs when each sinoatrial beat is followed by an extrasystole and a *trigeminal rhythm* when two sinoatrial beats are followed by an extrasystole

In the **supraventricular extrasystoles**, the impulse, due to its site of origin, propagates to the ventricles according to the typical sinus beats sequence, for which the QRS complex has a normal appearance. A supraventricular extrasystole is then recognized for its early appearance, the presence of non-compensatory postextrasystolic pause and the normal morphology of the QRS. The P wave can be irregular or absent (not visible).

The supraventricular extrasystoles that arise in the AV node are indicated as *nodal extrasystoles* and are subdivided into high, medium and low nodal extrasystoles. In all three types, the impulse propagates retrogradely to the atria and in anterogradely to the ventricles.

In the *high nodal extrasystoles,* the impulse arises in the upper part of the AV node. Retrograde atrial activation causes the P wave to be upside down and, due to the proximity to the atrium of the point of onset of the impulse, it precedes the QRS complex (Figure 10.27). In the case of the *middle nodal extrasystole*, the impulse originates in the middle part of the AV node and the activation of the atria is more delayed so that the wave P seems absent because it is masked by the rapid ventricular complex. Finally, in the *lower nodal extrasystole*, the inverted P wave follows the QRS complex

In the **ventricular extrasystoles,** the impulse can arise from any cells located in one of the two ventricles below the bifurcation of His bundle. Most often it arises in a peripheral fiber of the conduction system, or even in the working myocardium. Because of the place of origin of the extra-systolic pulse, the sequence of ventricular activation is different from that of sinoatrial beats and supraventricular extrasystoles. The QRS complex, therefore, has an irregular shape and a longer duration and higher amplitude than normal. To these first and immediate signs of recognition is added the absence of P wave

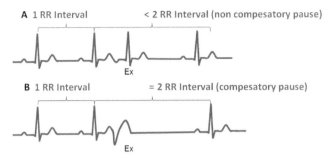

Figure 10.27 Supraventricular extrasystole with non-compensatory postextrasystolic pause (A). Ventricular extrasystole with compensatory postextrasystolic pause (B). See explanation in the text.

before the ventricular complex and the compensatory postextrasystolic pause (Figure 10.27).

Note that the morphology of a supraventricular extrasystole can be abnormal if it occurs too early and finds the ventricles only partially repolarized, therefore, the impulse travels along an anomalous pathway.

On the other hand, a ventricular extrasystole that originates exactly at the level of the bifurcation of the Hiss bundle may have a QRS morphology that resembles that of a regular beat. Therefore, in this case, the compensatory or not-compensatory postextrasystolic pause can be very useful for understanding the origin of the extrasystole.

Box 10.7 Why the Postextrasystolic Pause can be Compensatory or Non-compensatory?

The *postextrasystolic pause* is *non-compensatory* after supraventricular extrasystoles, while it is *compensatory* after ventricular extrasystoles (Figure 10.28). In part *A* of the Figure 10.28, it is schematically shown how in a sinus rhythm the impulse arising in the SAN, propagates through the atria, undergoes a slowing (delay) during the passage along the AVN to eventually propagate to the ventricles. The impulses occur at intervals of 800 ms from each other and the ventricles are activated with the same sequence and frequency but with a certain delay compared to the atria.

At a certain point in the atrium an extrasystole appears 300 ms after the previous normal beat (Figure 10.28) The extrasystolic impulse propagates not only to the ventricle through the AVN, but also to the

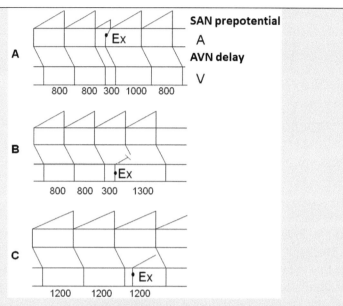

Figure 10.28   Spread of the excitation of a supraventricular (A), ventricular (B) and interpolated extrasystole (C). See explanation in the text.

atrium and the sinus node where it interrupts the formation of the sinus impulse. The ventricle is also activated 300 ms after it has been reached by the previous sinus beat. In the meantime, the sinus node has started to generate another impulse that will take place after 800 ms have elapsed since the moment when it was inhibited by the extrasystoles. Compared to this, the ventricles are activated after an extrasystolic pause equal to the sum of a sinus cycle (800 ms) and of the time taken by the extrasystole to inhibit the activity of the sinus node. If we imagine that this time is 200 ms, the postextrasystolic pause will be 1000 ms. It will not be compensatory since the interval between the preceding beat and the one following the extrasystole is lower (1300 ms) than two sinus cycles (1600 ms).

Even if it occurs in the AVN, a supraventricular extrasystole can spread within the atrial wall and reset the S AN; stopping the formation of an impulse (*the SAN does not "waste" time to reach the threshold for the generation of an impulse while the ventricle is already excited*).

Unlike the supraventricular extrasystoles, the ventricular extrasystoles do not normally come to inhibit the sinus node, which therefore succeeds in completing the formation of the impulse. As can be seen in part *B* of Figure 10.28, the extrasystole E is generated in the ventricles 300 ms after a regular beat. The impulse coming from the sinus node finds the ventricles in the refractory periods. Thus "wasting" time, and the subsequent impulse originating from the sinus node will thus activate the ventricle 1300 ms after the extrasystole. Therefore, a regular beat precedes the extrasystole by 300 ms and a regular beat follow the extrasystole by 1300 ms after (*the sinus node wasted time to reach the threshold for the generation of an impulse while the ventricle is already excited*). Thus we see that the RR interval between the R wave preceding the extrasystole and the R that follows is 1600 ms, that is, the duration of two normal sinus cycles.

In part *C* of Figure 10.28, we can see a ventricular extrasystole that falls between two sinus beats without altering either the formation or the propagation of the two regular beats. In this case, it is an *interpolated extrasystole* that appears between two sinus beats at a normal rate and is not followed by a postextrasystolic pause. Figure 10.28 illustrates how this extra-systole cannot block the sinus pulse due to the low heart rate. This is a real extra-systole (an additional beat). The others are premature beats but are usually called extrasystoles.

The extrasystoles can be distinguished according to various classifications. A criterion may be the duration of the interval that separates them from the regular beat that precedes them. Indeed, extrasystoles may have fixed or variable coupling intervals. If this interval is always the same every time the extrasystole appears, it is called *fixed-coupled extrasystoles*, while if it varies, it speaks of *variable-coupled extrasystoles or parasystoles*. Generally, while the latter is due to *ectopic firing*, the former is likely due to *reentrant mechanisms or early afterdepolarizations*. For more details on classification and mechanisms, the reader is referred to *Box 10.9 on Mechanisms of arrhythmias* and specialized treatises.

Here we spend a few words on the **hemodynamic classification of extrasystoles** that can be useful for the student to practice on mechanical aspects of heart contraction. Indeed, from a hemodynamic point of view, the extrasystoles are divided into *effective* and *ineffective*. The *effective extrasystoles* generate an SV, albeit reduced, since, compared to the previous sinus beat, they arise after a diastole time reduced, but long enough to ensure adequate ventricular filling, whereas the *ineffective extrasystoles* cannot generate any SV as they occur too early to allow adequate ventricular filling. Because of the long duration of the postextrasystolic pause, the SV of the beat that follows the extrasystole is greater than that of the other sinus beats. To generate this greater SV both a heterometric and a homeometric (inotropic) regulation of the force of contraction occurs. Indeed, the greater diastolic time besides allowing a greater filling (that it is a heterometric regulation of the force of contraction) it allows also the Ca^{2+} to flow into the terminal cistern from the point where it was pumped by the SERCA2 into the sarcoplasmic reticulum. Therefore, from the terminal cistern more Ca^{2+} comes out and the force of contraction increases (that is a homeometric-inotropic regulation of the force of contraction).

10.4.2.2 Tachyarrhythmias

The most frequently detected supraventricular tachyarrhythmias are *atrial flutter* and *atrial fibrillation*.

In the **atrial flutter**, a very high frequency (200–300 b.p.m.) of impulses occurs within the atria. These impulses are counted throughout the atrial tissue and determine synchronized fiber contraction. In the ECG the P wave is no longer represented but it is replaced by several waves of anomalous morphology called flutter waves or *F waves* (Figure 10.29), which resemble the teeth of a saw blade. Each of these waves is hemodynamically effective as they cause simultaneous contraction of atrial myocardial fibers. Due to the refractory period of the AVN, only some waves are conducted to the ventricles. The result is that the QRS complex, of normal morphology, appears only after some F waves. Often, in the atrial flutter, there is a conduction ratio that can be 2: 1, 3: 1, 4:1, thus every 2-3 or 4 F waves there is a single QRS (Figure 10.29). Thus, in the presence of an atrial rate of 300 b.p.m., if the conduction ratio is 2: 1 the ventricular rate will be 150 b.p.m., while if the conduction ratio is 3: 1, the ventricular rate will be 100 b.p.m. In both cases, the ventricles pulsate rhythmically. Usually, there is a *counterclockwise macro-reentry* around the atrioventricular annulus as the mechanism underpinning atrial flutter, but several sub-classifications are described in the clinics.

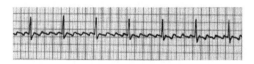

Figure 10.29 Atrial flutter. Note the atrio ventricular conduction ratio of 4: 1. For each ventricular cycle, a wave is hidden by the QRS complex.

A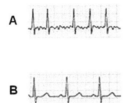

B

Figure 10.30 **A**: Atrial fibrillation; **B**: sinus rhythm of the same patient. The mean ventricular rate in atrial fibrillation is higher than in sinus rhythm: in the unit of time, more impulses cros s the AV conduction system.

Atrial fibrillation is the most frequent cause of *total arrhythmia of the ventricles*. Indeed, the AV node, continuously *"bombarded"* by very chaotic and high-frequency of impulses (400–1000/min), allows only a limited number of these impulses to pass through in a completely irregular manner. In this type of arrhythmia, the various atrial myocardial fibers are neither activated synchronously nor contracted simultaneously, *i.e.* we can say there is both a complete electrical and mechanical desynchronization of atrial activity. Because of this desynchronization, the atrial contraction does not participate in the ventricular filling. In the ECG the P waves are replaced by small and irregular waves, whereas the QRS may have normal forms but with a completely arrhythmic sequence (Figure 10.30). The average ventricular rate is generally quite high. The pathophysiology of atrial fibrillation recognizes several types of disturbances that promote *ectopic firing and reentrant mechanisms*, and include at least the following: *a)* autonomic neural dysregulation, *b)* Ca^{2+}-handling abnormalities, *c)* ion channel dysfunction and structural remodeling. Aging, diabetes mellitus, heart failure, hypertension, myocardial infarction, obesity, smoking, thyroid dysfunction, valve disease, and endurance exercise training all cause structural remodeling. Heart failure and prior atrial infarction also cause Ca^{2+}-handling abnormalities that lead to focal ectopic firing via delayed afterdepolarizations/triggered activity (see Box 10.9 on mechanisms of arrhythmias).

From a hemodynamic point of view in atrial fibrillation, depending on the interval that separates two consecutive ventricular beats, the SV can be more or less large. Differently from the flutter, in the fibrillation the atria do not participate with their contraction to the filling, as a consequence, the ventricular beats that follow short diastolic time intervals (*e.g.* 300 ms) are ineffective (no SV) in atrial fibrillation. Indeed, in atrial flutter, a 300 ms diastolic interval is accompanied by an SV, although reduced in volume.

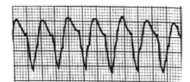

Figure 10.31   Example of ventricular tachycardia.

Atrial fibrillation very often alternates with the flutter. Flutter followed by fibrillation and fibrillation followed by flutter can be transient manifestations or last for months and years. An interrelationship between the two arrhythmias exists, suggesting that they are two sides of a coin: they likely reflect an important underlying similar pathophysiology and even similar mechanism(s) at the onset.

As far as cardiac output is concerned, atrial fibrillation with a low average ventricular rate can be tolerated by the patients, which can eventually refer to "thumping in the chest" and irregular heartbeat, whereas when the ventricular rate is excessive the patients may experience one or more of the following symptoms: dizziness, faintness or confusion, fatigue at rest or when exercising, rapid and irregular heartbeat, sweating, shortness of breath and anxiety. Nevertheless, atrial fibrillation predisposes to the formation of cloths within the atria and represents a condition with a *very high risk for thromboembolism*. Indeed, thromboembolism, ranging from stroke through mesenteric ischemia to acute limb ischemia, is a severe complication in atrial fibrillation. Therefore, therapy with warfarin or other anticoagulants is highly recommended.

Ventricular tachycardia resembles a rapid succession of ventricular extrasystoles (Figure 10.31). The frequency can sometimes be so high (above 180 b.p.m) to seriously compromise SV, CO, and arterial pressure. Ventricular tachycardia, when not interrupted by pharmacological or electrical interventions, may result in lethal ventricular fibrillation, which is favored by the hypoperfusion of the coronary bed because of the reduced arterial pressure. Indeed, coronary disease and myocardial ischemia are often at the basis of ventricular tachycardia and subsequent fibrillation. Ischemia much shorter than the time of about 30 min required for an infarction to occur may be responsible for fatal fibrillation. The cause of ventricular tachycardia usually resides in serious myocardial diseases. *Catecholaminergic polymorphic ventricular tachycardia* (CPVT) is a recently described stress-related, bidirectional ventricular tachycardia and atrial tachyarrhythmia in the absence of either structural heart disease or prolonged QT interval. CPVT is a channelopathy in which tachycardia is triggered by an adrenergic stimulus in a structu rally normal heart and at rest normal ECG.

Among the mechanisms of ventricular tachycardia ectopic foci and delayed or early afterdepolarizations may be included (see BOX 10.9 on mechanisms of arrhythmias)

In *ventricular fibrillation*, there is both a complete electrical and mechanical desynchronization of ventricular fiber activity, so that the development of a ventricular pressure is impossible. Arterial pressure drops rapidly and if the caregiver does not intervene immediately with defibrillation, the patient dies (*ventricular fibrillation is the vestibule of the death*). The sudden deaths due to ventricular fibrillation are generally attributed to an unidentified cardiac arrest. In the ECG, ventricular fibrillation is characterized by bizarre and chaotic waves with variable amplitude and frequency without the appearance of any appreciable PQRST complexes. (Figure 10.32B). Usually, the amplitude of the bizarre and chaotic waves is reduced as time passes and fibrillation becomes irreversible.

Often, the transition from tachycardia to ventricular fibrillation is observed when a progressive deterioration of the interventricular conduction occurs. In these cases, during t achycardia the QRST complexes, in addition to being more frequent, become progressively more enlarged (Figure 10.32A). Under these conditions, each *depolarization occurs before the repolarization of the previous beat is completed*, so that the ventricular fibers are no longer activated synchronously and the ventricular fibrillation is established (Figure 10.32B).

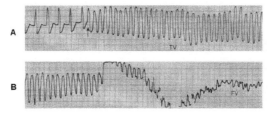

Figure 10.32 Increased frequency of ventricular tachycardia (A) with fibrillation outcome (B). The arrow in A indicates the increase in frequency and the enlargement of the ventricular complex. In B the oscillations of the trace are due to movements of the dying subject and electrocardiographic cables.

Box 10.8 The Defibrillation

While atrial fibrillation is compatible with life and can sometimes be reversible spontaneously, ventricular fibrillation is among the principal causes of sudden death. In ventricular fibrillation, the sinus rhythm can be restored with electrical defibrillation. While in ventricular fibrillation the defibrillation technique is mandatory, it can also be used in atrial fibrillation or flutter and other forms of tachycardia.

The device used to deliver this therapeutic shock to the heart is called a *defibrillator*. Different types of defibrillators are used nowadays and these include external defibrillators, transvenous defibrillators, and implantable defibrillators.

External defibrillation is done thanks to the use of a couple of paddle electrodes of the dimensions of the heart. An electric discharge of about 200-400 Joule, lasting about 5 ms, is applied on the chest with these two electrodes. In this way, all the myocardial fibers are suddenly depolarized and therefore they are all in the same electrical situation for a while. The heart stops, and after a few seconds, if its conditions are good enough, it may start to pulse with sinus rhythm.

Also, an *implanted defibrillator* may deliver a dose of electric current to the heart whenever necessary. Usually, implantable defibrillators are designed to deliver a shock within about 10 seconds after ventricular fibrillation or tachycardia is detected. Implantable defibrillators deliver up to 40 Joule of energy to the heart within a few milliseconds. This current depolarizes a large amount of the heart muscle, ending the arrhythmia. After a few seconds, the body's natural pacemaker in the SAN of the heart or an implanted artificial pacemaker can re-establish a normal rhythm.

Box 10.9 Mechanisms of Arrhythmias

Mechanisms of arrhythmias can be divided into two types: focal mechanisms and re-entry.

Focal mechanisms:

1. *Abnormal automaticity (ectopic focus)*: in the presence of cellular suffering (*e.g. ischemia, hypokalemia, and/or hyperkalemia*) any

region of the myocardium may present anomalous diastolic depolarization (partial depolarization of the membrane) and constitutes an "ectopic focus". Unlike the mechanism responsible for normal automaticity, the incoming current that supports the anomalous diastolic depolarization is primarily the *calcium current* (ICa). Since this current is augmented by beta-adrenergic stimulation, abnormal automaticity is favored by sympathetic activation. Beats that originate in a ventricular focus (a secondary pacemaker) may be responsible for ventricular parasystoles. In this case, the focus represents a parallel pacemaker, with its frequency, which competes with the physiological heart rhythm. When a variable exit block of the focus is present, parasystoles may appear at a rate that is related to a multiple or submultiple of parasystole typical frequency. Abnormal automaticity is likely at the origin of complex forms of arrhythmia, such as extrasystoles, whereas its role in the genesis of sustained ectopic tachycardia or fibrillation is likely to be infrequent. Indeed, besides the hypothesis that considers disturbances of the intramyocardial conduction as a frequent cause of fibrillation (see re-entry below), also the theory of single or multiple ectopic foci has been proposed. According to this theory, under pathological conditions, single or multiple sites of automatic activity at high frequency can arise in the myocardium, dividing it into various units capable of functioning independently of each other without any synchronization.

2. *Triggered activity:* is sustained by *"post-potentials"*, that is, abnormal oscillations of the membrane potential that follow an action potential. Unlike automatism, the post-potentials depend on the previous action potential (that is the "trigger") and the resulting arrhythmia maintains a relationship with it. We can distinguish two types of "post-potentials":

   (a) **Early afterdepolarizations (EADs):** EADs arise during the repolarization of the action potential (phases 2 and 3) when various incoming currents (ICa, INa) and outgoing currents (IK, IK1) contribute to determining the pattern of the potential. The EADs reported in Figure 10.33 are induced by *decreased conductance (gK) of the IK channel*, that is the cause of frequent clinical finding. Hypokalemia and

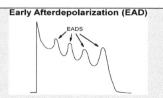

Figure 10.33   Early afterdepolarization. EADs are favored by conditions that prolong the action potential duration.

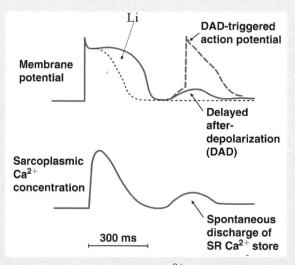

Figure 10.34   Delayed afterdepolarization. If Ca^{2+} is replaced by Li, DAD disappears.

bradycardia, which reduce the K^+ currents, favor the genesis of EADs.

The EADs are inhibited by all the conditions that can decrease the inward currents (*e.g.* the block of INa or ICa) as well as the conditions that increase the outward currents (*e.g.* tachycardia, activators of K^+ channels).

(b) *Delayed afterdepolarizations (DADs)*: DADs arise during electrical diastole, under conditions of *intracellular Ca^{2+} overload (high $[Ca^{2+}]i$)*. Indeed, DADs (Figure 10.34) are due to the activation of depolarizing currents by abnormal fluctuations of the intracellular concentration of Ca^{2+} ($[Ca^{2+}]i$). These fluctuations are generally promoted by the previous action potential (*i.e.* the trigger) and are favored by tachycardia that increases $[Ca^{2+}]i$. The calcium overload *per se* is not responsible for DADs, but triggers *activation of $3Na^+/1Ca^{2+}$ exchanger* (NCX) responsible for the depolarizing current.

Ischemia and digitalis intoxication is among the most typical causes of DADs, which also appear in other conditions of metabolic distress (*e.g.*, ischemia) associated with intracellular Ca^{2+} overload, especially if associated with adrenergic activation.

The triggered activity can support simple and complex arrhythmia forms, including ectopic tachycardias. A special case is represented by ventricular tachycardia with "***Torsades de pointes***". François Dessertenne called "*Torsades de pointes*" a ventricular arrhythmia with an ECG pattern of continuous changes in the morphology of the QRS complexes, which appeared to rotate around an imaginary line. *Torsades de pointes* is characteristically associated with all the conditions favoring the EADs and, therefore, it is commonly attributed to "triggered activity". It is, however, possible that this arrhythmia is supported by re-entry (see below), of which the EADs represent the triggering mechanism. Several pathological conditions, including LQT syndrome and drug intoxications (amiodarone, sotalol or quinidine) which prolong ventricular repolarization may favor the *Torsades de pointes* which appears on ECG as a polymorphic ventricular tachycardia with a characteristic of a twisting QRS complex: it seems that peaks of QRS, which are pointing up, appear to point down in subsequent complexes (Figure 10.35).

Re-entry circuit
The genesis of a re-entry circuit (Figure 10.36) requires an activation dissynchronism, such that within the same cardiac cycle, in which zones coexist still in activity and areas that have already passed the refractory

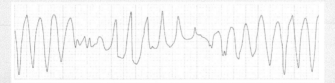

Figure 10.35 *Torsades de pointes.* Polymorphic ventricular tachycardia in which the QRS complexes "twist" around the isoelectric line.

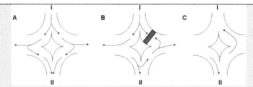

Figure 10.36   Return of excitement (re-entry) in the genesis of extrasystole. A: normal condition; B: unidirectional block of excitement in a branch of the myocardial syncytium; C: return of excitement immediately after block resolution [the red part has a short refractor period and unidirectional block and/or the impulse (arrows) travels at low velocity].

period. Therefore, the latter zones can be reactivated by the zones still in activity. This condition is favored by a reduction in conduction velocity (CV), a shortening of the refractory period (RP), and a presence of unidirectional conduction.

The genesis of the reentry circuits is favored by the inhomogeneity of myocardial conduction and repolarization. This inhomogeneity may have an anatomical and permanent basis (*e.g.* connective infiltration) or functional and temporary (*e.g.* regional differences in action potential duration also due to EADs).

The higher the perimeter (real Wall Length) of the circuit, the higher the probability of re-entry and frequency of arrhythmia. Because of myocardial non-uniformity, the real perimeter of a circuit is always greater than its *theoretical Wall Lenght* (tWL = CV × RP); there is, therefore, an *excitable gap* (EG) between the head and the tail of the circuit. For instance, if the conduction velocity CV is 1 cm/s and refractory period RP is 1 s , the tWL will be 1 cm. If the real length of the ring is longer than 1 cm an EG is present and re-entry can occur. If, due to the increase in CV and/or RP (along the circuit), this interval is canceled (no EG is present) and the circulating/re-entering impulse tends to disappear.

Circuits can be functional (due to variation of CV and/or RP) or anatomic (due to structure variations). Circuit whose path is not defined by fixed obstacles (*functional circuits*) have generally short EG and variable sites, those delimited by connective areas or anatomical structures (*anatomic circuits*) have a fixed site and, generally, larger EG.

A phenomenon similar to the one that characterizes the return of the excitement is represented in Figure 10.37. In this case, the re-entry is explained as a consequence of a *specific region* (not the circuit) with a longer refractory period. In this case, the impulse will not pass from this

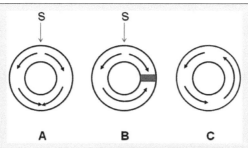

Figure 10.37 Establishing circular re-entry. A: normal situation; B and C: consequences of a difference in refractoriness between one part and the other of a myocardial ring. The arrows indicate the stimulus S, while the red part indicates the point of the ring where the refractory period is longer.

region the first time but will pass after a while and will return to the point of origin from which it will reactivate the ring for an unknown number of times.

Re-entry may be at the basis of *fixed-coupled extrasystoles (e.g.* the ectopic beats tend to occur just after a *fixed time* concerning the P wave of a sinusal beat). The re-entry is the most common mechanism underlying the sustained ectopic tachycardia and is certainly involved in the genesis of flutter and fibrillation. *Short-circuit and long-circuit , as well as variable-sites circuits (functional-type), are needed to explain fibrillation.* A natomical and functional circuits with variable EG are at the basis of the trigger and sustainment of fibrillation, which is characterized by desynchronization of the excitement of the various myocardial fibers, which will not allow an efficient contraction.

The late sodium current (INaL)
DADs and especially EADs are increased in the case of anomalous INaL. INaL is a persistent component of voltage-dependent sodium current (INa). It is normally very thin (negligible). However, anomalous INaL appears in some pathologies (ischemia or hereditary) and consists of an anomaly of the sodium-dependent voltage-current (INa) of the myocardium that increases in its persistent component (Figure 10.38).

The increase in INaL slows the repolarization speed with a tendency to EADs; it increases intracellular Na^+ concentration. Consequently:

(a) the expulsion of Ca^{2+} is reduced through $3Na^+/1Ca^{2+}$ *exchanger* (NCX) leading to Ca^{2+} overload;

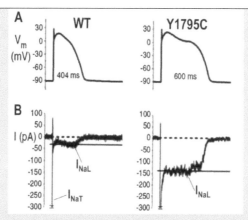

Figure 10.38 Action potentials (A) and relative sodium currents (B) in a normal cardiomyocyte and a cell with a mutation (Y1975C) in a sodium channel. The sodium current (INa) has a transient component (INaT) responsible for phase 0 of the action potential, and a late component (INaL). INaL is negligible in the normal cell but particularly marked in the mutated cell that prolongs the action potential and predisposes to a type 3 LQTS.

(b) sodium and calcium overload tend to increase ATP consumption;

(c) the expulsion of H^+ is also reduced through sodium/hydrogen exchanger (NHX) leading to intracellular acidosis;

(d) the ability to generate ATP is reduced in acidosis;

(e) in acidosis the lusitropism is reduced and the RyR are destabilized with consequent DADs.

Box 10.10 Acute Myocardial Infarction

In the first few hours after the beginning of acute myocardial infarction (AMI), the ECG alteration can be crucial for the diagnosis of AMI. However, the diagnosis of AMI is not only based on the ECG.

Indeed, a myocardial infarction is defined by several signs and symptoms. Detection of elevated serum levels of enzymes (i.e. CKMB or Troponin T) dismissed by ischemic cardiac cells is necessary for the diagnosis of AMI but cardiac enzymes can only be detected in the serum 5–7 hours after the onset of the chest pain that may accompany myocardial infarction. Therefore, especially for early diagnosis, the ECG plays a pivotal role to start therapeutic approaches.

Evolution of STEMI

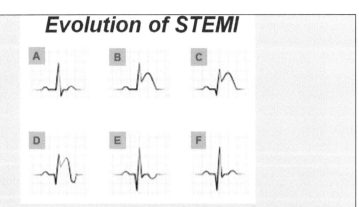

Figure 10.39   Evolution of ST-elevation myocardial infarction (STEMI). **A**: Normal; **B**: Begin of STEMI; **C**: a few minutes/hours after; **D**: a few hours/days after; **E**: weeks/months after; **F**: months/years after; note the *deep Q wave* (> 25% of the depth of QRS complex) in **E** and **F**, which sometimes can be the only sign of prior myocardial infarction.

The ECG signs are several, besides all kinds of arrhythmias and blocks, the ECG shows typical alteration of the ST segment. Indeed, elevation and/or depression (in opposite leads) of ST segments are described as typical signs of myocardial ischemia. Subsequently, pathological Q waves and T wave inversion develop on the ECG (Figure 10.39)

Necrosis causes a *deep Q wave*: this sign appears after months and remains for life. Not all infarcts have a Q wave. Indeed, based on the presence or absence of the Q wave infarct are classified as Q or not Q. The Q wave is a product of *endocavitary potentials* (electrical potentials of the subendocardial cells). Usually, these electric potentials are not seen by the exploring electrode. It is said that after an AMI there is an "electric hole" that allows *endocavitary potentials* of alive tissue to pass through the entire thickness of the dead wall and then "seen"/recorded by the exploring electrode as a negative field.

The ST segment alterations may be of different morphology and the mechanisms that sustain their appearance are very often not clear. However, a hypothesis based on the dipole theory sustains that it is not the ST segment that is altered, but the PQ segment that is elevated or depressed. As a result, when the PQ is elevated, the ST is depressed and *vice versa*. Indeed, at the time of PQ, all ventricular myocardium should be repolarized and then positive outside all myocardial fibers. That is, the ventricles are at rest and without the presence of a dipole. Instead, *the ischemic zone has an excess of negative charges on the outside of the fibers* which then creates a dipole whit the healthy (positive) ventricular myocardium, which is at rest. Thus, the *anomalous dipole* in some ECG leads depresses and in others elevates the PQ segment. When the healthy part depolarizes is negative as the ischemic part: no dipole is there and the J point can be considered the "real" zero. The ST seems elevated because the PQ is below the J point, or *vice versa* the ST seems depressed because the PQ is above the J point. Nevertheless, these modifications in the clinic are described as elevation and depression of the ST segment, respectively, considering the PQ as a reference zero line.

Indeed, ECG allows distinguishing between STEMI (ST-elevation myocardial infarction) and non-STEMI infarct. A STEMI infarct is usually caused by a sudden complete occlusion of a coronary artery. A non-STEMI is often caused by a not completely occluded coronary artery. Usually, a patient having a STEMI needs to be treated immediately and this usually involves angioplasty and stent placement; non-STEMI usually does better with medical management. However, a patient with a non-STEMI also often should undergo programmed cardiac catheterization, angioplasty and stent placement.

A condition that can sometime be mistaken with AMI is the socalled *Takotsubo syndrome* (TTS), also called "stress cardiomyopathy" or "broken heart syndrome". It is characterized by thoracic pain, ST segment elevation and cardiac enzymes and biomarkers elevation. The syndrome was initially recognized in the Japanese population and has recently been described in the United States and Europe. It can present as an acute heart failure or transient ventricular dysfunction which is often triggered by acute physical or emotional stress, although in some patient has no identifiable antecedent triggers. The syndrome has a surprising predilection for postmenopausal women, although it can also be found in younger patients and men. The incidence of TTS is around 5-10% of women who present with a suspected acute coronary syndrome. There are currently no accepted criteria for the diagnosis of TTS. As said, TTS patients often experience chest pain, and have dynamic and aspecific ECG changes and elevated cardiac biomarkers of injury. The elements that allow to distinguish TTS from an acute myocardial infarction are: ventricular dysfunction involving more than one vascular territory, the absence of plaque rupture and coronary thrombosis, and, sometimes, a rapid and spontaneous recovery of ventricular systolic function. In fact, the prognosis is generally favorable, but the condition is not to be considered benign, as it has been associated with serious complications such as thromboembolic events, cardiogenic shock, and potentially lethal arrhythmias.

Box 10.11 Wolff-Parkinson-White Syndrome

Wolff-Parkinson-White (WPW) syndrome is a pre-excitation syndrome. The disorder's estimated prevalence is about 1-3 per 1,000 people. It depends on the fact that in fetal life as an electrical point of contact between atria and ventricles there is not only the bundle of His but there are at least 3 accessory bundles: Manheim, Kent, and James bundles. The accessory bundles are reabsorbed with the maturation and growth of the heart and only the bundle of His remains. If instead of disappearing, one of these remains, more frequently the bundle of Kent, the AV node/His bundle is no longer the only point of the passage of electricity between atria and ventricles, so that the impulse can pass through different points. The Kent bundle allows the passage of the impulse between atria and ventricles and *vice versa* (bidirectional passage). Circuits can be established that generate very high frequency (paroxysmal tachycardia) in some circumstances of excitement.

The ECG can show three signs (Figure 10.40):

- short PR interval (<0.12 s)
- a ventricular pre-excitation (*delta wave* at the beginning of QRS)
- wide QRS complex > 0.12 s between the beginning of the *delta wave* and the end of QRS.

These ECG abnormalities correspond to an abnormal bundle that directly connects the atria and a ventricle. The syndrome may be silent

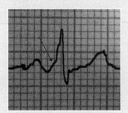

Figure 10.40 Typical ECG of Wolf-Parkinson-White syndrome. The arrow indicates the *delta wave*.

or may cause paroxysmal tachycardia and sudden deaths. An ablation of the Kent bundle by radio frequency which cauterizes the bundle can be recommended especially in young people. The diagnosis may be occasional in otherwise healthy adults.

In a variant of the WPW syndrome, the anomalous bundle connects the atria with the His bundle. In this case, the PR is shorter but the QRS has a normal duration. This condition is called Lown-Ganong-Levine syndrome.

Box 10.12 Electrocardiographic Alterations in Athletes and Heritable Heart-Muscle Disorder (Brugada Syndrome, Cpvt and Arvc)

We decided to treat together these completely different clinical conditions (ECG alteration in athletes and muscle disorder) to underline their subtle differences that sometimes only experienced cardiologists can distinguish.

Physical conditioning induces numerous cardiovascular adaptations, including *vagotonia* (high vagal tone) and increased cardiac volume and mass; these are features of the so-called *"athlete's heart"* which may be also characterized by functional modifications without a necessary presence of anatomical and/or structural appreciable changes. The importance of the ECG exam for athletes is undisputable. It is extremely necessary to understand electrocardiographic changes deriving from high-performance training. These changes are defined as *"athlete's ECG"* and are considered as part of the "athlete's heart". Common ECG changes which can be related to training and can be observed in an athlete are the following:

a. sinus bradycardia and sinus phasic or respiratory arrhythmia;
b. change in ventricular repolarization, of the early repolarization type, prominent U waves, and increased QT intervals;
c. First-degree AV and/or second-degree AV block, Mobitz I type;
d. right bundle-branch conduction delay.;
e. ventricular overload pattern by voltage criteria only: the so-called Sokolow-Lyon criteria, which is considered positive when the sum of the amplitude of the S wave in the V1 lead with the amplitude of the R wave in the V5/V6 leads is greater than 35 mm (see Figure 10.41).

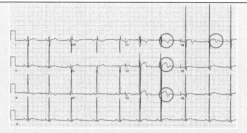

Figure 10.41 Normal ECG in an athlete with *aspecific anomalies* of the repolarization phase. Compare the parts circled with the ECG of fig 10.42.

Sometimes is difficult to differentiate a normal "athlete's ECG" from the athletic patient with a diseased heart (compare Figures 10.41 and 10.42). Therefore, further evaluations are necessary, especially when the athlete also presents with abnormalities of the cardiovascular physical examination, including murmurs and hypertension and/or there is a familiarity for cardiovascular diseases or sudden death.

Brugada syndrome is a channelopathy expressed by a defect in sodium channels in the right ventricle epicardium. It is responsible for up to 4% of all sudden cardiac deaths worldwide and up to 20% of sudden cardiac deaths in patients with structurally normal hearts at autopsy. This syndrome affects mainly men with a family history of sudden death. It is transmitted by autosomal dominant inheritance and is genetically heterogeneous, involving at least 13 genes. On the ECG, it is characterized by a J point elevation in the V1 and V2 leads, with three possible presentations:

a. Type 1 pattern: *coved ST-segment elevation* = 2 mm (J point), followed by T-wave inversion, with a significantly slow R' decline in V1 and V2.

b. Type 2 pattern: *saddle-shaped ST elevation* with a = 2 mm peak (J point) and = 1 mm base (ST terminal portion) and a positive or biphasic T wave in V2.

c. Type 3 pattern: *saddle-shaped ST elevation* with a = 2 mm peak (J point) and < 1 mm base (ST terminal portion) and a positive T wave in V2 (Figure 10.42).

We discuss here this syndrome because to distinguish between a Brugada pattern (Type 2 or 3) and one of many "Brugada-like" patterns presents challenges, especially in athletes.

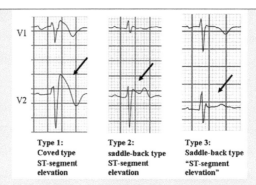

Type 1:
Coved type
ST-segment
elevation

Type 2:
saddle-back type
ST-segment
elevation

Type 3:
Saddle-back type
"ST-segment
elevation"

Figure 10.42   Typical ECG of Brugada syndrome.

Brugada, long and short QT syndromes and CPVT are *channalopathies* often associated with threatening arrhythmias in young people and athletes.

Also, the *arrhythmogenic right ventricular cardiomyopathy* (ARVC) is a heritable heart-muscle disorder included in the leading causes of arrhythmic cardiac arrest in young people and athletes. ARVC predominantly affects the right ventricle with a progressive replacement of cardiomyocytes by electroneutral fibrotic and adipose tissues. ECG hallmarks of the ARVC are negative T waves and the presence of so-called *epsilon wave* (delayed and fragmented depolarization of parts of the right ventricle) in the late part of QRS in right precordial leads. ARVC was initially designated as dysplasia, the discovery that the disease is caused by a *genetic defect in the cardiac desmosomes* has led to its classification as cardiomyopathy. Other rare and congenic cardiopathies associated with dangerous arrhythmias are hypertrophic, dilated, and restrictive cardiomyopathies. *For these pathologic conditions and all the above-described arrhythmias, the reader is redirected to specialized books for a more in-depth description.*

11

Vascular Hemodynamics

11.1 Introduction

Ventricular ejection occurs against the *impedance* present in the aorta which depends mainly on *aortic pressure, elastic properties of the arteries,* and *vessel resistance.* Impedance is modified instant by instant by the ejected blood itself. Indeed, the study of *vessel characteristics*, of *arterial pressure*, and the *work of the heart*, as well as that of *venous return* and *cardiac output* in a rather unified way would better describe the effect of cardiac activity on the vessels and, *vice versa*, the effects of the vessels on the heart. *Like a kick with adequate strength and a soccer ball with adequate weight are needed to get a good shot with a good chance of scoring a goal, in the same way, the heart and the vascular system must be coupled with the appropriate strength of the cardiac contraction and impedance of the vascular system to achieve optimal cardiac work.* Therefore, a distinction between *cardiac hemodynamics* and *vascular hemodynamics* is conceptually rather arbitrary. However, there are some peculiarities of blood movement in the vessels that lead us to dedicate a chapter to the so-called *vascular hemodynamics*.

In Chapter 1, we mentioned the difference between flow and blood velocity, and it was reported that, at a constant flow rate, in each section of the cardiovascular system, an increase in the cross-sectional area involves a decrease in fluid velocity. We must keep in mind this simple concept from now on for a clearer understanding of what will be explained.

11.2 The Total Energy of Fluids

If we do not specify, when we say "pressure" we intend the *lateral pressure* (see below) exerted by the blood on the vessel wall and measurable with the sphygmomanometer. It is commonly said that a liquid moves from a point with greater lateral pressure to a point with less pressure, *i.e.,* according to

Poiseuille's law, the flow is determined by a pressure gradient. This statement does not take into account the role that *position energy* and *kinetic energy* have in the movement of a fluid. In the case of blood and circulation in humans, the statement on the pressure gradient is valid if the subject is in a *clinostatic position* (lying), while it is not correct if the subject is in an *orthostatic position* (standing). Indeed, the lateral pressure in the arteries of the lower limbs to a standing subject is much higher than that present in the aorta, even though the blood moves from the aorta to the lower limbs. Moreover, in a subject in the orthostatic position, the pressure is always a little greater in the ventricles than in the atria but in the filling phases, the blood moves from the atria to the ventricle thanks to the position energy.

We can say that in the orthostatic position the blood flow against pressure gradient can occur due to gravity. Although this statement is acceptable, it is necessary to consider that moving blood spends energy. Therefore, we can say that *blood flow is established between a point where the blood has a greater total amount of energy and a point where the blood has a less total amount of energy.* Indeed, this difference in total energy between the two points is due to energy dissipation, thus it is better to say that *a liquid flows along a gradient of total energy.* For simplicity, we usually say that in a lying person a liquid moves along a pressure gradient.

The pressure itself is not a form of energy. It can, however, be included in a form of *potential energy* if it is multiplied by a volume, in particular by the volume of blood V whose flow is studied. The potential energy depending on the lateral pressure P of this blood is, therefore, P multiplied V, PV.

If we indicate as E_t the total amount of energy that a volume of liquid has at a certain point of a hydraulic system, we see that

$$E_t = PV + 1/2mv^2 + mgh,$$

that is that E_t is given by the sum of the potential energy PV impressed by the heart, of the *kinetic energy 1/2 mv²* of the volume of blood in movement and by the *position energy, mgh,* that the same volume has at that particular height. Obviously, *m* indicates the mass of the blood, *v* the velocity, *g* the acceleration of gravity, and *h* the difference in level between two points considered.

If, instead of referring to the total energy, we prefer to continue using the concept of pressure without taking into account the volume and mass of the blood moving, we can use the following equation:

$$P_t = P + 1/2\rho v^2 + \rho gh,$$

where P_t indicates the sum between the lateral pressure P and the components $1/2\rho v^2$ and $\rho g h$, in which the mass m has been replaced with the density ρ. In this way the expressions $1/2\rho v^2$ and $\rho g h$ has the units of pressure values. In fact, since the density is the mass of a unit volume, ρ is equal to g*/cm³, where g* indicates the mass expressed in grams. Since velocity v is expressed in cm/s, the acceleration of gravity g is cm/s² and the difference in level h is cm, with the appropriate substitutions we can see that all these components can be expressed as dyn/cm² that indicates pressure.

Whether we prefer to consider the total energy of the fluid E_t or the sum P_t of three pressure components, we see that in the aortic arch of a subject in an orthostatic position the blood has total energy greater than that which it has in the arteries of the lower limbs. For the same reasons, the blood can flow in diastole from the atria to the ventricles because in the atria it has total energy greater than that present in the ventricles. The above seems in contrast to Bernoulli's theorem which states that the sum of the pressure energy, kinetic and potential energy per unit mass is a constant, for the laminar flow along a tube. However, Bernoulli's theorem is valid for an ideal, *non-viscous* fluid. Since *blood is a viscous fluid*, along vessel energy is dissipated as heat and to overcome the viscous resistance.

11.2.1 The Lateral, Head and Tail Pressures (Figure 11.1)

Besides explaining the apparent paradox that in a person in an orthostatic position the blood moves from a point with less lateral pressure to another one with higher lateral pressure, the total energy calculation allows also to understand how the blood can circulate in the *vascular arches*, or rather in those *inosculation anastomoses* in which two branches detached from the same artery continue one into the other forming an arch, or by-pass, that runs

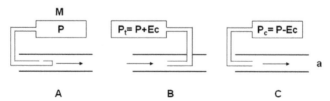

Figure 11.1 In A, the lateral (P) is recorded (the pressure sensor is placed at 90° concerning the direction of the moving blood (arrow)). In B and C, the head (Pt) and tail (Pc) pressure are obtained respectively by adding and subtracting the kinetic energy (Ec) of the moving blood to the lateral pressure. M: pressure gauge; a: artery.

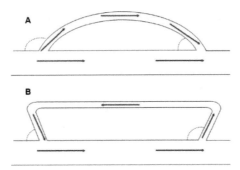

Figure 11.2 Anastomosis by inosculation forming arches (by-pass) parallel to the main vessel. The arrows represent the direction of blood flow. See the explanation in the text.

parallel to the vessel from which they originate (Figure 11.2). Inosculation anastomoses are present, for example, in the mesenteric circulation.

To easily understand how circulation occurs in one of these arches or anastomoses, besides the concept of *lateral pressure* it is helpful to introduce the concepts of *head pressure* and *tail pressure*. It is noteworthy that with the sphygmomanometer it is possible to measure only the lateral pressure. To record the head and tail pressures we need to introduce a catheter into a vessel and to connect it to a manometer (Figure 11.1).

Imagine a horizontal tube in which a liquid flows. Being the tube horizontal, we can neglect the position of energy. A catheter connected to a manometer can be introduced into the tube both in the direction of the current and countercurrent. When the tip or the pressure sensor of the catheter "sees" the flow coming towards them, the manometer registers the *head pressure, i.e.* the sum of the lateral pressure plus the kinetic component of the liquid (Figure 11.1B). When the tip of the catheter "sees" the flow moving away from it, the manometer registers the *tail pressure, i.e.* the lateral pressure minus the kinetic component (Figure 11.1C). When the sensor is perpendicular to the direction of the flow we record the *lateral pressure* (Figure 11.1A).

We can now see how the circulation takes place in a vascular arch or bypass (Figure 11.2). From a horizontal tube in which a liquid flows from left to right, an arch takes origin parallel to the main vessel. At the point of origin, the arch forms an *obtuse angle* concerning the direction of flow in the main tube; the arch then returns to the main tube forming an *acute angle* concerning the flow direction (Figure 11.2A).

From Figure 11.2A, it is easy to understand how the energy at the beginning of the arch is greater than at the end so that the liquid flows from left to right as in the main tube.

In Figure 11.2B the origin and end of the arch form with the main tube angles opposite to those we saw in Figure 11.2A. It is clear that in these conditions the energy from the right side of the arch is greater than that from the left side, so the blood now flows in the opposite direction to the main tube.

The pressures at the beginning and the end of the arches do not correspond exactly to the lateral, head, or tail pressures. The arch branches detach with an angle that is not 90° concerning the main tube, and a portion of kinetic energy must be added or subtracted at the beginning and the end of the arches to the lateral pressure value. The value of kinetic energy added or subtracted is, in fact, a function of the *cosine* of the angle between the arch branches and the main tube. Since the cosine can vary between 0 (90° angle) and 1 (0° angle) or −1 (180° angle), the kinetic component can be 100% "appreciated" by a pipe placed at 0 degrees, i.e. "in front" of the mainstream, that is head pressure. Instead, the kinetic component cannot be at all recorded by a 90° pipe, which is lateral pressure. At 180 degrees we subtract the 100% of kinetic component, which is the tail pressure (Figure 11.1). Intermediate angles add or subtract intermediate values of kinetic energy depending on cosine. For instance, if a branch detaches with a 45° angle from the mainstream, the cosine is 0.5, therefore, half kinetic energy is added, *i.e.* an intermediate value between lateral and head pressure is recorded.

11.3 The Blood Viscosity

The word viscosity derives from the Latin word "*viscum*", that is the *mistletoe* containing a glutinous fluid. Newton defined viscosity as "*defectus lubricitatis*" or lack of slipperiness. The blood has an absolute viscosity and an apparent or relative viscosity.

The *absolute viscosity* depends on the composition of the blood in the water, macromolecules, and corpuscular component, while the *apparent viscosity* also depends on the flow modalities. When the viscosity of distilled water is equal to 1, the absolute viscosity of the plasma is about 1.6 owing to the presence of proteins, while that of the blood rises to 4–5 owing to the presence of the corpuscular elements. Given the prevalence of red blood cells between the corpuscular elements, the absolute viscosity of the blood increases with hematocrit. As can be seen in Figure 11.3, the increase is not linear but is steeper for values higher than the normal value, which is 45%. It follows that, when the hematocrit falls below 45%, the reduction in viscosity is modest, while its increase is noticeable when the hematocrit rises above 50%. The increase in hematocrit takes place in *polycythemia* of

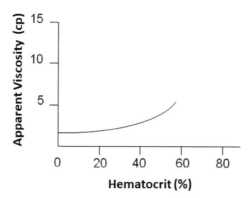

Figure 11.3   Relation between the apparent viscosity of the blood and the hematocrit value. cp: centipoise. See the explanation in the text.

various kinds as well as in *dehydration* where there is a greater concentration of red blood cells. Also, the auto-transfusion used by athletes may increase the hematocrit, ideally less than 50%, to improve tissue oxygenation but not increase appreciably the blood viscosity.

The viscosity of a flowing liquid is its internal friction which is opposed to sliding. The *apparent or relative viscosity* (aV) depends on the flow velocity and the diameter of the vessel in which the blood flows. It can be considered the friction between two layers of the flowing liquid and according to Newton's Law, apparent viscosity (AV) can be defined as the *ratio of shear stress (SS) to shear rate (ShR).*

$$aV = SS/ShR.$$

Indeed, shear stress is viscosity multiplied for shear rate:

$$SS = aV \times ShR.$$

The **shear rate** can be defined as the change in liquid velocity per unit distance across the tube. Indeed, the shear rate is the velocity of the upper lamina (in meters per second) divided by the distance between the two laminas (in meters) whose unit is 1/s or reciprocal second s^{-1}. If the sliding velocity increases, the velocity of splitting between the various laminas or *shear rate* increases (see below the velocity profile of laminar flow).

The **shear stress** can be defined as the tugging force by one layer on the other (the last stationary layer is tugging on the endothelial surface) and has the unit N/m^2 as it depends on the axial pressure gradient, related to the kinetic component of pressure.

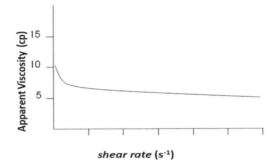

Figure 11.4 Relationship between apparent blood viscosity and *shear rate*. cp: centipoise; s^{-1}: indicates the energy gradient expressed in (m/s)/m. See the explanation in the text.

When blood velocity increases shear rate increases more than shear stress and viscosity decreases. Indeed, if the liquid moves in a laminar way (see Paragraph 11.4), the red blood cells stay in the central strand rather than in the peripheral lamina. Therefore, the viscosity and the shear rate influence the shear stress and *vice versa*.

In other words, an increase in shear rate is accompanied by a decrease in relative viscosity as the red blood cells align better along the central strand avoiding the tendency to sedimentation that would increase it (Figure 11.4). In this way, we understand how the blood has a higher relative viscosity in the veins than in the arteries. If then the blood moves very slowly, *stackings or rouleaux* (clumps of red blood cells that look like stacked dishes) may be formed, which *in vitro* favor red cells sedimentation.

However, relative viscosity does not diminish indefinitely with an increase in blood velocity. I f this rises too much the flow becomes turbulent (see Paragraph 11.4), and the viscosity increases suddenly due to the reciprocal collision of the corpuscular elements. Despite the decrease in viscosity, shear stress increases with flow velocity and increases even more with the appearance of turbulence. However, in turbulent flow shear stress direction changes continuously. This affects endothelial cells' physiology and pathophysiology (see Chapter 13.2).

In vessels of a very small radius, the relative viscosity decreases with the diameter of the tube. This apparent paradox is known as the *Fåhraeus-Lindqvist phenomenon*, for which in very small radius vessels, which generally depart from a larger vessel, there is a very low hematocrit, as they are perfused by blood poor in red blood cells, that is the *marginal plasma layer*, which in the main vessel flows along the more peripheral laminae

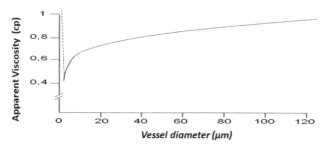

Figure 11.5   Relation between the apparent viscosity of the blood and the diameter of the vessel (*Fåhraeus-Lindqvist phenomenon*). In capillary size tubes (about 6 μm), blood viscosity reaches a minimum value; this explains the low viscosity of blood *in vivo*. The dotted line is the apparent viscosity in *in-vitro* tubes smaller than capillaries. cp: centipoise. See the explanation in the text.

(Figure 11.5). If, on the other hand, the collateral branch has a larger diameter it receives blood from the laminae that contain more red blood cells.

The viscosity of the blood should not be confused with the *density* which is about 1055 kg/m^3 (distilled water density is 1000 kg/m^3).

11.4 The Laminar Flow and the Turbulent Flow

In Chapter 8.1, in the study of the Poiseuille device, the flow of the liquid in the horizontal tube takes place according to a laminar flow. A liquid flowing in stationary mean velocity in a tube with a circular section and a constant diameter along its entire length can be considered as a system formed by concentric cylindrical *laminas* that move at increasing velocity from the periphery to the center. The most peripheral lamina in contact with the wall of the tube is considered immobile, while in the center, where there is no longer a lamina, but an axial or central strand, the liquid moves at the highest velocity.

In a liquid that flows with laminar flow, it is possible to draw the velocity profile of the various laminae and the axial strand (Figure 11.6A). The profile has the appearance of a parabola in that the relationship between the velocity and distance of each lamina from the axis of the tube is parabolic. The different velocity of the various laminas indicates that the resistance that a liquid encounters in flowing in a tube is due to the friction of the various laminas between them. It can be calculated that the average flow velocity of the liquid is equal to half the sliding velocity of the central strand.

The ideal conditions for the laminar flow are those of a liquid that moves in a stationary way, in a rigid tube with a constant section, however, it is

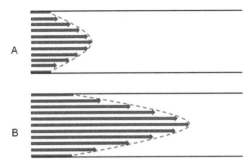

Figure 11.6 Laminar flow. Notice how the velocity increase (B) makes the profile of the parabola more pointed. It is increasing the shear rate. See the explanation in the text.

also possible a laminar flow in the arteries in which the blood proceeds in a pulsating manner in tubes with non-constant section and with a variously distensible wall.

If the velocity increases, the outer lamina continues to be stationary while the velocity of the other sheets and that of the axial strand increases. As can be seen in Figure 11.6 B, the parabolic velocity profile becomes sharper and more pointed. If then the velocity exceeds a certain *critical value*, the flow is transformed from laminar to turbulent (Figure 11.7). In these conditions, the laminas dissolve so that the various fluid particles end up forming vortices. Since this type of flow causes more friction between the various parts of the liquid, the viscosity increases, and the resistance to flow increases.

The passage of flow from the laminar to the turbulent state was studied by *Sir Osborne Reynolds* who caused the flow rate of a liquid to vary in a transparent tube. After adding a dye to the flowing liquid, he could see that, when the velocity was not excessive, the colored outline of the advancing liquid had a parabolic shape. This shape became more pointed as the velocity increased, until, after exceeding a critical velocity value, the parabolic profile disappeared revealing the passage to turbulent motion.

Based on the data obtained, Reynolds expressed the following formula to calculate the *critical velocity*:

$$V = \frac{R\eta}{\rho r}.$$

When this *critical velocity* V is exceeded the flow becomes turbulent.

The critical velocity expressed in cm/s, the viscosity η of the liquid is expressed in poise (or milliPascal), the density ρ is expressed in g/cm^3, r the radius of the tube is expressed in cm and R or *Reynolds' number* is a

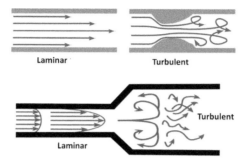

Figure 11.7   Passage from linear to turbulent flow. See the explanation in the text.

dimensionless constant whose value is 1000 (2000 if, instead of the radius, we consider the diameter of the tube). The formula also shows that

$$R = \frac{V\rho r}{\eta}.$$

In other words, for certain given values of density and viscosity of the blood and the tube's radius, the critical velocity is the flow velocity that causes R to rise to about 1000. Besides velocity also an increase in radius may cause an increase of Reynolds' number and thus turbulence, as can occur in aneurysms (Figure 11.7). For further details on aneurysms see Box 6.1 on Laplace's Law and the Bernoulli's theorem explain the varices and aneurysms, in Chapter 6.

Physiologically, Reynolds' number reaches about 2300 value in the aorta root of a normal adult, where the radius is 1.25 cm and peak flow velocity 70 cm/sec, blood density and viscosity are 1.06 g/cm^3 and 0.04 centipoise, respectively. That is why sometimes the turbulence may become audible as an *"innocent" systolic ejection murmur.*

Since in the presence of turbulence there is no longer a fixed/stationary lamina in contact with the tube wall, the friction between the liquid and the wall is particularly important in contributing to a dramatic increase of the resistance to flow.

The value equal to 1000 of the Reynolds' number applies only to the constant flow in tubes that do not have any restrictions. If along the course of a tube there is shrinkage, not only the velocity of the liquid can increase at that point, but the turbulence may appear even before R has reached the value of 1000. Moreover, turbulence appears when the Reynolds' number is lower than 1000 because blood flow is pulsatile and the vessels are neither straight nor uniform in size. In those circumstances, moreover, the turbulence extends also to the section of the tube immediately downstream of the narrowing.

In the cardiovascular system, the flow is turbulent in the ventricles and the initial part of the aorta and pulmonary artery. In other arteries, turbulence may appear when they are partially occluded by an atherosclerotic plaque (Figure 11.7). The flow becomes turbulent also in the section of the humeral artery that is compressed by the *Riva-Rocci* sphygmomanometer cuff when the arterial pressure is measured. These turbulences are responsible for the *Korotkov* sounds or Korotkoff's sound (see Chapter 8.12).

When the turbulences reach certain intensity, the value of which also depends on the characteristics of the vessel wall, the latter can become the site of audible vibrations responsible for the so-called *murmur (bruit or noises)*. These can occur in the presence of congenital heart disease, acquired valve defects, as well as in the presence of arteriovenous anastomoses. In addition to these anatomical situations, reductions in blood viscosity due to a reduction in the number of red blood cells may generate audible murmurs in systole. These are the so-called *anemic or functional murmurs* that appear when the number of red blood cells per mm^3 of blood falls below 3,000,000 and blood velocity increases.

11.5 The Visco-Elastic and Contractile Features of Vessels

The vessel walls have viscous, elastic, and contractile characteristics. The large arteries, such as the aorta and its main branches, which contain many elastin fibers and a few smooth muscle cells in their walls, have predominantly elastic walls and, due to their large diameter, offer little resistance to blood flow. Of course, the collagen fibers and smooth muscle cells contribute to the *visco elastic* properties of the large arteries.

Unlike the large arteries, the arteries of small diameter, and in particular the precapillary arterioles, contain in their wall many smooth muscle cells. For this reason, they can vary the resistance depending on the degree of contraction of the smooth muscle. The set of precapillary resistance of the systemic circulation is responsible for the total peripheral resistance, TPR, which we have seen to be one of the mechanical determinants of arterial pressure, which can be finely regulated. These smooth muscle cells are multiunit muscle fibers controlled by autonomic nerves, hormones and mechanical stimuli as stretching.

The capillaries, lacking elastic fibers and smooth muscle cells, are modestly extensible and generally have walls permeable to water and solutes, but not to plasma proteins. The characteristics and function of capillaries will be discussed in greater detail in Paragraph 11.6.

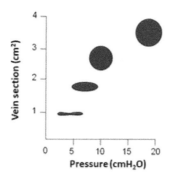

Figure 11.8   Relative variation of the inferior vena cava section due to an increase in transmural pressure. See the explanation in the text.

The capillaries are followed by venules endowed with a certain number of smooth muscle cells that regulate postcapillary resistance. The veins, located downstream of the venules, are also characterized by the presence of a certain smooth musculature which can make them vary in their capacity, thus changing the venous return to the heart. Their distensibility also contributes to the variations in the *capacity* of the veins, which can modify the section as can be seen in Figure 11.8. This figure illustrates how the increase in hydrostatic pressure, due to the passage from the clinostatic position to the orthostatic position, makes the section of the inferior vena cava vary. In particular, it can be observed how, with the increase in pressure, the vein section passes from an elliptical shape to a circular shape. Note that when the veins collapse they never close completely, but assume a binocular shape at low pressure.

To better understand the *visco-elastic properties of the vessels* it is appropriate to refer to *Hooke's law* and *Young's modulus* or *modulus of elasticity*. According to Hooke's law, the lengthening of spring with linear elasticity is directly proportional to the applied force. The relationship between strength and elongation or, more generally, between strength and deformation, is given by Young's modulus (Figure 11.9).

If we call E Young's module, we see that

$$E = \frac{F/A}{\Delta l/l}$$

where F is the force expressed in dyn, A the section surface of the spring, l its initial length and Δl the variation in length due to the application of F. The Young modulus corresponds to the force per unit area of section required to

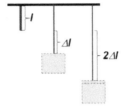

Figure 11.9 Young's law. The elongation (Δl) of the schematized spring (blue line) is proportional to the applied force (gray rectangle). See the explanation in the text.

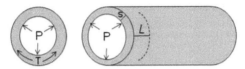

Figure 11.10 Application of Young's modulus to an elastic tube. See the explanation in the text.

stretch the spring by 100%. As the modulus increases with F, it is understood that the more rigid is a body the higher is its elasticity modulus.

For perfectly elastic bodies Young's modulus is constant for any degree of deformation of the body, while it increases with the deformation for bodies that are not perfectly elastic.

In the case of an elastic tube, the distension of the walls is due to the lateral pressure exerted by the liquid flowing inside it. As can be seen in Figure 11.10, the pressure in the tube wall produces a circumferential tension (T) which, at each point of the circumference, can be represented by two vectors facing centripetal concerning the point of application of the vector indicating the pressure. From this, it can be inferred that T tends to move away from each other two adjacent points of the vessel wall (tends to break the wall). If Young's modulus is applied to this particular case, we see that the force F is the circumferential tension T, while the Δl represent the variation of the initial length, while A is the product of the thickness (s) of the tube for the unit of length (L) of the tube itself. As can be easily understood, this product represents the section surface of a segment of the tube with unit length and is therefore analogous to the section surface of the spring considered above.

Hooke's law, valid for perfectly elastic bodies, foresees that these can be deformed linearly and without limits with the increase of the force acting on them. In nature, there are no perfectly elastic bodies. Furthermore, the blood vessels, in addition to not being perfectly elastic, do not even have a

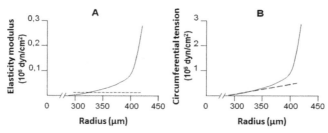

Figure 11.11   Increase in the modulus of elasticity (A) and the circumferential tension (B) concerning the radius increases. In black are the values of a small artery and in red those of a tube with perfectly elastic walls. Note the identical unit for the two parameters. See the explanation in the text. (from Hinke JA, Wilson ML, 1962)

homogeneous structure of the walls. Because of these characteristics, Young's modulus is not constant but increases with vessel distension. In other words, the more the pressure, radius, and circumferential tension increase, the more the vascular wall becomes less stretchable, which is more rigid. In Chapters 8.2 and 8.6 we have seen how the *windkessel effect* of large arterial vessels ceases resulting in increased pulse pressure due to lack of distension and subsequent elastic return of the walls. While in Chapter 8.6, rigidity was attributed to atherosclerotic processes, in Chapter 8.2 it was seen also as the consequence of an increase in diastolic pressure.

The non-homogeneous structure of the arterial walls means that the relationship between radius on one side and Young's modulus on the other is not a linear but an exponential function as shown in Figure 11.11A. This figure compares the progression of the *elasticity module (Young's modulus)* of a perfectly elastic imaginary tube with that of a small artery concerning increases in the radius of the lumen.

Another difference between a perfectly elastic tube and a small artery is given by Figure 11.11B in which we see the relationship between the radius and the circumferential tension. In this figure, the linear increase of the circumferential tension of the tube is simply the consequence of Laplace's law. Since Young's modulus is not constant also, in this case, the circumferential tension increases more with the increase in radius ($T = P \times r$). Note that the increase in radius is obtained by increasing the pressure.

If the pressure in the tube increases, the radius increases. According to Laplace's law, also the circumferential tension inc reases. The radius increases until the distension of the elastic structures of the tube produce a force such as to balance the circumferential tension. The radius that the tube assumes for each pressure increase is called the equilibrium radius for that pressure.

An anatomic-physiological consequence of Laplace's law is the effect of the different radius r of arteries that have the same pressure P inside them. Since $T = P \times r$, the arteries of greater radius develop a greater circumferential tension and would tend to break if they do not have a thicker wall than those with a smaller radius. The aorta which has a much wider radius has a much thicker wall than a carotid artery or a coronary artery, which has a similar pressure inside. The capillary which can have a very high P transiently (*e.g.* when standing up) does not break due to the very small radius, despite the very thin wall.

11.6 The Miogenic Regulation of the Radius of Resistance Vessels

The myogenic regulation of the vascular radius concerns vessels that contain smooth muscle cells in their walls. Therefore, it mainly concerns arterioles responsible for district resistance. In various vascular districts it can be observed that an increase in transmural pressure results in vasoconstriction, while a pressure reduction determines vasodilation. In both cases, the result is that the flow remains fairly constant despite changes in perfusion pressure, over a wide pressure range that differs slightly between the districts.

Figure 11.12 shows the myogenic regulation within the typical pressure range of the coronary circulation. It can be seen as initially for low average pressure values, the flow increases more than the pressure, as evidenced by the concavity of the curve to the left. This type of increase is due to the distension of the vascular walls not very tense yet, and therefore presenting a low Young's modulus. When the perfusion pressure exceeds 70 mmHg, the increase in flow due to increased pressure becomes very modest (almost absent) due to the increase in the contraction strength of smooth muscle cells.

Exceeding 120 mmHg, the flow starts to rise again, this time in linear relation with pressure since the upper limit of myogenic regulation has been reached and the radius of the vascular lumen no longer varies. It can, therefore, be said that, within the auto-regulation limits of the coronary circulation, *the equivalent radius* of the resistance vessels progressively decreases with the pressure increase from about 70 to about 120 mmHg and, *vice versa*, the radius increases with the return of the pressure from 120 to 70 mmHg.

If the myogenic auto-regulation fails at all pressure levels, as happens in the presence of maximum pharmacological vasodilation, the flow initially rises more and more rapidly, due to the progressive increase in the radius of

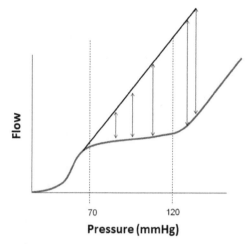

Pressure (mmHg)

Figure 11.12 Myogenic auto-regulation. Tre pressure values reported in the X-axis are typical of the coronary circulation autoregulation range. The red curve is the autoregulated flow, the black line is the non-autoregulated flow observed in a maximally dilated vascular bed. See the explanation in the text.

the vessels. When the wall of the vessels can no longer be extended further, the flow increases linearly with pressure

We can say that he myogenic response in the range of auto-regulation of vascular districts acts on the resistances in the opposite direction to the viscoelastic deformations (expansion or recovery) induced by the pressure variations.

Figure 11.13 illustrates how a sudden drop in pressure, based only on elastic recovery of the wall, would lead to an increase in resistance due to a decrease in the lumen of the vessels, with a consequent decrease in the flow to lower levels than those due to the simple reduction of the pressure of perfusion. However, the overlapping of the vasodilator myogenic response after a few seconds brings the resistance below the control value; inducing at the same time a partial recovery of the flow (the flow recovery is partial because the pressure is lower despite the vasodilation).

Myogenic responses are of particular importance in the auto-regulation of blood flow in some circulatory districts, such as, in particular, the coronary, cerebral, renal, and limb vascular districts.

Vasoconstriction due to increased transmural pressure is due to the contraction of smooth muscle cells stimulated by stretching. The mechanism of the myogenic response is still a matter of debate. Currently, concerning the

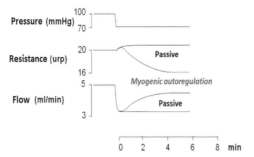

Figure 11.13 Myogenic auto-regulation. Following the fall in pressure to an increase in resistance and to a decrease in the flow due to passive collapse (in black) of the vessels, vasodilatation (in red) of myogenic nature is opposed which causes a reduction in resistance and an increased in flow. Of course, in the presence of autoregulation, the black curves (for resistance and flow) are not observed. See the explanation in the text.

sequence of events that couple changes in intravascular pressure or stretch with alterations in vascular smooth muscle activation, there are several broad hypotheses. These hypotheses consider:

(a) modifications of membrane properties leading to activation of ion channels;
(b) activation or inhibition of biochemical cell-signaling pathways within the smooth muscle;
(c) length-dependent changes in contractile protein function; and
(d) endothelial-dependent modulation of vascular smooth muscle tone.

In particular, it has been suggested that the following stretching, Ca^{2+} is released from the intracellular deposits through the ryanodic channels (these are present on SR of smooth muscle and of cardiac muscle, see Chapter 4.3). It has been suggested that in some district *nitric oxide* may blunt the myogenic response. In this case, NO is released by an endothelial nitric oxide synthase (eNOS) *via* the phosphoinositide-3-kinase–protein kinase B/Akt (PI3K-PKB/Akt) pathway activation induced by vessel stretching.

11.7 The Flow of the Blood According to the Waterfall Model

In Chapter 15 we will see how some aspects of the pulmonary circulation can be explained with the waterfall water model. With this model, valid in general for many district circles, authors have also tried to explain the systolic

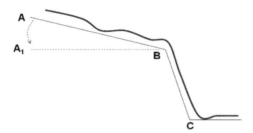

Figure 11.14   Water cascade model. See the explanation in the text.

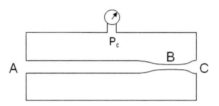

Figure 11.15   Collapse of a rubber tube to increase the chamber pressure (Pc). See the explanation in the text.

reduction of the coronary circulation. Yet, this model does not explain many aspects of coronary circulation (see coronary circulation, Chapter 14.1).

This model is described in Figure 11.14 where a leaning/sloping river-bed allows water to flow from point A to point C. Between two points there is a waterfall in B. The difference in height between the various points represents a difference in pressure (or energy). From the observation of Figure 11.14 we understand how the velocity of water flow is not due to the difference in level between A and C, but between A and B; if point A is lowered to the same level as B at the point indicated with A1, there is no more difference in level or pressure between A1 and B, consequently, the flow between the two points will cease and there will be no more water in C.

Now imagine a device (Figure 11.15) consisting of a collapsible tube that for most of its course is contained in a chamber in which it is possible to vary the pressure that the air exerts from the outside on the tube. The proximal and distal ends of the tube, both located outside the air chamber, are indicated by the letters A and C, respectively.

The flow will depend on the pressure gradient between A and C, as long as the pressure in B will not collapse the rubber tube. If the pressure inside the chamber is increased beyond the pressure in C, the rubber tube will begin to collapse at a point B. This point is far from A and near C due to the energy dissipation along the tube, in B the water pressure is lower than at the beginning of the tube. The tube collapses as the air pressure in the chamber increases. In the presence of collapse, the water flow depends on the pressure gradient between points A and B, rather than between A and C. Moreover, if the chamber pressure equals the pressure in A, the collapse in B completely obstructs the tube causing the cessation of the flow. In this model the pressure in B corresponds to the level of the jump B of the waterfall, this explains the name of the model.

In the circulatory system, points A and C of the waterfall model correspond to the arterial and venous pressure, respectively, at the two ends of a perfused organ. The organ tissue exerts from the outside a pressure B which tends to make the micro-vessels collapse. It is clear that, if the pressure from the arterial side goes down to the value of the pressure developed by the surrounding tissue, the flow stops. This is why, when blood pressure is reduced in a circulatory district, the flow stops before the head pressure has dropped to the value of the venous pressure. This arterial internal pressure is the *critical closing pressure* at which a blood vessel closes completely and blood flow is stopped. In many tissues, at resting condition, the critical closing pressure is around 20 mmHg.

A similar, but not the identical condition is the flow interruption and zeroing of the flow observed in the coronary circulation. This occurs when, with the same head pressure, the pressure in the tissue increases: i.e. when the systole increases the pressure exerted by the myocardial tissue on the coronary vessels that pass through it. This is only in part responsible for coronary flow reduction in systole, as we will see in Chapter 14.1. Indeed, the coronary flow in isovolumic systole not only can be stopped but can go backward and the waterfall model cannot explain this backflow.

11.8 The Wave of Flow Along the Arterial Tree

In Chapter 8.8 we have seen how the pressure curve varies as it proceeds from the aorta to the peripheral arteries. Figure 11.16 illustrates that in addition to the pressure curve, the flow curve is also modified. In the ascending aorta, the flow curve is the systolic flow curve, and the diastolic flow is zero. Indeed, after a wide oscillation, corresponding to the ejection phases with acceleration and deceleration, it presents a small backflow due to the attempted of the blood to return in the ventricle, stopped by the closure of the aortic valve. In diastole, there is no flow since in this phase of the cycle no blood comes out from the ventricle.

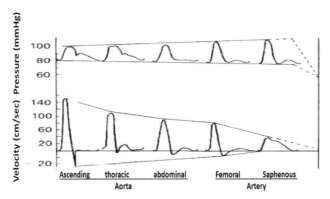

Figure 11.16   Fluctuations pressure (top) and in flow (bottom) starting from the aorta to the peripheral arteries. See the explanation in the text. (modified from DA McDonald, 1960)

If from the ascending aorta we move to the thoracic and abdominal aorta and, subsequently to the femoral artery, we notice a progressive reduction of the amplitude of the curve accompanied by reflux of longer duration followed by a second diastolic positive flow.

The changes in the curves are due to the *backward flow* which must be subtracted from the *forward flow*, which in diastole is favored by the elastic recovery of the large arteries. Note that while the backward pressure wave is added to the forward wave, the backward flow is subtracted from the forward flow.

Since part of the systolic flow does not proceed in the aorta but perfuses collateral branches that depart from it, in the upstream parts of the explored point, the mean amplitude of the flow velocity is reduced moving from ascending to the abdominal aorta. Since each point of the curve represents a velocity, the mean velocity (cm/min) multiplied for the sectional area (cm^2) of the vessel is equal to the flow (cm^3/min or mL/min). Therefore, "the absence" of part of the collateral flow can explain the reduction in the mean amplitude of the curve in the abdominal aorta compared to the thoracic aorta, as they have a similar cross-sectional area. However, the abdominal aorta and femoral artery have a very different sectional area, a different flow amount, but, by chance, a similar mean amplitude of the flow velocity curves.

11.9 The Microcirculation

The microcirculation is mainly made up of arterioles, capillaries, and venules. In some organs, such as the intestine and skeletal muscle, metarterioles and the so-called "thoroughfare channels" or "preferential channels" are also part of the microcirculation. In addition to providing exchanges between blood and tissues at the level of the capillaries, the microcirculation participates in the regulation of arterial pressure by variations in the vasomotor tone of the arterioles. The microcirculation functions are many, including important thermoregulation in the skin.

The arterioles, as already mentioned several times in this book, are resistance vessels that present in their wall abundant smooth muscle cells arranged in one or more layers. The increase in the smooth muscle tone causes an increase in resistance or *vasoconstriction*, while the decrease allows a reduction in resistance or *vasodilation*. The vasomotor tone is regulated by the sympathetic innervation, but also by humoral factors, autacoids, and/or by myogenic mechanisms. A few important districts, such as the salivary glands and the cavernous bodies of the genital organs, receive a parasympathetic innervation with a vasodilator action.

Metarterioles are small vessels that in microcirculation sometimes connect an arteriole with a venule, thus bypassing the capillaries and forming the *preferential channel* (Figure 8.8). The musculature of their wall becomes progressively sparser preceding from the arteriole to the venule.

The capillaries originate from the proximal part of the metarterioles. Capillaries after having nourished/cleaned the surrounding tissue terminate in the distal pre-venular part of the same channel. At the origin of the capillaries, the metarterioles have muscular rings or *precapillary sphincters*. If these are open, the blood enters the capillaries, while, if they are closed, the blood passes directly into the venule without passing into the capillaries. In Chapter 14.4 we will see this organization in the microcirculation of skeletal muscle.

The capillaries are exchange vessels placed between the arterioles and venules. Their diameter varies from 6 to 10 μm. They are formed by a layer of endothelial cells placed on a basement membrane. There are no smooth muscles in the capillary wall. Depending on the architecture, the capillary wall can be continuous, fenestrated, or discontinuous.

The *continuous-wall capillaries* have an uninterrupted basement membrane on which rather bulky endothelial cells are placed with tight junctions with each other. This type of capillary is mainly present in skeletal muscles.

The *fenestrated-wall capillaries* also have a basement membrane without interruption, while the endothelial cells, smaller than the previous ones, present scattered spaces or windows of 60–70 nm in diameter. Sometimes the windows have a very thin diaphragm. This type of capillaries mainly characterizes the *renal glomeruli*, the endocrine and exocrine glands, the gastrointestinal tract, and the gallbladder.

The *discontinuous-wall capillaries*, also called sinusoids, have discontinuities in the basement membrane that can reach a diameter of 30–40 μm, and that can also be crossed by the corpuscular elements of the blood. These capillaries have a diameter that can reach 100 μm and is therefore much greater than that of the two previous types, while their lumen is irregular due to the morphology of endothelial cells. This type of capillary is well represented in the liver, spleen, bone marrow, endocrine glands, and lymphoid tissue.

The permeability of discontinuous-wall capillaries is greater than that of fenestrated-wall capillaries and the permeability of this is greater than that of continuous-wall capillaries.

In addition to the capillaries above described, *arterial capillaries* and *venous capillaries* have been observed. The formers are interposed between the arterioles and capillaries, the latter between the capillaries and venules. They have an intermediate structure to that of the vessels between which they are interposed. In their wall, scant and discontinuous smooth muscle tissue is observed.

Exchanges through the capillary wall occur both by filtration and diffusion. In a capillary, the permeability is due to the presence of pores of about 90 Å in the capillary wall. *Filtration* means the passage of a liquid and the solutes contained in it from one side to the other of a permeable membrane, according to the gradient of pressure acting on the liquid.

Diffusion means the passage of solutes through a membrane by their gradient. Concentration gradient for solid substances or partial pressure gradient, in the case of gas. The diffusion is completely independent of the pressure gradient of the liquid, whatever it is.

Filtration: Generally, in the systemic circulation, blood pressure in the capillary is 38–25 mmHg from the arteriolar side and 8–14 mmHg from the venular side, being its average value along the capillary length of about 18–20 mmHg. Since the pressure in the interstices is equal to 0 mmHg, if not even negative, the blood pressure tends to cause filtration of liquid and solutes in the interstitial liquid. Given the pore size, only protein-free plasma can be filtered. Staying mainly in the capillary the proteins exert their oncotic pressure of about 25 mmHg (reflection coefficient is close to 1, see below). While capillary pressure falls progressively, plasma oncotic pressure usually does not change significantly because the filtration fraction is small along the systemic capillaries (filtration fraction is the ratio between the filtered liquid and the plasma flow). Since plasma oncotic pressure acts in the

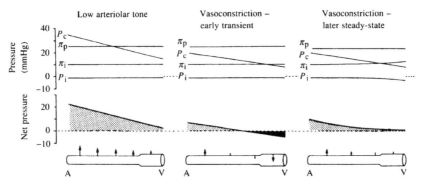

Figure 11.17 Variation of net filtration pressure and reabsorption in a capillary. The capillaries are prevalently filtrating (left and right panels). Indeed, the interstitial fluid pressure (Pi) and the oncotic pressure (πi) are re-adjusted after a period of transient reabsorption (middle panels) and the net pressure returns to be positive along all the capillary (right panels). See explanation in the text (*Levick JR* Experimental Physiology 1991; 76:825–857).

opposite direction to blood pressure, it slows the leakage of liquid from the arteriolar side and determines its return from the venular side (Figure 11.17). Proteins are also present in the interstitium and favor leakage of fluids. *Usually, in many vascular districts the net filtration pressure is positive and tends to release liquid and solutes along the entire capillary.* Transiently, when the vasomotor tone of arterioles increases, the net filtration pressure is negative on the venular side (middle panels of Figure 11.17) where the liquid can re-enter in the capillary (reabsorbed). This reabsorption will reduce the interstitial pressure, which becomes negative, and increases the interstitial oncotic pressures, thus favoring again the leakage of fluids (left panels of Figure 11.17).

From the above description, it is clear that the quantity of filtered liquid is greater than the amount reabsorbed in the microcirculation. The effects of this difference are attenuated by the greater development of the capillary wall area on the venular side. Despite this "correction", not all the filtered liquid is reabsorbed. The lymphatic circulation provides the removal of excess liquid from the interstitium (see Chapter 18).

The factors that regulate filtration and reabsorption through the capillary wall are summarized in the *Starling formula*:

$$J_v = K_f([P_c - P_i] - \int [\pi_c - \pi_i]),$$

where **Jv** is the net movement of the liquid between the two intra- and extra-capillary compartments, K_f is a filtration coefficient dependent on the type of capillary, **Pc** and **Pi** represent respectively the blood pressure in the capillary and the interstitial liquid pressure; $\pi_c - \pi_i$ are, respectively, the oncotic pressure in the capillary and interstitium and f is a *reflection coefficient*. It can theoretically vary between 1 and 0, concerning the permeability to proteins from non permeable to completely permeable. In liver it is 0.1 and in lungs 0.75. The mean reflection coefficient in our body is about 0.85.

The pressure values inside a capillary may vary depending on the variations in pre- and post-capillary resistance, in the sense that capillary pressure decreases if the pre-capillary resistance increases (middle panels of Figure 11.17) or if the post-capillary resistance decreases. Of course, capillary pressure increases if pre-capillary resistance decreases or post-capillary resistance increases. Moreover, the net filtration and reabsorption pressures vary not only depending on the pressure gradient of the liquid on the two sides of the membrane but also the oncotic pressure gradient. These mechanisms can be used to explain the formation of *edema or ascites* as well as the recovery of part of the volume of circulating fluid after hemorrhage. For example, ascites in cirrhotic patients is favored by the low level of plasma proteins and the high pressure in the capillaries of the enteric and portal circulation (see also Chapter 14.3.3).

Diffusion: The spread of respiratory gases depends on their partial pressure gradient. In arterial blood, the O_2 that arrives in the systemic capillaries has a pressure of about 95–100 mmHg, while in tissues it is estimated to have an average pressure of only 40 mmHg. The result of this gradient is a passage of O_2 from the blood to the tissues.

In the tissues, there is a CO_2 pressure that on average is calculated to be 46 mmHg, while in arterial blood it is about 40 mmHg. In this case, CO_2 passes from the tissues to the blood that removes it through the lungs. The fact that oxygen with carbon dioxide crosses the capillary wall with a different pressure gradient depends on the greater diffusibility of CO_2. Transit time is the time it takes for blood to travel through a capillary. At rest, the gases complete the exchange and reach equilibrium in the first quarter/third of the alveolar capillary, i.e. in less than 1/3 of the transit time. Blood velocity can increase three to four times and blood gas pressures can still reach alveolar pressures.

Venules are the last part of the microcirculation and they receive the blood from the capillaries. Since their walls contain smooth muscle cells that can be contracted by the sympathetic action and by humoral factors, venules can show a certain resistance (post-capillary resistance) to the blood flow. Due to the reduced smooth musculature with a low basal tone, venule resistance is much lower than that of arterioles.

12

Nervous Control of the Cardiovascular System

12.1 The Innervation of the Cardiovascular System

The regulation of the cardiovascular system by the autonomic nervous system concerns the reflex control of both the heart and the vessels. While the heart is subjected to the efferent control of both the sympathetic and the parasympathetic/vagal innervation, the vessels receive almost exclusively a sympathetic innervation, being the parasympathetic innervation reserved only to a few vessels, such as the vessels of the salivary glands and the cavernous bodies of the genital organs. The afferent limb of the reflex involves sensor inputs that are described in Paragraph 12.6.

Sympathetic fibers directed to the heart are postganglionic fibers that arise from the *superior, middle,* and *inferior cervical sympathetic ganglia* and in small quantities from the *first thoracic ganglion*. In mammals, the inferior cervical ganglion is usually fused with the first thoracic ganglion to form the *stellate ganglion*. These ganglia receive preganglionic fibers coming from the first five segments of the *intermediolateral column (IMLC)* of the thoracolumbar medulla. The IMLC extends from the 1st thoracic (T1) to the 2nd lumbar (L2) segment and gives rise to the entire sympathetic innervation of the body. Postganglionic fibers arising from the *superior, middle, and stellate ganglia* are grouped in the *superior, middle, and lower cardiac nerves*.

The mediator released by preganglionic fiber is *acetylcholine (ACh)*, which acts on *nicotinic receptors* on the postganglionic cells.

Acetylcholine nicotinic receptors are linked to cation channels and are present in the ganglia of the two divisions of the autonomic nervous system. While these autonomic receptors are blocked by *hexamethonium*, the *nicotinic receptors* found on the skeletal muscle are blocked by curare. Indeed, *Acetylcholine* is the chemical mediator released by both the motor-neuron innervating the muscle fibers of skeletal muscle and by the preganglionic

261

fibers of the sympathetic and parasympathetic system. For *acetylcholine muscarinic receptors* present on the heart and smooth muscle see below (Paragraph 12.4.2)

The right and left *postganglionic sympathetic fibers* from the *superficial and deep cardiac plexuses*. The sympathetic fibers of these plexuses innervate the sino-atrial node (SAN), the atrio ventricular (AV) conduction system, the atrial and ventricular myocardium, as well as the coronary vessels. The right sympathetic fibers innervate mainly the SAN, the atria, and part of the right ventricle, while those on the left innervate the left ventricle and part of the right one.

The main chemical mediator of the sympathetic action on the heart and vessels is *noradrenaline (NA),* also called *norepinephrine(NE).* In this book, both terms are used indifferently. NA acts both on *alpha- and beta-adrenergic receptors*. More details on these receptors will be given in *Paragraph 12.5.*

Parasympathetic fibers directed to the heart are preganglionic fibers belonging to the *vagal nerves*. The parasympathetic system usually has very long preganglionic fibers that make synaptic contact with very short postganglionic fibers in ganglia close to or within the target organ. The preganglionic vagal fibers directed to the heart come from the *medulla oblongata* where they originate in the *ambiguous nuclei* and the *dorsal motor nucleus*. In the thorax, the two vagal nerves give rise to the groups of *superior, middle, and low cardiac vagal branches.*

After forming the cardiac plexus with the sympathetic fibers, the preganglionic vagal fibers make synaptic contact with the short postganglionic fibers that innervate the SAN, the atrial myocardium, the atrio ventricular node (AVN) and the upper parts of the ventricles and AV conduction system. In particular, the SAN and the right atrium are mainly innervated by fibers dependent on the right vagus, while the left atrium, the AVN, and the remaining parts of the conduction system receive mainly fibers of the left vagus. A few postganglionic parasympathetic fibers also innervate the ventricular myocardium and the coronary vessels. These few ventricular fibers have an uncertain functional meaning.

The main chemical mediator of the parasympathetic action on the heart is *ACh* which acts on *muscarinic cholinergic receptors*. Of note, a high density of acetylcholine muscarinic receptors has been described both in the atria and ventricles.

While the heart is normally under the prevailing control of the vagus that already in basal conditions slows down the intrinsic heart rate (HR), the vessels are subjected almost exclusively to the *sympathetic vasoconstrictor tone* which guarantees partial vasoconstriction called *"basal vasomotor tone"*. As mentioned above, the sympathetic pregangliar fibers originate from the *lateral horns or intermediolateral nucleus of the spinal cord* of the thoracolumbar sympathetic column (IMLC) which extends from T1 to the L2 segment. These fibers are usually short and end in the paravertebral ganglia from which the longer postganglionic fibers depart. The postganglionic fibers are directed to the artery and veins, especially to resistance vessels, where they innervate the smooth muscle cells. The splanchnic nerve fibers are an exception to this scheme. These are rather long preganglionic fibers with a cholinergic transmission that terminate in the adrenal medulla where they induce the release of *noradrenaline (norepinephrine) and adrenaline (epinephrine) hormones*. We can say that cells of the adrenal medulla have a function analogous to postganglionic fibers.

As mentioned in other chapters, in addition to the sympathetic vasoconstrictor, there is also a *sympathetic vasodilator*. These are fibers that anatomically pertain to the sympathetic system, whose postganglionic fibers, instead of norepinephrine, release ACh as their chemical mediator. They innervate some smooth muscle cells of skeletal muscle resistance vessels. The cholinergic fibers, such as those contained in the celiac and superior mesenteric plexuses that dilate the pancreas vessels, as well as some fibers innervating the salivary glands, can also be considered as vasodilator sympathetic fibers. Some authors consider the *sympathetic vasodilator* as normal adrenergic fibers, whose mediator (noradrenalin) acts on the β-receptors; rather than cholinergic fibers acting on muscarinic receptors (see also the Chapter 8.5).

12.2 Action of the Sympathetic and Vagal Nerves on the Heart

While the increase in sympathetic activity on the heart exerts several "positive" effects, the increase in vagal activity causes "negative" effects. Indeed, the sympathetic stimulation determines *positive chronotropic* (increase in HR), *dromotropic* (increase in conduction speed) *bathmotropic* (increase in cell excitability), and *positive inotropic* (increase in the force of contraction) *actions*. On the contrary, the vagal stimulation determines atrial *negative chronotropic, dromotropic, bathmotropic, and inotropic effects on atria*. While sympathetic stimulation determines also a *positive lusitropic action* (increases relaxation velocity), the vagal effect on *lusitropy* is not clear.

For the anatomic reasons described above, the "positive" sympathetic actions are exerted both on the atria and the ventricles and the AV conduction system, the "negative" effects of the vagus are exerted mainly on the atria and the AV conduction system.

The heart of mammals is simultaneously subjected to both the sympathetic and the vagal tone. In large mammalians, including the humans, *vagal tone is prevalent*. Indeed, if the vagal fibers directed to the heart are pharmacologically blocked (*e.g.* with *atropine*), HR can double, while if the sympathetic system is blocked (*e.g.* with *beta-blockers*), the HR decreases only by 20–30%. Actually, ***in many physiological conditions, the increases in HR are due to a decrease in vagal activity rather than an increase in sympathetic activity***. As will be seen in Paragraph 12.6, the prevalence of the parasympathetic tone on the heart is attributed to a reflex sustained by the continuous stimulation of the arterial baroreceptors by pulse pressure.

12.2.1 Action of the Sympathetic Nerves on the Heart

As has already been said, the sympathetic nerves act on the heart mainly through the release of *noradrenaline*. On the myocardium, on the SAN and the conduction systems, noradrenaline acts by binding to adrenergic receptors. In the myocardial tissues, the prevalent adrenergic receptors are the *beta-1 adrenoreceptors*.

Figure 12.1A shows the action of norepinephrine on the membrane potentials of the SAN. The increase in chronotropism takes place through an increased velocity of *diastolic depolarization* (*prepotential*). Therefore, the threshold value for the triggering of the action potential is reached in advance compared to a control situation.

The cascade of signals that determines the higher HR includes activation of *adenylyl cyclase* (AC) and a *protein kinase A* (PKA) (Figure 12.2). While PKA opens the sarcolemmal channels for Ca^{2+}, the cAMP directly opens the so-called *funny channels* for cations. The increase in currents for Na^+ and Ca^{2+} accelerates the development of diastolic depolarization.

The cascade of signals that follows the binding of norepinephrine to these receptors mediating the increase in inotropism has already been discussed in Chapter 7.1.3. Briefly, β 1-receptors act on a Gs protein which in turn activates AC which determines the formation of cyclic AMP (cAMP) from the ATP. The cAMP then activates a PKA that phosphorylates calcium channels and phospholamban. Thus, determining an increase in Ca^{2+} channel opening and removing the inhibitory action that phospholamban exerts on SERCA2. In this way, there is a transient increase in the intracellular concentration of

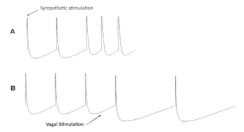

Figure 12.1 Action of sympathetic activity mediated by norepinephrine (A) and vagal activity mediated by acetylcholine (B) on the action potential (AP) of the Sinus Atrial Node (SAN). **A:** A few beats after the application of norepinephrine there is a more rapid ascent of the prepotential to the threshold value to trigger an AP. These changes are responsible for an increase in the SAN discharge rate. The effect occurs after a certain number of beats, due to the time required by the second messengers to determine the effect of norepinephrine. **B:** After the application of acetylcholine there is an immediate hyperpolarization due to the rapid opening of potassium channels activated by muscarinic receptors (muscarinic potassium (K_{ACh}) channels) and then a subsequent slowdown in the development of the prepotential due to cyclic AMP level reduction. Therefore, the reaching of the threshold value for the genesis of AP is delayed, and the SAN discharge rate is reduced. When the prepotential fails to reach the threshold value, the heart is stopped with a possible subsequent *escape from the vagus* (not visible in this figure). For further explanation, see the text. See Chapter 4.1.1.

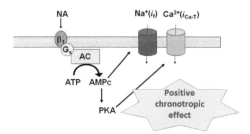

Figure 12.2 Positive chronotropic action of noradrenaline (NA) on the sinoatrial node. β1: type 1 beta-adrenergic receptor; Gs: stimulating G protein; AC: adenylyl cyclase; PKA: protein kinase A; *If*: funny sodium current; iCa-T: transient calcium current.

Ca^{2+} and a more rapid recovery in the sarcoplasmic reticulum (Figure 7.5). While the transient calcium increase is the main responsible for the positive inotropic effect, the rapid recovery is responsible for the positive lusitropic effect.

There may also be a few β2-receptors on cardiomyocytes that act similar to β1-receptors, as well as few α-receptors which also exert a modest positive inotropic action, by increasing the intracellular concentration of Ca^{2+}. The α-receptor signaling considers the activation of a Gq protein-coupled to the receptor, which activates a phospholipase C (PLC) and determines the formation of inositol-triphosphate (IP3) which opens the RYR channels of the sarcoplasmic reticulum.

As a result of the increased inotropism and lusitropism the ventricle contracts and relaxes more quickly and more completely. This leads to an increase in stroke volume. The increase in ejection is because the more energetic contraction also uses a part of the systolic residue, which is thus decreased, the ventricular filling is also favored by faster ventricular relaxation so that even the end-diastolic ventricular volume is preserved despite the increase in HR. Figure 4.10 illustrates the left ventricular pressure curve following sympathetic stimulation. It can be seen how, compared to the basal situation, during contraction the pressure reaches higher levels in a shorter time, thus increasing the +dP/dT ratio. Since relaxation is also faster, the –dP/dT ratio is also increased.

The increases in dromotropism and bathmotropism are also mediated by the greater opening of the Na^+ channels due to the action of cyclic AMP and PKA.

12.2.2 Action of the Vagus on the Heart

The most obvious action of the vagus on the heart is the slowing of the HR, easily detectable by placing the fingertips on an artery, during carotid sinus massage triggering a baroreflex (see Paragraph 12.6). By electrocardiographic examination, we can appreciate the slowing of atrio ventricular conduction by the lengthening of the PR interval.

The right vagus has a prevailing action on the HR dictated by the SAN and the left vagus acts mainly on the speed propagation of the impulse along with conduction system. Prolonged and intense stimulation of the right vagus can lead to the arrest of the whole heart if the activity of the SAN ceases. Only the ventricles stop if the atrio ventricular conduction is interrupted by intense stimulation of the left vagus. In both cases, while the stimulation continues, the ventricles, however, can resume pulsatile activity with a very low *idioventricular rhythm* (20–30 b.p.m.). This is the so-called *escape from the vagus*, due to the onset of the automatic activity of vicarious centers located in the ventricular conduction system (*i.e., Purkinje fibers*).

The signaling pathway evoked by the activity of the vagus considers ACh binding to M2 *muscarinic receptors*, thus opening K_{ACh} channels for K^+ and leading to an immediate hyperpolarization of the membrane (Figure 12.1B). Furthermore, acetylcholine exerts, through a Gi protein, an inhibitory action on AC that leads to a reduction of cAMP concentration and to effect opposite to those of norepinephrine.

12.3 Action of the Sympathetic Fibers on the Vessels

It is necessary to reiterate the concept that *total peripheral resistance are regulated solely by the thoracolumbar sympathetic system which has a vasoconstrictor action. Indeed, vasodilation affecting total peripheral resistance is due to the withdrawal of sympathetic tone on the majority of vascular districts.* Of course, the fibers that regulate vascular resistance are mainly directed to the precapillary arterioles, although sympathetic fibers also supply arteries and veins of a greater caliber.

As said, the sympathetic also innervates venules and veins. The greater or lesser constriction of the venules, situated immediately after the capillaries, can regulate the *postcapillary resistance* and therefore the pressure in the capillaries. Capillary pressure increases when venular tone increase and *vice versa.* The constriction of the veins determines the decrease of the *capacity* of these vessels and therefore favors the return of blood to the heart.

Due to the innervation of the sympathetic to vessels of resistance and capacity, *a generalized activation of the sympathetic nervous system determines both an increase in total peripheral resistance and venous return.* Since a sympathetic discharge also determines an increase in cardiac inotropism and lusitropism, we understand how in the states of excitement (*fight or flight response*) there is an increase in venous return, in stroke volume, and arterial pressure. To the increase in pressure contributes also an HR increase, which is mediated mainly by a reduction in vagal tone (see also Chapter 16.2), but also by an increase in sympathetic tone.

The action of the sympathetic on the vessels is mainly due to the release of noradrenaline from the post-ganglionic endings. As we will see below, norepinephrine causes vasoconstriction by acting on the numerous $\alpha 1$-receptors. However, catecholamines can also cause vasodilation acting on $\beta 2$ dilator receptors present on vascular smooth muscle (VSM) of a few vascular districts. Of note, norepinephrine is not the only sympathetic mediator released by sympathetic fibers (see paragraph 12.4).

The stimulation of the sympathetic nerve fibers directed to the heart causes a very modest and transient coronary vasoconstriction followed by marked and persistent vasodilation. In fact, on the VSM of the coronary vessels are both $\alpha 1$ constrictor and $\beta 2$ dilator receptors. While the former is in an intrasynaptic position and is then activated in the initial phase of stimulation, the latter have an extrasynaptic position and are activated with a few seconds of delay when they are reached by the norepinephrine "escaping" from the synaptic space. However, more than this direct mechanism, *coronary vasodilation due to sympathetic stimulation is due to the increased myocardial metabolism induced by the simultaneous increase in HR and contractility mediated by β1-receptors on cardiomyocytes.*

Beta-2-adrenergic dilator receptors are also present on the VSM of vessel resistance of the skeletal muscle, where they cause vasodilation in synergy with the sympathetic cholinergic vasodilator. As already mentioned (see also Chapter 8.5), in some animal species, the precapillary arterioles of skeletal muscles are also innervated by a *sympathetic vasodilator* which has as chemical mediator ACh, although they pertain to sympathetic system from an anatomical viewpoint.

Parasympathetic fibers innervate only a few vascular districts, such as the salivary glands and the cavernous bodies of the genital organs. Also, the vagus does not affect the general circulation as its action on the cardiovascular system can be considered limited to the heart and, to a very small extent, to the coronary circulation. Indeed, the vagal nerves have direct and indirect actions on the coronary vessels. Direct action is responsible for modest vasodilation (approximately 10% resistance reduction), while the indirect action determines vasoconstriction due to a reduction in myocardial metabolism caused by the decreased HR (see also Chapter 14.1). In experimental models, we can observe the direct vasodilator action of the vagus, if the heart is driven at a constant HR, or the ventricles are fibrillating and coronary artery perfused by extracorporeal circulation. In these conditions, the vagus cannot have any influence on HR and metabolism.

Some salivary gland and pancreatic vessels also receive cholinergic dilator fibers pertaining anatomically to the sympathetic system (these are fibers similar to those of the sympathetic vasodilator system).

Of note, the few vagal and non-vagal parasympathetic fibers innervating vessels have no effects on total peripheral resistance variations during cardiovascular reflexes (see also below the discussion on baroreflex).

12.4 The Sympathetic and Parasympathetic Receptors of the Cardiovascular Apparatus

We have already seen how the action of the sympathetic and parasympathetic on the cardiovascular system is due to the release of NA and ACh, which act by binding to specific receptors.

12.4.1 Adrenergic Receptors

Adrenergic receptors are divided into α- and β-receptors with various subtypes (α1A, α1B, α1D; α2A, α2B, α2C; β1, β2, and β3). There is no α1C receptor. All of the adrenergic receptors belong to the family of G-protein-coupled receptors (GPCRs) that link to heterotrimeric G-proteins. All β adrenergic receptors couple to Gs (G stimulator) and activate AC and increase cyclic AMP. However, $\beta2$ and $\beta3$ can also couple with Gi (G inhibitor) which inhibits AC and thus decreases cyclic AMP. Activation of α adrenergic receptors also couples to a variety of isoforms, such as Gi ($\alpha2$-receptors) and Gq ($\alpha1$-receptors) proteins.

In addition to the norepinephrine released by the sympathetic endings, also the norepinephrine and the epinephrine of adrenal origin bind to the adrenergic receptors. In particular, while ***noradrenaline acts preferably on α-receptors, adrenaline preferably acts on β-receptors.***

The identification of types and subtypes of adrenergic receptors can be recognized by specific agonists and blockers. In particular, the α-receptors are activated by the agonist *phenylephrine* and blocked by *phentolamine*, while the β-receptors are activated by *isoproterenol* and blocked by *propranolol*.

12.4.1.1 Alpha-receptors

Generally have a vasoconstrictor action. The $\alpha1$-*receptors* are located on the membrane of smooth muscle cells of the majority of vessels. They are located at the post-ganglionic sympathetic endings in an intrasynaptic position and are therefore immediately exposed to the action of norepinephrine. The $\alpha2$-*receptors* are instead in a postsynaptic and presynaptic position. The *postsynaptic $\alpha2$-receptors* are also located on VSM, outside of synapsis membrane on which sympathetic varicosity appears. Their action is also a vasoconstrictor. The *presynaptic $\alpha2$-receptors* are instead placed on the post-ganglionic sympathetic fibers shortly before their endings. When activated, they reduce the release of norepinephrine from the ending, so it can be assumed that they are part of a negative feed-back mechanism capable of limiting an excessive release of the chemical mediator.

For instance, once stimulated by the ligand, the VSM α-receptors, through a membrane protein Gq, activate a phospholipase C (PLC). This, acting on the phospho-inositol-diphosphate (PIP2), determines the formation of diacyl-glycerol (DAG) and inositol-triphosphate (IP3) (Figure 12.3). While DAG opens non-selective sarcolemmal cation channels that allow Ca^2 and Na^+ to enter the cell, IP3 mediates the release of Ca^2 from the sarcoplasmic reticulum. Calcium causes also the opening of channels that allow the Cl^- exit from the cell.

The entry of Ca^2 and Na^+ and the exit of Cl^- depolarize the membrane with possible development of the action potential by opening the long-lasting

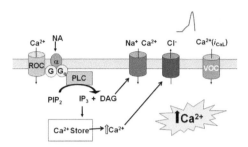

Figure 12.3 Increase in intracellular calcium concentration in response to adrenergic α-receptor activation in smooth muscle cells: pharmacomechanical and electromechanical coupling. G and Gq: membrane proteins; PLC: phospholipase C; PIP2: phosphoinositol diphosphate; IP3: inositol triphosphate; DAG: diacylglycerol; ROC: receptor-operated channels; VOC: voltage-operated channels. See the explanation in the text.

calcium channels, or L-type channels. For the mode of their opening, these channels are also called *voltage-dependent or voltage-operated channels* (VOC).

The increase in intracellular Ca^{2+} allows the formation of the complex Calcium-Calmodulin which in turn activates the myosin light chain kinase (MLCK) which determines the phosphorylation of myosin light chain (MLC) and contraction of smooth muscle cells (Figure 12.4). Due to the intervention of a variation of the membrane potential, the path to contraction just described is called *electromechanical coupling. Pharmacomechanical coupling* occurs when contraction of smooth muscle cells is due to the action of agents that raise intracellular Ca^{2+} mainly *via* the IP3 mechanism.

The α2-like receptors would be found on the endothelial cells from which they would determine the release of NO which induces smooth muscle relaxation. Indeed, NO *via* cGMP activates myosin light chain phosphatase (MLCP) which removes the phosphorylation from the myosin heads.

Although the *presynaptic α2-receptors* and the *endothelial α-2-like receptors* do not have a vasoconstrictor action, the overall effect of stimulation of alpha receptors by the noradrenaline released by the sympathetic system results in an increase in vasomotor tone due to vasoconstriction induced by postsynaptic α1-receptors. This vasoconstriction occurs for both pharmacomechanical and electromechanical coupling. As mentioned, *pharmacomechanical coupling* is a smooth muscle contraction due to a chemical agent that leads to the sarcoplasmic Ca^{2+} release *via* IP3 mechanism, without the necessity of change in membrane potential.

The stimulation of sympathetic division may cause vasoconstriction even in the presence of an α-receptor blockade. This is because, in addition to

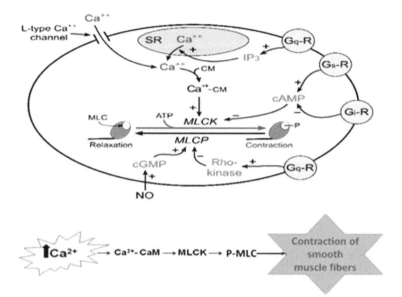

Figure 12.4 Contraction and relaxation of smooth muscle cells due to phosphorylation and de-phosphorylation of myosin light chain (MLC), respectively, at the end of complex pathways. In the lower panel are reported the main steps leading to smooth muscle contraction following Ca^{2+} concentration increase. Gq- Gi- and Gs-R: different isoforms of G protein-coupled receptors; IP3: inositol triphosphate; CM: calmodulin ; cAMP: cyclic adenosine monophosphate; cGMP: cyclic guanosine monophosphate; MLCK: myosin light chain kinase; MLCP: myosin light chain phosphatase; NO: nitric oxide.

noradrenaline, sympathetic postganglionic fibers can also release two co-transmitters, *ATP and neuropeptide Y*. ATP is produced in the post-ganglion sympathetic ending and released with norepinephrine, while neuropeptide Y is synthesized in the cell body of the postganglionic neuron and reaches to the nerve ending moving along the axon, where it accumulates in vesicles. When nerve impulses arrive, both co-transmitters are released with norepinephrine. Binding to the *P2x purinergic receptors*, ATP causes the opening of aspecific cationic channels responsible for a brief depolarization with contraction of smooth muscle cells. The neuropeptide Y, on the other hand, acting on its receptors, determines a depolarization of longer duration and sensitizes the smooth muscle membrane to noradrenaline. It serves as a strong vasoconstrictor and also induces the growth of fat tissue.

As mentioned above, some α1-receptors are present on the cardiac cells, where they can cause very modest increases in HR and contractility.

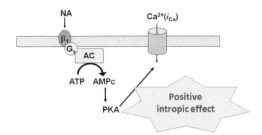

Figure 12.5   Main steps leading to *positive inotropic action of noradrenaline* (NA) on the myocardium. β1: type 1 beta-adrenergic receptor; Gs: stimulating G protein; AC: adenylyl cyclase; PKA: protein kinase A; iCa: calcium current. AMPc: cAMP or cyclic adenosine monophosphate. See the explanation in the text.

12.4.1.2 Beta-receptors

Beta-receptors are present on the cardiac cells and all the VSMs. Besides working cardiomyocytes, β1-receptors are present on the slow and rapid cardiac conduction system. A small number of β2-receptors may also be present on the cardiomyocytes. Nevertheless, *the activation of β1-receptors is responsible for the typical "positive" actions of the sympathetic system on the heart.*

When the beta-receptor is bound to norepinephrine, it activates the membrane Gs protein, which in turn activates the AC enzyme that transforms ATP into cAMP. While cAMP can directly increase the probability of opening funny channels for Na^+, the PKA can increase the opening probability of Ca^{2+} channels, thus mediating a positive chronotropic/dromotropic action in SAN and NAV (Figure 12.2). In the myocardium, PKA determines the phosphorylation and the opening of both the L-type Ca^{2+} channels (Figure 12.5) and of the subtypes of Ik channels for the K^+. While the opening of the channels for the Ca^{2+}, determines an increase in the intracellular concentration of the ion and therefore of the contractility, the opening of the channels for the K^+ allows more rapid repolarization with shortening of the action potential plateau, thus allowing a greater number of action potentials in the unit of time.

As said, PKA also phosphorylates and inhibits *phospholamban*. Since phospholamban normally inhibits SERCA2 by attenuating its functioning, its inhibition "frees SERCA2 activity", accelerating the recovery of Ca^{2+} in the sarcoplasmic reticulum and thus making its relaxation faster (*lusitropic action*) (Figure 12.6). Of note, the positive lusitropic effect is linked to the positive inotropic effect, as the greater re uptake of Ca^{2+} will favor a greater

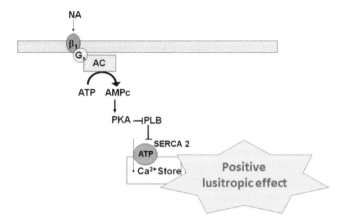

Figure 12.6 Main steps leading to *positive lusitropic action of noradrenaline* (NA) on the myocardium. $\beta1$: type 1 beta-adrenergic receptor; Gs: stimulating G protein; AC: adenylyl cyclase; PKA: protein kinase A; PLB: phospholamban; SERCA: sarcoendoplasmic reticulum Ca^{2+}-ATPase. GMPc: cAMP or cyclic adenosine monophosphate. See the explanation in the text.

release of Ca^{2+} in the subsequent beat by the calcium-induced calcium release mechanism.

The $\beta2$-receptors are located on the VSM of some vessels, including skeletal muscle and coronary circulation resistance vessels. They are located in an extra-synaptic position and mediate a vasodilator response. Although they are preferably stimulated by adrenaline, they are also activated by norepinephrine released by the sympathetic endings due to their extra-synaptic position. Although the affinity of norepinephrine for α-receptors is greater, when the release of the ligand is sufficient it activates also β-receptors. The chronology of receptor activation can be highlighted with the stimulation of the sympathetic directed to the heart: after a transient phase of coronary constriction due to $\alpha1$-receptor activation, vasodilation due to $\beta2$-receptors occurs. As said, then will prevail the *metabolic vasodilation* mediated by $\beta1$-receptor activation.

On VSM, the $\beta2$-receptors are coupled to Gs membrane proteins, which, through the activation of AC lead to the formation of cAMP and therefore to the activation of PKA (Figure 12.7). Unlike, however, what we have seen for cardiac receptors, PKA does not determine an increase in the intracellular concentration of Ca^{2+}, but rather its decrease. In fact, in smooth muscle cells, PKA activates Ca^{2+}pumps, both the sarcolemmal pump for Ca^{2+} and

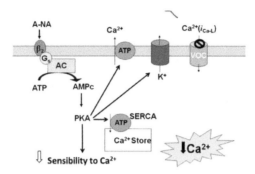

Figure 12.7   Reduction of intracellular Ca^{2+} concentration and reduced sensibility to Ca^{2+} by myofilaments in response to activation of $\beta2$ adrenergic receptors by adrenaline (A) and/or noradrenaline (NA) in smooth muscle cells. Gs: stimulating G protein; AC: adenylyl cyclase; PKA: protein kinase A; VOC: voltage-operated channels; SERCA: sarcoendoplasmic reticulum Ca^{2+}-ATPase. AMPc: cAMP or cyclic adenosine monophosphate See the explanation in the text.

SERCA pump with the consequent passage of the ion to the outside of the cell and in the sarcoplasmic reticulum, respectively.

Beta3-receptors have been described on myocardial fibers. They increase in the number of heart failure and have negative inotropic effects. The meaning of this action in heart failure might be protective. In these conditions, the heart is subjected to an increased adrenergic action with a positive inotropic effect. If in the short term the adrenergic action of stimulating contractility can improve cardiac output and therefore the patient's hemodynamic situation, in the long term it may have a maladaptive effect as it ends up worsening the conditions of the myocardium. A negative inotropic action can, therefore, attenuate the harmful effects of excessive activity of the catecholamines on the $\beta1$-receptors.

To sum up, the increased activity of sympathetic system on the vessels determines vasoconstriction via $\alpha1$-receptor, whose final effects are a reduction of local blood flow, a reduction of capillary pressure, a reduction of blood volume contained in the veins and, if the increase in sympathetic outflow is generalized, an increase in total peripheral resistance. Beta1 effects on the heart determine an increase in HR and contractility whose final effect is an increase in cardiac output. Vasodilatation via $\beta2$-receptor is present only in some districts, such as skeletal muscle vasculature. We can say that the regulation of blood pressure is the most important function of the sympathetic vasomotor system. While $\alpha1$- and $\beta1$-receptor stimulation leads

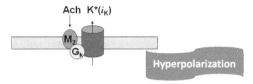

Figure 12.8 Action of acetylcholine (ACh) on the sinoatrial node. M2: type 2 muscarinic receptor; Gk: G protein isoform; iK: potassium current.

to an increase in blood pressure, the stimulation of β2-receptor improves the perfusion of active skeletal muscle limiting an excessive increase in pressure.

12.4.2 Acetylcholine Receptors

In the effector organs of the cardiovascular system, the cholinergic receptors, also called parasympathetic receptors, are muscarinic (M) receptors. The M receptors are divided into five subtypes from M1 to M5 and are all blocked by *atropine*.

While the M1, M4, and M5 receptors are present in structures other than the cardiovascular system (*e.g.* salivary glands, exocrine pancreas, and gastrointestinal mucosa), the M1, M2 and M3 are typical the cardiovascular system. The M2 receptors on the heart produce negative chronotropic, inotropic, and bathmotropic effects. The M1, M2, and M3 receptors on the vessel can mediate vasoconstriction or vasodilation.

The cholinergic M2 receptors on the heart are found on the SAN and the AVN. They are also present on the myocardium, especially on atrial tissue, and to a lesser extent on ventricular tissue. While the atrial myocardial receptors are in contact with the vagal endings, and therefore more exposed to the action of ACh, the ventricular ones are stimulated only pharmacologically or in the event of arrival to the ventricles of ACh released in excess from the vagus. It follows that usually ventricular inotropic and bathmotropic function is little or nothing influenced by the vagal nerves.

On the ASN and AVN cholinergic receptors are responsible for chronotropic and dromotropic negative actions, respectively. Activation of M2 receptors leads to hyperpolarization of the cell membrane as a result of the opening of K_{ACh} channels through a membrane Gk protein (Figure 12.8).

The effects of ACh on the vasomotor tone are vessel and specie specific. In some vascular districts, ACh may cause vasodilation *via* a direct hyperpolarization of the VSM. In other districts, such as the aorta or coronary vessels, it can induce vasoconstriction *via* a direct depolarization of the

VSM. Whether ACh of neural origin can induce the release of endothelial NO and/or endothelial-derived hyperpolarizing factors (EDHF) is a matter of controversy. Nevertheless, ACh applied directly on the endothelium (from the luminal side) acting om M3 receptors may induce the release of either NO and/or EDHF which mediate vasodilation. The damage or removal of endothelial cells was seen to replace vasodilation with ACh-dependent vaso-constriction mediated by M2 receptors on VSM. In cerebral arterioles, ACh causes endothelium-dependent relaxation via the M3 receptor and directly constricts VSM *via* the M1 receptor.

The parasympathetic fibers of the vagus do not dilate other vessels besides the coronary ones, on which the action is also modest and counteracted by the metabolic effect on the myocardium. The nerves of the sacral spinal part of the parasympathetic system are directed to some vessels of the lower intestinal tract and the genital erectile tissue.

In some districts, parasympathetic vasodilation persists partly after block-age of muscarinic receptors with atropine. It is likely that, in addition to ACh, other substances with vasodilating action can be released. Together with cholinergic fibers, the parasympathetic nerves may contain the so-called *non-adrenergic and non-cholinergic (NANC)* fibers. Most of these fibers secrete the *vasoactive intestinal peptide* (VIP), others release *substance P* and others *NO*. Therefore, NO can be secreted directly by specific NANC fibers or by the action of ACh on endothelial cells. Even the vasodilator action of the VIP seems to be mediated by its action on the endothelium, which, stimulated by VIP, releases NO.

To sum up, due to the scant and mixed effect of ACh on the vasomotor tone of few vascular districts, it is reasonable to say that the vasomotor tone is mainly controlled by the sympathetic vasoconstrictor system. The parasympathetic tone is prevalent in the control of HR.

12.5 The Nervous Control Centers of the Cardiovascular Apparatus

The main nerve centers controlling the cardiovascular system are located in the *medulla oblongata* (also called simply medulla or bulb) and the hypotha-lamus. On them act other centers located in various areas of the cerebral cortex, in the pons and the cerebellum.

The centers in the *medulla* control both the sympathetic and the vagal activity. The old classification considered a *pressure area* located more

Figure 12.9 Bulbar pressor areas (squared) and depressor areas (horizontal dashed) in the brainstem.

laterally and upper part, and a *depressor area* located more medially and in the lower part of the medulla (Figure 12.9).

Subsequently, it has been seen that the sympatho-excitatory neurons are located in a rather small area, called *rostral ventrolateral medulla* (RVLM). Its neurons project to the *intermedio-lateral nucleus* of the thoracolumbar medulla (ILM), where the cellular bodies of sympathetic preganglionic neurons are contained. RVLM neurons act tonically on ILM neurons in an excitatory sense, through the release of noradrenaline. The RVLM is spatially organized; therefore the sympathetic efferent activity can increase at a certain district and simultaneously decrease at another district. For example, during the *fight or flight reaction,* it is possible to lower the sympathetic vasoconstrictor tone in the skeletal muscle and to increase it in the skin territory. Indeed, the RVLM receives afferent activator stimuli from the *posterior hypothalamus* and inhibitor stimuli from the *caudal ventrolateral medulla* (CVLM). The latter corresponds more or less to what was once classified as a depressor area (Figure 12.10).

The nuclei of the vagus are also found in the bulb, namely the *ambiguous nucleus* and the *dorsal motor nucleus.* The fibers directed to the heart come mainly from the ambiguous nucleus.

An important nucleus present in the bulb is the *nucleus of the solitary tract* (NTS) to which arrive the afferent fibers of the *glossopharyngeal nerve* (IX pair of cranial nerves) *and vague* (X pair of cranial nerves) whose terminals constitute the arterial baroreceptors (Paragraph 12.6). Since these afferent fibers eventually cause a reduction in HR and pressure, the nerves to which they belong (IX and X pair) are called *buffer nerves.*

The glutamate releasing fibers of NTS activates both the vagal nuclei and the CVLM. After being activated, the CVLM sends inhibitory impulses to the RVLM through fibers that release the ã-aminobutyric acid (GABA). Thanks to these connections, in the end, the NTS stimulation reduces sympathetic

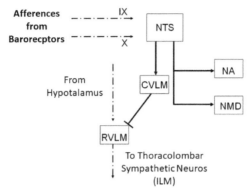

Figure 12.10   Baroreceptor reflex pathways. NTS: nucleus of the solitary trait; NA: ambiguous nucleus; NMD: dorsal motor nucleus of the vagus; CVLM: caudal ventrolateral medulla; RVLM: rostral ventrolateral medulla. See the explanation in the text.

activity and increases parasympathetic activity. Therefore, stimulation of NTS by baroreceptor activation leads to a reduction of peripheral resistance, cardiac contractility, and HR. This is how the baroreflex correction of a pressure increase occurs. NTS also has connections to the cerebellum involved in the regulation of cardiovascular responses during exercise.

The NTS also projects to a depressor area and the *supraoptic nucleus of the hypothalamus*. From this connection, a regulation of ADH (vasopressin) production occurs. The damage of the NTS determines the appearance of hypertension.

Neurons are present in the *perifornical region of the hypothalamus* and the *periaqueductal region of the pons,* which directly and/or through the inhibition of NTS cause activation of RVLM and a reduction in vagal activity, with subsequent increase in HR and arterial pressure. This is how the *fight or flight reaction* takes place (see Chapter 16.2).

The hypothalamus intervention in the *fight or flight reaction* is also regulated by the *cerebral cortex*, in particular by the motor and premotor cortex, by the orbital and temporal cortex and by the *limbic system* as a whole. Also, the stimulation of the *thalamic nuclei* of the midline can cause tachycardia.

The stimulation of some nuclei of the *cerebellum* can determine various responses affecting blood pressure and HR. Cerebellum connections with the hypothalamus and NTS may be responsible for these effects. It is generally a physical exercise response accompanied by the stimulation of the *cerebellar vermis*, which causes vasodilation in the skeletal muscles and vasoconstriction of the renal circulation. It has also been observed that in mammalians the lesion of the *fastigial nucleus* reduces the increase in HR and in blood pressure that occurs during exercise. Cerebellum connections with the NTS may be responsible for the so-called *baroreflex reset* (no HR reduction regardless of the increase in aortic pressure) observed during exercise.

12.6 The Nervous Reflex Control of the Cardiovascular Apparatus

When we talk about nerve reflex control of the cardiovascular system, we refer to reflexes starting from arterial and cardiac pressure receptors or baroreceptors, from peripheral chemoreceptors, as well as to less specific reflexes starting from various parts of the body and to local reflexes, such as axonal reflex.

12.6.1 The Baroreceptors

We have mentioned several times the baroreceptors and the reflexes evoked by them. In short, the baroreceptors are stimulated by the increase in arterial pressure and give as a reflex response a decrease in the HR due to vagal activation and a reduction in cardiac contractility and vascular peripheral resistance by sympathetic inhibition. Baroreflex has been discussed in the short term control of blood pressure (Chapter 8.10.1). Here, we see some more details of this important reflex.

As has already been seen, the afferent pathways from the baroreceptors, or buffer nerves, act on the NTS, which in turn stimulates the ambiguous nucleus of the vagus and the CVLM. While the vagus slows down the HR, CVLM inhibits the tonic discharge of RVLM on the sympathetic neurons and therefore attenuates the vasomotor tone and cardiac contractility. Figure 12.11 illustrates the contemporary effect of stimulating baroreceptors on vagal and sympathetic efferent tone following progressive pressure increases.

The baroreceptors of the systemic circulation are located in the *brachiocephalic artery*, in the *arch of the aorta*, near the origin of the *left subclavian artery* and the *carotid sinuses*. The latter consists of the arterial expansion which extends from the bifurcation of the common carotid artery to the initial section of the internal carotid artery (Figure 12.12). In particular, the baroreceptors, namely the nerve endings, are located between the middle layer and the external part of the arterial wall of the arteries (between tunica media and adventitia).

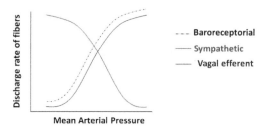

Figure 12.11   Action of the mean pressure on discharge rate of the afferent baroreceptor fibers, and sympathetic and vagal efferent fibers.

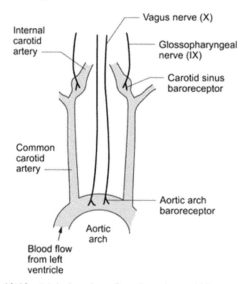

Figure 12.12   Main location of aortic and carotid baroreceptors.

Baroreceptors are also present in the arterioles afferent to the renal glomerulus. Sensitive to local pressure increases, they also participate in the control of blood volume as will be seen in Paragraph 12.7.

High-pressure aortic baroreceptors send their afferent fibers to the NTS through the vagal nerves. In particular, on the left side, the fibers reach the vagus through the *Cyon depressor nerve*. *Carotid baroreceptors* send the afferent fibers to the NTS through the glossopharyngeal nerves, to reach which they pass short nerves known as *carotid sinuses or Pagano-Hering nerves*.

Besides the above described high-pressure baroreceptors also *low-pressure baroreceptors* are present in our body. Receptors of the pulmonary

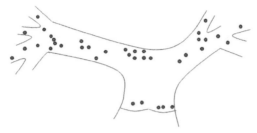

Figure 12.13 Low-pressure baroreceptors in the pulmonary artery and its branches.

artery and its branches are stimulated by increases in pressure within the lower levels exiting in this circulation. The responses to stimulation are however the same for both high-pressure and low-pressure baroreceptors. In the pulmonary artery, they are located near the insertion sections of the pulmonary valve leaflets and in the distal section of the bifurcation of the vessel where the head pressure is felt. (Figure 12.13). The other receptors are found along the course of the right and left branches and in the subsequent subdivisions. A particular type of low-pressure baroreceptors has been described in the atria (see below).

In Chapter 8.10 we have seen the role of baroreceptors in controlling blood pressure, we must consider some characteristics such as *range, maximum gain, and adaptation.* While the range is from about 40 to about 200 mmHg, the maximum gain corresponds to an average pressure of 100–110 mmHg, and the adaptation is completed in a few days. Of note, the baroreceptor system is a short-term mechanism for controlling blood pressure as it achieves the maximum gain in a few seconds.

Since the baroreceptors are subject to adaptation, one wonders if they are better stimulated by the pulse pressure (PP) rather than the mean aortic pressure (MAP). Integration between the level of MAP and PP takes place. Baroreceptors stimulation by a pulsatile pressure elicits a greater baroreceptor discharge compared to a stationary pressure at the same MAP. Figure 12.14 illustrates the experiment conducted by Erich Neal already in 1954. He had highlighted the integration between MAP and PP in the stimulation of baroreceptors. Therefore, both MAP and PP levels are important.

In Figure 12.14, we see a representation of the action potentials recorded from a fiber of the aortic nerve. At normal blood pressure, the baroreceptor discharge is observed during the systolic oscillation: the discharge rate is higher during the pressure rise phase and lowers during the descent phase. By blood withdrawal, MAP progressively decreases, while PP decreases

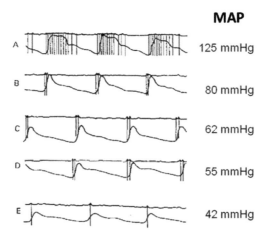

Figure 12.14   Recording of arterial pressure and baroreceptor discharges in an anesthetized animal in the presence of a progressive reduction in mean pressure (from A to E) following blood withdrawal. We can see that the lowering of the mean arterial pressure (MAP) causes a decrease of discharges along the afferent fibers by the baroreceptors, even if the pulse pressure is scarcely reduced. The baroreceptor discharge is indicated by vertical lines (from Neil E., 1954).

less. Nevertheless, the rate of baroreceptor discharge is reduced. The correspondence of baroreceptor discharge with the oscillations indicates the importance of the PP, while its reduction with MAP lowering reveals the importance of the latter. Of note, a certain activation of baroreflex leading to a certain degree of inhibition of sympathetic discharge and activation of parasympathetic discharge already occurs for normal MAP and PP values. Therefore, at rest, a reflex parasympathetic tone prevails over sympathetic activity.

12.6.2 The Ventricular Receptors

The Ventricular Receptors are located in the myocardial wall of the two ventricles, are connected with type C unmyelinated fibers, and are stimulated by increased ventricular pressure. Their discharge rate increases with the rate of pressure rise and therefore with the increase in contractility. In the case of the left ventricle, their most evident stimulation was experimentally obtained with the extreme distension of the ventricle by preventing the efflux of blood in the aorta. The effect of this stimulation is remarkable bradycardia and marked vasodilation in the skeletal muscle and a modest one in the skin.

This is vasodilation mainly due to the reduction of the sympathetic tone. The afferent information of these baroreceptors also travels along vagal nerves and reaches the NTS. In addition to the receptors described above, cardiac chambers receptors have also been described. These are receptors located on the inner surface of atria and ventricles that are connected to myelinated fibers and can give the hypotensive responses common to the other baroreceptors.

The *"Bezold and Jarisch reflex"* is a reflex described by Albert von Bezold and Adolf Jarisch Junior.It is due to a sort of antagonism between arterial and ventricular baroreceptors. In the upright position, which in certain subjects sharply limits venous return to the heart and decreases cardiac output and arterial pressure, we would expect an increase in HR due to reduced arterial baroreceptor activity. In the studied subjects, Bezold and Jarisch have instead observed paradoxical bradycardia with hypotension. They hypothesized that the increase in contractility, due to the greater sympathetic discharge from reduced stimulation of arterial baroreceptors, stimulates the ventricular receptors by evoking a baroreflex resulting in bradycardia and pressure drop. *The Bezold-Jarisch reflex seems due to an improper excitement of the baroreceptors located at the level of the left ventricle when ventricles are particularly empty.* Therefore, the induced reduction in HR aggravates the pre-existing arterial hypotension. The phenomenon is commonly referred to as the *vaso-vagal reflex*. Bradycardia with a fall in pressure occurs even if these receptors are stimulated with *veratrin* and *veratridine*.

12.6.3 Atrial Receptors are of Different Types

For simplicity, we can consider them as low- or high-threshold receptors. Some atrial receptors are comparable to the ventricular ones, but have a very high threshold and are stimulated only by extreme distension of the atria. They also discharge along type C fibers. They form a network of fine fibers in the atrial wall. The responses to stimulation are bradycardia and peripheral resistance reduction. In the atria, however, *Paintal receptors* are also present. These receptors are also called volume-recpetors and are characterized by a much lower threshold than the previous ones and can trigger tachycardia instead of bradycardia. This is *Bainbridge's reflex* that will be discussed in Paragraph 12.7.

12.6.4 Chemoreceptors

Reflex nervous control of the cardiovascular system also includes the contribution of aortic and carotid chemoreceptors (see Chapter 8.10.1.3). Unlike the baroreceptors that are located in the walls of the arteries, the chemoreceptors are located in formations, known as *carotid body or glomus*, outside the arteries.

The *carotid body* is nodules formed by loops of capillaries adjacent to chemosensitive cells which are in turn in contact with fibers afferent to the vagus or glossopharyngeal nerves depending on whether they are aortic or carotid glomus, respectively. While the *aortic glomi* is located below the concave part of the arch of the aorta, the carotid ones are located between the external and internal carotid artery immediately after the bifurcation of the common carotid artery. They receive capillaries deriving directly from the artery where they are located. The decrease in pO_2, the increase in pCO_2 and the reduction of pH in arterial blood are the stimuli sensed by chemoreceptors. Experimentally, chemoreceptors can also be stimulated by the alkaloid *lobeline*.

In contrast to the baroreceptors, ***chemoreceptor stimulation determines an increase in HR and peripheral resistance.***

The decrease in pO_2 is the most active chemoreceptor stimulus. Indeed, the decrease in pO_2 occurs rapidly when breathing low-oxygen gas mixtures or when we are exposed, as happens in high mountains, to low barometric pressures. The increase in pCO_2 instead occurs very slowly following hypoventilation or alteration of the uniformity of the ventilation-perfusion ratio. Therefore, chemoreceptors are stimulated by the rapid decrease of the pO_2, rather than by the slow increase in pCO_2.

Of note, chemoreceptor intervention takes place in emergency conditions when protection against serious falls in blood pressure is required (see also Chapter 8.10). Indeed, chemoreceptors are stimulated when blood pressure is lower than 70 mmHg and the carotid body is under-perfused and, thus, under-oxygenated.

12.6.5 The Axonic Reflexes

Axonal reflexes are reflexes that occur without the involvement of the central nervous system. They are partly responsible for that spreading flare that occurs in a point of the skin where a noxious stimulus has been applied. These reflexes occur in the peripheral bifurcation of slow-conduction type C afferent pain fibers (Figure 12.15). Downstream of the bifurcation, a branch acts as a nociceptive receptor, while the other innervates a precapillary resistance arteriole. An eventual painful stimulation sends afferent impulses which, upon reaching the bifurcation, activate both the part of the fiber directed to the

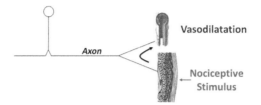

Figure 12.15 Scheme of axonic reflex. See explanation in the text.

cellular body located in the dorsal ganglia of the spinal cord and the branch that innervates the arteriole, which is likely dilated by the release of ATP, CGRP or substance P.

Failure to involve the central nervous system can be evidenced by sectioning the fiber upstream of the bifurcation in order to prevent the arrival of impulses to the ganglion cell. Under these conditions, the axonic reflex remains until the two branches undergo degeneration due to their disconnection from the cell body.

The sensory C-fibers axon reflex can be part of the so-called *Lewis triple response.* This is a skin response that occurs following stroking of the skin, which produces an initial red line, followed by a flare around that line, and finally, a bulge called a wheal.

12.7 The Control of the Blood Volume and the Bainbridge Reflex

Stimulation of baroreceptors found in the aorta, carotid, and afferent arteriole to the glomerulus determines a reflex reduction of sympathetic discharge. The baroreceptors also moderate the release of antidiuretic hormone (ADH) from the *supraoptic nucleus* of the hypothalamus, to which fibers arrive from the NTS.

An increase in osmolarity is an adequate stimulus to increase ADH release. In the presence of blood loss or when the volume of circulating fluid is reduced and the arterial pressure decreases, the reduced stimulation of baroreceptors determines an increase in ADH release. Therefore, ADH is released when arterial pressure decreases. ADH is released and renal water reabsorbed, even if plasma osmolarity is decreased when the arterial pressure decreases by about 30%.

When baroreceptors are less stimulated, the increase in sympathetic discharge also determines constriction of the glomerular afferent and efferent arterioles by α-adrenergic stimulation, with consequent reduction of glomerular filtration and of the filtered Na^+ load. The reduction of the tubular Na^+ load stimulates the juxtaglomerular apparatus to secrete a greater amount of *renin*. The *renin-angiotensin-aldosterone system* (RAAS) is thus activated, which results in increased reabsorption of Na^+ and, therefore, of water.

Activation of the juxtaglomerular apparatus and the RAAS also occurs by increasing the sympathetic discharge on β-adrenergic receptors of afferent arterioles. Furthermore, sympathetic discharge improves the reabsorption of Na^+ by acting on the α-receptors of the proximal renal tubule.

Bainbridge's Reflex. In Paragraph 12.6 we mentioned the Bainbridge reflex evoked by the stimulation of Paintal receptors (atrial type B receptors or volume-receptors). These are located at the venoatrial junctions of both atria. The distension of the atria due to an increase in venous return or, in pathological conditions, such as, for example, stagnation of blood due to ventricular failure, causes *an increase in HR and saline diuresis (natriuresis)*. These Bainbridge's effects occur even if arterial blood pressure is not increased.

The increase in HR has been attributed to a reflex increased discharge of sympathetic fibers directed to the SAN and a direct effect due to the distension of pacemaker cells in the SAN.

The increase in diuresis has a multifactorial origin. The following factors have been suggested: *a)* inhibition of the hypothalamic-pituitary secretion of ADH; *b)* inhibition of the renin-angiotensin-aldosterone system; *c)* an increased release of the *atrial natriuretic peptide* from the stretched atrial walls; *d)* the production of a non-identified natriuretic hormone, different

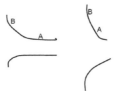

Figure 12.16 Location of Paintal receptors A and B in the atrium: in the absence of dilation (left) the A receptor is in the large vein. Following distension of the atrium by an increase in blood volume (right) both A and B receptors are in the atrial cavity. See further explanation in the text.

from the atrial natriuretic peptide: it can be the *adrenomedullin* produced by the adrenal medulla and/or the *brain natriuretic peptide* produced by the brain or cardiac ventricles; *e)* renal vasodilation due to the reduction of sympathetic discharge toward the kidney.

Initially, the Paintal receptors were divided into A and B (Figure 12.16). Under normal atrial filling conditions, receptors A discharge at the atrial systole, while B sends impulses concurrently with the V wave of the atrial pressure curve. Only when the atrium is stretched by an accumulation of blood, both receptors A and B discharge the degree of distension regardless of the phase of the cardiac cycle.

It has been suggested that the difference between the two types of receptors lies solely in their position in the atria. While the receptors of type A are found in the last part of the large veins, and therefore they are stimulated only when the wave of atrial contraction arrives, those of type B are found inside the atria in an easier position to allow the blood that accumulates to stretch, that is when the V wave of the atrial pressure curve is recorded. However, if the atria are pathologically dilated, both A and B receptors are located in the atrial cavity and subjected to the same stimuli (Figure 12.16).

13

Humoral Control of the Cardiovascular System

13.1 Catecholamines

Catecholamines are *adrenaline (epinephrine), noradrenaline (norepinephrine), and dopamine*. Although they are adrenal medullary hormones, noradrenaline, produced only in small quantities by the adrenal gland, is mainly the neurotransmitter of adrenergic sympathetic fibers. Catecholamines are produced in chromaffin cells according to the following scheme:

L-tyrosine $\Rightarrow$ L-DOPA $\Rightarrow$ dopamine $\Rightarrow$ noradrenaline $\Rightarrow$ adrenaline (Figure 13.1).

While adrenaline and the small amount of noradrenaline released by the adrenal gland act at a distance conveyed by blood, the action of noradrenaline produced by the sympathetic postganglionic neurons occurs locally on special effectors. In the case of the cardiovascular system, the targets and effectors are the VSM and all the functional tissues of the heart. The different localization and composition of target organs determine the modalities of catecholamines catabolism. In this process mainly two classes of enzymes are involved, the *monoamine oxidases* (MAOs) and the *catechol-ortho-methyltransferases* (COMTs). While MAOs are found linked to the outer mitochondrial membrane, COMTs are widespread in cells. The final product of catecholamine catabolism is mostly *vanilmandelic acid*. The half-life of catecholamines is very short. Therefore, the determination of urinary elimination of *vanilmandelic acid and metanephrine* is used in the clinic to ascertain the extent of catecholamine production by a hypertensive patient. For example, these catabolites increase in the case of catecholamine-producing tumors such as the *pheochromocytomas* of the adrenal gland.

Intravenous administration of *norepinephrine*, which acts mainly on α-receptors, may be used, with the due caution, in cases of acute critical

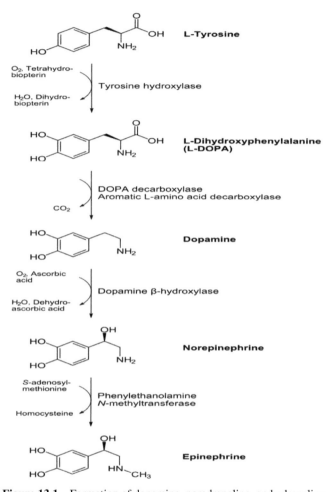

Figure 13.1   Formation of dopamine, noradrenaline, and adrenaline.

hypotension. The pressure can be brought back to the normal value or even at higher levels for the concomitant arteriolar vasoconstriction, venoconstriction, which improves the venous return to the heart and increases in contractility thus causing an increase in cardiac output. Increased pressure can induce bradycardia by stimulating baroreceptors. The bradycardia response is not observed if epinephrine is used instead of norepinephrine. Indeed, adrenaline acts mainly on β-receptors, inducing vasodilation in some districts and, therefore, only a moderate increase in pressure by the other positive effects on HR, contractility, and venous return (prevalence of α-receptors on the veins).

Catecholamines, preferably adrenaline, can be administered within myocardium in case of cardiac arrest for asystole. However, if the cause of cardiac arrest is ventricular fibrillation, it is impossible to restart the heart with catecholamines only. Yet, even in the case of asystole, adrenaline or noradrenaline restarting the heart activity can induce arrhythmias and ventricular fibrillation.

13.2 Endothial Factors

Various factors released by the endothelium of the vessels are active on the cardiovascular system. Here we recall the so-called *endothelial-derived relaxing factor* (EDRF) which was found to be *nitric oxide* (NO), the *endothelial-derived hyperpolarizing factors* (EDHFs), and *endothelial-derived contractile factors* (EDCFs) among which *endothelins* are of particular importance.

13.2.1 Nitric Oxide

The name of EDRF was initially attributed to an unidentified substance released by vascular endothelium, necessary for ACh to exert its vasodilating action. The obligatory role of endothelial cells in the relaxation of arterial smooth muscle by ACh was established in 1980. Subsequently, the substance was identified with NO.

It was then discovered that in addition to endothelial cells, NO is produced by cells of different tissues and is therefore currently considered to be an almost ubiquitous substance. In the case of the cardiovascular system, it is produced not only by vascular endothelium but also by the endocardial endothelium, by cardiomyocytes, by smooth muscle cells and by *non-adrenergic and non-cholinergic* (NANC) fibers of the autonomic system, which can determine vasodilation.

The synthesis of NO takes place by some NOS that converts L-arginine into L-citrulline incorporating molecular oxygen. The enzymatic catalysis of l-arginine requires dimerization of enzyme monomer units and several essential cofactors, such as tetrahydrobiopterin (BH_4), calmodulin (CM), heme, flavins, and NADPH. There are three types of NO-synthases: a NOS-1 or neuronal (n-NOS), a NOS-2 or inducible (i-NOS), and a NOS-3 or endothelial (e-NOS).

The NOS-1 and the NOS-3 are constitutive, *i.e.* already present pre-formed in the cells of many tissues, while the NOS-2 is inducible in many tissues, and as such it is produced following stimuli by of other substances. It follows that, while in the case of constitutive NOS the action of NO occurs almost immediately after the stimulus, in the case of i-NOS the action of NO appears belatedly (at least 3 hours after the initial stimulus) and can last over 24 hours. The delayed appearance of the NO by i-NOS support the idea that this enzyme is not pre-formed, but must be synthesized *de novo*, in many tissues. The activation of the NOS-2 and the corresponding production of NO occur massively in septic shock and can lead to the death of the patient following a serious and irreversible fall in blood pressure (*septic shock*).

The nomenclature with numbers is preferable as the other name is not correct: the inducible form (NOS-2) in some tissue, *e.g.* gastrointestinal mucosa, is always present (does not need to be induced). Also, despite the terms, n-NOS and e-NOS suggest a well-localized presence of the two enzymes, in neurons and endothelium, respectively, it has been demonstrated that both the n-NOS and the e-NOS can be present in other tissues. For instance, in cardiomyocytes all three types of enzymes have been described: NOS-1 and NOS-3 are involved in cardiac contractility regulation and NOS-2 in delayd cardioprotection.

The action of NO-synthases can be inhibited by L-arginine analogs such as L-monomethyl-arginine (L-NMMA), LN-nitro-arginine (L-NNA) and LN-nitro-arginine-methyl ester (L-NAME). These substances are used experimentally to study the role of NO in many physiological and pathological conditions.

Already *in vitro* endothelial cells present a basal production of NO in very small quantities. *In vivo*, in basal conditions, the quantity released slightly increases. Even when it is produced by the endothelium in basal conditions, NO prevents platelet aggregation and moderates the vasoconstrictor tone.

Stimulation of NOS-3 of endothelial cells and the relative production of NO can increase due to mechanical and chemical stimuli. *Mechanical stimuli* consist of an acute increase in *shear stress* and *pulse pressure*. Shear stress depends on the friction that the flow of blood exerts on the inner vessel wall, namely on the endothelium. It increases with the speed of viscous blood flow (see Chapter 11.3). The increase in *pulse pressure* is less effective than shear stress in stimulating NOS-3. Usually, the two stimuli are accentuated in a physical exercise where, by increasing the production of NO, they contribute to the dilation of the coronary and skeletal muscles. It has been demonstrated that if, at the same mean aortic pressure, pulse pressure increases the coronary flow increases and it increases even more in the presence of low concentrations of adenosine. Adenosine and NO may have synergistic vasodilator effects in the coronary circulation.

Chemical stimuli that determine the increase of NO production from the endothelium are several, including acetylcholine, bradykinin, catecholamines, thrombin, substance P, as well as serotonin, ATP, ADP, and adenosine. Acetylcholine activates e-NOS via M3 muscarinic receptors, while bradykinin acts through B2 receptors. The release of NO by catecholamine action appears to be mediated by α2-like endothelial receptors.

The mechanical and chemical factors that determine the production of NO increase the concentration of Ca^{2+} in endothelial cells. Unexpected complexity in endothelial calcium influx pathways has been described. For example, different pathways mediate acetylcholine and thapsigargin induced calcium influx, which results in the production and release of NO: *i.e.* the STIM/ORAI/TRPC channels are integral components of the calcium influx pathway activated by ACh and thapsigargin. Yet, TRPC4 contributes to ACh but not to thapsigargin elicited store-operated calcium entry. TRPC is a family of about 30 types of *transient receptor potential channels* located on the plasma membrane of numerous cell types.

Calcium entry activates the e-NOS which, acting on the L-arginine, causes the synthesis of nitric oxide. As said, for NOS to synthesize NO, it is necessary to have some cofactors among which a major role is played by *calmodulin (CM)* and *tetrahydrobiopterin* (BH_4) which renders constitutive NOS Ca^{2+}-dependent. In the absence of BH_4, various reactive species, including superoxide anion and hydrogen peroxide are formed instead of nitric oxide, by NOS.

After being produced in endothelial cells, NO can be poured into the vascular lumen (*luminal secretion*) where it prevents platelet aggregation, or it can diffuse into the vascular wall (*abluminal secretion*) where it causes the relaxation of *smooth muscle cells*. Nitric oxide action both on the platelets and on the smooth muscle cells is mediated by the activation of the *soluble guanylyl cyclase* which determines the transformation of *guanosine-triphosphate* (GTP) into cGMP. The latter activates the *protein kinase G* (PKG) that reduces the intracellular VSM Ca^{2+} concentration by the closure of the L-type channels/VOC (Figure 13.2). Indeed, VSMs depend mainly on extracellular Ca^{2+} for contraction. In these cells the sarcoplasmic reticulum (SR) is a rudimentary organelle; however, in some smooth muscle cells also the ryanodic channels of the rudimentary SR can be closed by PKG. In cardiac muscle the majority of Ca^{2+} required for contraction comes from

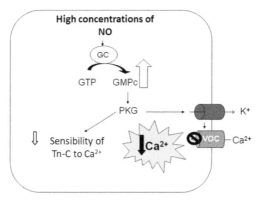

Figure 13.2　Vasodilation or reduced myocardial contractility in response to a high concentration of nitric oxide (NO). GC: guanylyl cyclase; PKG: protein kinase G; VOC: voltage-activated calcium channels; Tn-C: Troponin-C (in cardiac muscle only; smooth muscle does not contain the Ca^{2+}-binding protein troponin). GMPc = cGMP or cyclic guanosine monophosphate. See the explanation in the text.

the SR; therefore, the PKG formation plays a major role in determining the negative inotropic effects of NO closing ryanodic channels and the sensibility of Troponin-C to calcium (see below and Figure 13.2).

Nitric oxide also prevents the adhesion of granulocytes on the vessel wall where it hinders the formation of atherosclerotic plaques.

The NO half-life is about four seconds. However, it can be removed from the plasma by hemoglobin and cholesterol, which plays the role of scavenger towards it. While the constancy of hemoglobin concentrations in a subject's blood should not affect the presence of NO, changes in cholesterol can modify its concentration.

Based on the action of NO on vessels, vasodilator stimuli can be classified into endothelium-dependent or independent. Endothelium-independent stimuli, such as those of some metabolic factor, although starting vasodilation independently of the production of NO, it determines a subsequent production of NO, due to the greater shear stress determined by the initial increase in flow. This incoming NO leads to longer duration of vasodilation. In fact, in the coronary circulation, it has been observed that reactive hyperemia, evoked by metabolite accumulation, has a reduced duration if e-NOS is inhibited (see coronary reactive hyperemia in Chapter 14.1).

On the *myocardium,* NO induces a positive inotropic action in small concentrations and a negative inotropic action at high concentrations. A said above and in Chapter 7, the mechanism of *negative inotropic action* is

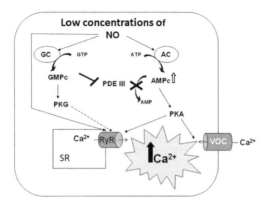

Figure 13.3 Vasoconstriction or increase in myocardial contractility in response to a low NO concentration. GC: guanylyl cyclase; PKG: protein kinase G; AC: adenylyl cyclase; PDE III: phosphodiesterase III; PKA: protein kinase A; RyR: ryanodic channels; SR: sarcoplasmic reticulum; VOC: voltage-activated calcium channels. AMPc: cAMP or cyclic adenosine monophosphate. GMPc: cGMP or cyclic guanosine monophosphate See the explanation in the text.

similar to that of vasodilation (Figure 13.2), including the closure of ryanodic channels and the lower sensitivity of the Troponin-C to Ca^{2+} induced by PKG. The mechanisms underlying *positive inotropic action* consist of increasing the concentration of intracellular Ca^{2+} (Figure 13.3). It has been suggested that a low concentration of cGMP determines the opening of the Ca^{2+} ryanodic channels, while the increase in cAMP *and* PKA leads to the activation of VOC/L-type calcium channels. The increase in cAMP *and* PKA are determined by AC activation and phosphodiesterase III (PDE3 or PDE III) inhibition by NO. In this regard, it is interesting how a very low NO concentration (sub physiological concentrations) can determine constriction, rather than dilatation also in isolated vessels. Compartmentalization of various NOSs and PDEs may also explain the different effects of different concentrations of NO.

Nitric oxide is also responsible for protecting the myocardium from ischemia and reperfusion injury (see Chapter 17, in particular Paragraphs 17.2 and 17.3).

13.2.2 Endothelial Derived Hyperpolarizing Factors

The existence of *endothelial-derived hyperpolarizing factors (EDHFs)* has been hypothesized following the observation that, if NOS is blocked with analogs of L-arginine, some substances (*e.g.* acetylcholine or bradykinin) that induce NO-dependent vasodilation can still cause vasodilation and/or negative inotropic effects, though to a lesser extent. Hyperpolarizing factors capable of increasing K^+ permeability would be the mediators of this residual vasodilation. Hyperpolarization limits the entry of Ca^{2+} into smooth muscle cells. The nature and the effects of EDHF on smooth muscle and myocardium are not clear yet.

It has been suggested that ACh stimulates the release of EDHF by acting on M3 receptors (as in the case of NO) of endothelial cells. It seems that the release of these factors is also mediated by an increase in Ca^{2+} concentration in endothelial cells. Hyperpolarizing factors can compensate for the lack of NO.

Among the substances proposed as EDHF there are *prostacyclins*, or PGI1, various isomers of *epoxy-eicosa-trienoic acids* (EETs), *superoxide anion* ($O_2{}^-$) and *hydrogen peroxide* (H_2O_2). EETs are produced by the action of cytochrome P-450, activated by acetylcholine or bradykinin, on arachidonic acid. EETs are, therefore, a non-prostanoid product of arachidonic acid.

Superoxide anion and peroxynitrite ($ONOO^-$) are also produced by the e-NOS in the absence of the L-arginine substrate and/or the BH_4 cofactor. Apart from their characteristic of oxidizing substances, superoxide anion and $ONOO^-$ can determine a modest degree of vasodilation through hyperpolarization of the smooth muscles. The superoxide anion-induced vasodilation, however, may be limited by the fact that it, by joining with NO, can remove the latter by causing the formation of $ONOO^-$ which may have modest vasodilator or even vasoconstrictor effects.

The endothelium has been shown to produce H_2O_2 in response to ACh. In recent years, considerable emphasis has been placed on H_2O_2 among hyperpolarizing endothelial factors. This seems to have particular importance in the dilation of the coronary and mesenteric circles. In coronary it can derive also from myocardial mitochondria (see coronary circulation regulation in Chapters 14.1). Its hyperpolarizing action is attributed to the opening of various K^+ channels, including those K^+ channels activated by Ca^{2+}.

Strictly speaking, even NO could be considered a hyper-polarizing factor. Parkingtob and coll. (1993) have highlighted how the application of ACh to vascular endothelium causes two phases of hyperpolarization of smooth muscle, namely a shorter early phase followed by a longer delayed phase. Of these phases, only the second would be due to NO.

13.2.3 Endothelial Contraction Factors

Endothelial contraction factors include *prostaglandins* H_2 and F_{2a}, *thromboxane* A_2, *peroxynitrite*, and *endothelins*. All these factors act through the increase in the intracellular concentration of Ca^{2+} in VSM.

Peroxynitrite vasoconstriction may be because it is related to the removal of NO by O_2^-. Therefore, peroxynitrite can be either vasodilator or vasoconstrictor.

The *endothelins (ETs)*, peptides formed by 21 amino acids, are endowed with a very high vasoconstrictor power. Endothelin-1 (ET-1), endothelin-2 (ET-2), endothelin-3 (ET-3) and endothelin-4 (ET-4) have been described. Although all four isoforms are present in the human genome, ET-1 is certainly present in vascular endothelium.

Endothelins are the ligands of two types of receptors for endothelins, *Ea receptors*, present on smooth muscle cells, and cardiomyocytes, while *Eb receptors* are also present on endothelial cells. The binding of endothelins with Ea and Eb receptors of smooth muscle cells causes vasoconstriction, whereas the link with endothelial Eb receptors causes transient vasodilation due to the endothelial release of NO. This is the reason why the application of endothelins to the vessels determines vasoconstriction preceded by transient vasodilation. In smooth muscle cells, the binding to the Ea and Eb receptors, coupled with a Gq membrane protein, determines the activation of a phospholipase C (PLC) with the production of DAG. This, through the PKC, activates the H^+/Na^+ exchanger which, by increasing the intracellular concentration of Na^+, activates the Na^+/Ca^{2+} exchanger with less Ca^{2+} eliminated by the cell, or even an entry o Ca^{2+} into the cell, when the Na^+/Ca^{2+} exchanger works in *reverse mode*. At the same time, in the myocardial fiber only, the reduction of the intracellular concentration of H^+ increases the sensitivity of the *troponin C* to Ca^{2+} (Figure 13.4). Smooth muscle does not express troponin, therefore calmodulin regulates the contraction in these cells.

Acting on the endothelin receptors of cardiomyocytes, the ETs cause an increase in contractility with the same mechanism with which they cause vasoconstriction.

However, the binding to the endothelial Eb receptors, coupled with a Gi protein, activates the e-NOS *via* PI3K-PKB/Akt pathway and induces vasodilation (Figure 13.5).

Although it has been observed that there is a basal secretion of endothelins which, like the basal secretion of NO, can increase following shear stress, it is important to remember some of the main pathophysiological

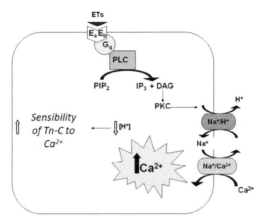

Figure 13.4 Endothelins (ETs) induce vasoconstriction acting on smooth muscle and increase contractility acting on myocardial fibers. Ea and Eb: endothelin receptors; Gq: membrane proteins; PLC: phospholipase C; PIP2: phosphoinositol diphosphate; IP3: inositol phosphate; DAG: diacylglycerol; PKC: protein kinase C; Tn-C: troponin C (in cardiac muscle only; smooth muscle does not contain the Ca^{2+}-binding protein troponin). See the explanation in the text.

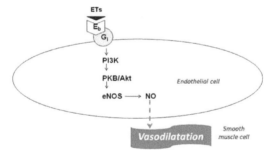

Figure 13.5 Endothelial vasodilation by endothelins (ETs). Eb: endothelin receptors; Gi: membrane inhibitory G protein; PI3K: phospho-inositol-3-kinase; PKB/Akt: protein kinase B/Akt; eNOS: endothelial NO-synthase. See the explanation in the text.

situations characterized by an increased concentration of these vasocon- strictors. They are essential hypertension, pulmonary hypertension, heart failure, and myocardial infarction. All these conditions can be characterized by *endothelial dysfunction* (decrease in NO and increase in ETs produc- tion, see below). Whether endothelins are responsible for these pathological conditions is a matter of controversy.

13.2.4 Endothelial Dysfunction

Endothelial dysfunction is observed when the endothelium shifts toward a phenotype characterized by impaired vasodilation, prothrombic, and proinflammatory status. In general, endothelial dysfunction is characterized by reduced NO production with increased production of reactive oxygen species (ROS) and ETs. It can be acute or chronic. Endothelial dysfunction can precede hypertension or can be a consequence of hypertension. The vessels become less compliant and will favor the establishment of higher blood pressures.

Acute endothelial dysfunction can be observed in the coronary circulation during reperfusion which follows myocardial ischemia. This dysfunction together with the production of ROS aggravates the vessels and the infarct damage. Chronic endothelial dysfunction may involve all vascular districts and can be considered the "initial answer" to the risk factors of atherosclerosis, including smoking. Acute endothelial dysfunction is discussed in Chapter 17.1 considering the role of the endothelium in ischemia/reperfusion injury. Here we only recall how in the post-ischemic reperfusion of the coronary circulation, the interaction of granulocytes and platelets with the endothelium, platelet aggregation, and vasoconstriction leads to the *non-reflow phenomenon* (absence of flow in a previously re-perfused fabric). All these events have been attributed to the post-ischemic NO deficiency.

In chronic endothelial dysfunction the activation of *adhesion molecules* (see Chapter 17.1), due to the phenomena of adhesion and aggregation, as well as the lack of NO, is favored by the presence of low density oxidized lipoproteins (ox-LDL). Increased platelet aggregation results in intravascular coagulation and migration of monocytes into the wall with the onset of atheromatous plaque formation. When there is an increased expression of specific leukocyte adhesion molecules and other soluble products by the endothelium some authors speak of *endothelial activation*. In this condition, nuclear factor-kappa B (NF-κB) is likely responsible for the transcription of various endothelial genes.

The lack of NO, possibly accompanied by a greater presence of ETs in the plasma, may be the basis of some cases of the so-called "essential hypertension". Given that often the increase in blood pressure is accompanied by hypercholesterolemia, it is necessary to keep in mind the action of cholesterol as a *scavenger* of NO. In this case, hypertension can sometimes be treated with dietary interventions or pharmacological therapies aimed at reducing plasma cholesterol concentration.

Endothelial dysfunction can favor vasospasm, or thrombosis, and is associated with the development of atherosclerosis and restenosis, and vascular complications observed in many cardio-metabolic diseases, including diabetes.

As already mentioned in this Chapter, a very low concentration of NO, and therefore of cGMP, can determine both an increase in cardiac contractility and vasoconstriction. A low concentration of cGMP can lead to the Ca^{2+} channels opening of both dihydropyridine receptor, DHP, and ryanodic receptors, RYR, on the cell membrane and SR respectively.

Box 13.1 Nitric Oxide and Erectile Dysfunction

The functional state of the penis, flaccid or erect is governed by smooth muscle tone. Sympathetic contractile factors maintain flaccidity whilst parasympathetic factors induce smooth muscle relaxation and erection. Indeed, smooth muscle relaxation allows chambers inside the penis (two *corpora cavernosa* and a *corpus spongiosum*) to fill with blood so the penis becomes erect. It is generally accepted that NO is the principal-agent responsible for the relaxation of penile smooth muscle. Nitric oxide is derived from two principal sources: directly from NANC fibers and indirectly from the endothelium lining *corporal/cavernosal sinusoids* and blood vessels in response to cholinergic stimulation. The generation of NO from L-arginine is catalyzed by either constitutive NOSs. There has been controversy over the relative prevalence of NOS-1 or NOS-3 within the penis of different animal species. *Sildenafil* and *tadalafil* used as a treatment for erectile dysfunction, are specific *phosphodiesterase type 5* (PDE5 or PDE V) inhibitors that enhance NO-mediated vasodilation in the *corpora cavernosa* by inhibiting cGMP catabolism. Since PDE5 is widely expressed in the cardiovascular system, it has been hypothesized that sildenafil could enhance NO-mediated vasodilation in other vascular beds and improve endothelial function. Moreover, cGMP and PKG, play an important role in counterbalancing maladaptive remodeling in the heart. It has been also suggested that the suppression of cGMP catabolism may have a positive impact on a variety of cardiac disease conditions. It seems that inhibitors of PDE5, such as *sildenafil* and *tadalafil*, ameliorate cardiac pressure and volume overload, ischemic injury, and may limit cancer drug cardiotoxicity.

13.3 Other Humoral Factors That Act on the Cardiovascular Apparatus

Humoral factors may be divided into local and inflammatory factors from one side (*e.g.* serotonin, prostaglandins, adenosine, histamine, and kinins) and hormones (*e.g.* adrenaline, angiotensin II, natriuretic peptides and vasopressin) form the other. The latter is more important in the control of blood pressure. Several humoral factors have acute effects on the cardiovascular system. The effects of many humoral/hormonal factors are evidenced in chronic conditions and/or when the neural control is impaired, as in transplanted hearts. Some of the hormonal factors, *e.g. adrenaline*, have been already described. *Acetylcholine* is the chemical mediator of the parasympathetic and cholinergic sympathetic action on the cardiovascular system (see Chapters 12.2 and 12.3). It cannot be considered a humoral factor. Nevertheless, besides being a vasoconstrictor, it acts as a vasodilator by inducing the release of humoral factors such as NO and EDHF already described above (see above and below). In this Chapter, some aspects of already discussed and new factors are described.

13.3.1 Serotonin

Serotonin, or 5-hydroxytryptamine (5-HT), is the product of the hydroxylation of the tryptophan, introduced with food, to 5-hydroxytryptophan by the tryptophan-hydroxylase. The subsequent decarboxylation of the 5-hydroxytryptophan by a decarboxylase leads to the formation of 5-HT. In addition to the serotonergic neurons of the central nervous system, present mainly at the hypothalamic and mesencephalic level, serotonin is produced by the entero-chromaffin cells of the intestinal glands.

The catabolism of serotonin ends with the formation of *tryptamine-O-sulphate* as a consequence of the action of an MAO and subsequently of a sulfotransferase. For serotonin, seven different 5HT receptors were identified. In particular, 5-HT1D receptor agonists are used in the treatment of migraine.

In the blood, serotonin is present mostly in platelets. Apart from the great variety of its actions (antidepressant, immune, thermoregulatory, memory-promoting, stimulation of pulmonary ventilation and intestinal motility, induction of platelet aggregation, etc.), serotonin actions on the cardiovascular system are quite complex.

Given parenterally, the substance produces a three-phase variation in pressure. In the first phase, the pressure decreases due to a decrease in peripheral resistance and heart rate. It is believed that this first phase is due to the triggering of the *Bezold and Jarisch reflex* (see Chapter 12.6.2), as suggested by the fact that the hemodynamic changes of this phase are absent after cutting the vague. In the second phase, the pressure increases due to an increase in peripheral resistance and cardiac output. It has been hypothesized that in this phase both direct action on VSM and a reflex action due to the stimulation of aortic and carotid chemoreceptors are involved. It is also possible that an increased venous return produced by the constriction of veins contributes to the increase in cardiac output. In the third phase, there is a reduction in pressure attributed to a fall in peripheral resistance due to the dilation of arterioles of the large vascular bed of skeletal muscles. The VSM of these arterioles would respond to serotonin with relaxation rather than contraction.

13.3.2 Prostaglandins

Prostaglandins (PG) are almost ubiquitous substances derived from arachidonic acid that can be taken by food or formed from the metabolism of linolenic acid. Various *prostanoids* such as prostaglandins, thromboxanes, prostacyclines, and leukotrienes can be derived from arachidonic acid, via the enzyme *cyclo-oxygenase*. Prostanoids are formed during inflammatory processes in which they contribute to vasodilation, increased vascular permeability, and chemotaxis, as well as to fever and pain. The production of prostanoids can be inhibited by anti-inflammatory drugs, such as NSAIDs (non-steroidal anti-inflammatory drugs; FANS in Italian: farmaci anti-infiammatori non-steroidei).

The action that the various prostaglandins (PGE1, PGEe, PGE3, PGF1, PGF2, PGF3, etc.) exert on the vessels depends not only on the type of prostaglandin but also on the animal species. It has been observed that, while PGE1 has vasodilator and hypotensive action on most species, PGF2 causes an increase in blood pressure in dogs and rats, probably due to a venous constriction that favors the venous return of blood to the heart. Other prostaglandins also appear to reduce renal hypertension (hypertension observed in some renal diseases, *see also below*).

13.3.3 Plasma Kinins

The active *plasma kinins* on the cardiovascular system are *bradykinin (Bk)* *and callidin (Kd)*. While Bk, a peptide of 9 amino acids, is produced by the action of *liver kininogenase kallikrein* on *high molecular weight kininogen*, Kd, a peptide of 10 amino acids, is the result of the action of *tissue kallikrein* on *low molecular weight kininogen*. Locally activation of *tissue kallikrein* is the result of trauma, sunburn, insect bites, or ischemia.

Bradykinin and callidin have B1 and B2 receptors in common. The *B1 receptors*, coupled with membrane G proteins, are present on the VSM of the small vessels where their expression increases in the presence of inflammatory phenomena. The *B2 receptors*, considered in some sense more important and more numerous than the previous ones, are associated with the membrane proteins Gq and Gi. While the Gq protein stimulates phospholipase C causing an increase in the intracellular concentration of Ca^{2+}, Gi inhibits the activity of adenylyl cyclase.

Plasma kinins are responsible not only for dilation but also for increased vascular permeability and pain, as in inflammation. If their effect is not only local but also general, they can cause a decrease in resistance capable of reducing arterial pressure up to shock. To this hypotensive effect, they also contribute to their natriuretic effect.

Neither the stimulation of PLC, due to the Gq protein, nor the inhibition of adenylyl cyclase (AC) due to Gi explains the vasodilator action of plasma kinins. Instead, they are more adapted to explain the contraction of non-vascular smooth muscle cells for the action of kinins. The vasodilator action is likely mediated by the activation of NO produced by the action of Bk and Kd on NOS-3 via the B2 receptor. For this reason, the vasodilator action of plasma kinins is considered endothelium-dependent. Antagonists of Bk are sometimes used in the treatment of local swellings of traumatic origin. BK is metabolized by the angiotensin-converting enzyme (ACE). Therefore, ACE inhibitors, widely used anti-hypertensive drugs, may increase BK accumulation in the lung interstitium, where BK can induce an irritating cough. ACE is an enzyme present in several cells, but especially on the membrane of endothelial cells of the pulmonary and renal circulations where it functions as an *exo-enzyme* (an enzyme that functions outside of the cells).

13.3.4 Histamine

Histamine is a compound that causes vasodilation and increased capillary permeability to white blood cells. Due to these characteristics, it is considered one of the factors that trigger inflammatory responses. Produced in basophilic granulocytes and mast cells by the action of the L-histidine decarboxylase enzyme on the amino acid histidine, histamine performs its action by binding to 4 subtypes of H1, H2, H3, and H4 receptors.

The H1 receptor is responsible for vasodilation as well as for itching and pain. It is interesting how this receptor, which releases the VSM of the vessels, causes contraction of the bronchial smooth muscles. The H2 receptor in addition to performing a strong vasodilating action seems to increase cardiac contractility. At the level of the parietal cells of the gastric glands, it stimulates the secretion of hydrochloric acid. The H3 receptor is present mainly in the nervous system where it restrains the release of neurotransmitters. Finally, the H4 receptor, present above all in basophilic granulocytes, promotes chemotaxis.

13.3.5 Antidiuretic Hormone

The antidiuretic hormone (ADH), also called *vasopressin*, is a polypeptide of 9 amino acids produced by the magnocellular neurons of the *supraoptic nucleus* and partly also by the *paraventricular nucleus*, of the *hypothalamus*. It is then transported along nerve fibers to the *posterior pituitary gland* where it is released into the bloodstream.

The production of ADH is mainly regulated by the osmolarity of the extracellular fluid: it is stimulated by the increase and inhibited by the decrease of the osmotic concentration (Osm/l or Osm/kg) of the extracellular fluid. Given the sensitivity to the osmolarity of the environment in which they are immersed, the hypothalamic cells that secrete ADH are called *osmoreceptors*. In the presence of a normal osmolarity of about 290 mOsm/kg, there is a basal hypothalamic production of the hormone, which allows water retention by the kidney. ADH production ceases when the osmolarity falls below 280 mOsm/kg.

Receptors for ADH, or V receptors, are three subtypes: V1A, V1B, and V2. The binding of ADH to the V1A receptors, present on the VSM of the resistance vessels, causes vasoconstriction by acting on a membrane protein Gq. The binding of ADH to the V2 receptors in the distal convoluted tubule and collecting duct of the kidney, through a *Gs protein/cAMP/PKA pathway,* allows the expression of aquaporin 2 (AQP2), which is shuttled from intracellular vesicles to the plasma membrane. At the end AQP2 allows the reabsorption of interstitial water, thus regulating the volume of the circulating liquid. The V1B receptor contributes to the *fight-or-flight response* (see Chapter 16.2). This receptor is present in the brain where it acts on a membrane Gq protein.

When the concentration of solutes increases in the interstitial fluid, the increased production of ADH increases the reabsorption of water by the kidney, thus limiting the increase in osmolarity. Diuresis from water ingestion occurs limiting ADH production. In addition to the lower osmolarity, hypothalamic osmoreceptors are also inhibited by alcohol. This explains why the introduction of wine or beer results in a diuretic effect higher than that induced by the ingestion of an equal quantity of water only. ADH production is also influenced by baro-reflexes that allow ADH production when blood pressure is reduced: a 30% reduction in blood pressure (or blood volume) can increase ADH production despite a reduction in plasma osmolarity (see below and regulation of blood pressure in Chapter 12). After a hemorrhage when one begins to recall/retain water the osmolarity decreases and the individual will continue to recall/retain water until the blood volume and pressure are too low.

High concentrations of ADH cause strong vasoconstriction in many vascular districts. Coronary circulation after transient vasoconstriction responds to ADH with NO-dependent vasodilatation. NO-dependent vasodilation induced by ADH has been observed also in the cerebral circulation.

13.3.6 Natriuretic Factors

Natriuretic peptides are hormones that induce natriuresis. Among natriuretic factors are included *1)* the *atrial natriuretic peptide* (ANP), produced by the atrial wall; *2)* the *brain natriuretic peptide* (BNP), produced mainly in cardiac ventricles, although initially isolated from brain tissue; *3)* the *type C natriuretic peptide* (CNP), which is widely expressed throughout the vasculature and in particular in the endothelium. Also *urodilatin*, produced in the distal convolute renal tubule and collecting duct, is included among natriuretic factors.

The ANP, is a peptide hormone consisting of 28 amino acids. It is produced in response to the distension of the atrial myocardial fibers as can occur in ventricular insufficiency. ANP is also produced by stimulation with angiotensin II or ETs, as well as by adrenergic sympathetic stimulation.

The BNP consists of 32 amino acids and is synthesized by the distended ventricular myocardium (in pathological or experimental conditions) as well as by some parts of the brain. Both hormones are degraded by endopeptidases, including *neprilysin*.

ANP and BNP activate the transmembrane guanylyl cyclase, *natriuretic peptide receptor-A* (NPR-A). CNP activates *natriuretic peptide receptor-B*

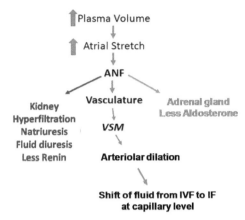

Figure 13.6   The principal effects of atrial natriuretic factors (ANF). IF: interstitial fluid; IVF intravascular fluid volume; VSM: vascular smooth muscle; See text for explanation. (*Modified from Maak T, Kidney International 1996*).

(NPR-B). Both receptors activate a guanylyl cyclase that catalyzes the synthesis of cGMP, which mediates the most known effects of natriuretic peptides. A third natriuretic peptide receptor, *natriuretic peptide receptor-C* (NPR-C), clears natriuretic peptides from the circulation *via* receptor-mediated internalization and degradation. However, a signaling function for NPR-C has been suggested as well. NPR-C is coupled to a Gi protein which leads to inhibition of AC or activation of PLC and regulates various cellular functions.

Natriuretic factors act directly and indirectly on the cardiovascular system (Figure 13.6). Directly they cause dilation both at the level of the vessels of capacity and of those of resistance. While dilation of the capacity vessels decreases central venous pressure and ventricular preload, dilation of the resistance vessels contributes to the reduction of arterial pressure. Natriuretic factors also increase capillary permeability favoring the shift of fluid from the vessels to the interstitium.

Indirectly, natriuretic peptides are responsible for an increased elimination of sodium and water following an inhibitory action on renin secretion by the renal juxtaglomerular apparatus with consequently reduced angiotensin formation and less aldosterone release. Together with the reduction of peripheral resistance, the elimination of sodium and water (reduces intravascular fluid volume) plays a very important role in reducing blood pressure. ANP inhibits ADH and aldosterone production.

The effects of natriuretic peptides, therefore, antagonize the activity of the renin-angiotensin-aldosterone system (RAAS) (see Paragraph 13.4). In the case of heart failure, the two systems tend to create a balance between retention and elimination of salt and water. While the RAAS tends to increase the volume of circulating liquid and to favor the ventricular Starling mechanism, natriuretic hormones tend to prevent this increase from leading to excessive dilation of the ventricular cavities, which triggers maladaptive cardiac modifications.

In summary, natriuretic peptides cause a reduction in expanded extracellular fluid volume by increasing renal sodium excretion. Therefore, they participate in the regulation of circulating volume and pressure. Indeed, natriuretic peptides increase in hypertensive subjects, in whom they could play a limiting role in the pressure rise.

Neprilysin (NEP) is the main enzyme involved in the degradation of natriuretic peptides. Various peptides including glucagon, enkephalins, substance P, neurotensin, oxytocin, and bradykinin are also degraded by NEP. Given its main role in the degradation of natriuretic peptides, the inhibition of NEP has been suggested as a good strategy to increase their plasma concentration, thus increasing their beneficial effects such as vasodilation, and natriuresis. The study of drugs capable of blocking the pathophysiological action of NEP has been considered as a therapeutic option in heart failure. However, NEP blockers alone or in association with ACE inhibitors or Angiotensin II receptor blockers (ARB) have some side effects (*i.e.* angioedema), which limit their use in the clinic. Among the angiotensin receptor–neprilysin inhibitors (called ARNi) there is the combination sacubitril/valsartan, which combines an ARB with the NEP inhibition. The hope is that these molecules will be useful in heart failure and hypertension.

13.3.7 Orexins or Hypocretins

The *orexins or hypocretins* are peptides secreted by hypothalamic cells, which, in addition to participating in the awakening reaction, coordinate cardiovascular and respiratory responses as well as lowering of sensitivity to pain, which occur in the *fight-or-flight response* (see Chapter 16.2). Orexin A, or hypocretin 1, consists of 33 amino acids, while orexin B, or hypocretin 2, consists of 28 amino acids. Orexins A and B, derived from the same precursor protein. Indeed, the two substances are identical for about 50% of the molecule. Intracerebroventricular administration of orexins stimulates food consumption in a dose-dependent manner. Within the hypothalamus, orexin nerve fibers and orexin receptors are found extensively in the paraventricular nucleus, whose neurons project directly to the sympathetic preganglionic neurons in the spinal cord and control sympathetic outflow. Therefore, it seems that orexins affect cardiovascular function *via* their action on the central nervous system. In particular, orexin-A may be involved in the control of multiple homeostatic functions, including the control of appetite, stress reactions, body fluids, heart rate, and blood pressure through sympathetic nerve activity.

13.4 Renin-Angiotensin-Aldosterone System

The renin-angiotensin system (RAS) has already been discussed in Chapters 7 and 8 and mentioned in Paragraph 13.3. Here we recall classical and new concepts that underline the key role this system plays in hypertension, heart failure, and myocardial fibrosis.

Renin is mainly produced by the *granular cells* of the juxtaglomerular apparatus of the kidney in the presence of lower Na^+ concentration in the tubular liquid that arrives at the *macula dense*. Renin is in turn responsible for the transformation of angiotensinogen into angiotensin I (Ang I), which will, in turn, be converted to angiotensin II (Ang II) by the angiotensin-converting enzyme (ACE), an exo-enzyme situated on the cell membrane, mainly at lung capillary level.

Renin production is also induced by a lower glomerular filtration and an increase in sympathetic stimulation due to reduced baroreceptor stimulation, secondary to a blood pressure decrease. It follows that the increase in renin production is the result of lower systemic pressure. However, we must remember that locally the filtration pressure can be reduced by renal artery stenosis, which determines a reduction of the renal pressure distal to the stenosis. In this case, the systemic pressure is increased (*Goldblatt hypertension*) by the activation of RAAS and systemic actions of Ang II, so that the renal pressure and renin production can return to normal values. As systemic blood pressure increases, also sodium excretion by the intact contralateral kidney rises (*pressure natriuresis*); therefore, there is no sodium retention in *Goldblatt hypertension*.

In addition to exerting a vasoconstrictor action, *Ang II* stimulates the production of *aldosterone* by the cells of the glomerular area of the adrenal cortex. Aldosterone then stimulates sodium reabsorption and potassium elimination by the distal convoluted tubule and kidney collecting duct. Aldosterone causes the reabsorption of sodium to be followed by that of water. Since excessive production of aldosterone causes hypertension, spironolactone is used as a diuretic and antihypertensive drug. Antidiuretic hormone

ADH is often released simultaneously with aldosterone, which may also favor the release of ADH.

In the contractile insufficiency of the heart, the first attempt of compensation is due to Starling's mechanism. This is favored by a greater increase in ventricular end-diastolic volume due to the increase in the systolic residue. If the cardiac output does not return stable to normal values, the reduced blood pressure is responsible for a lower glomerular filtration. Under these conditions, stimuli from the *macula dense* start the renin-angiotensin-aldosterone secretion with increased sodium and water reabsorption. There is, therefore, an increase in the circulating liquid volume which can favor a further dilatation of the cardiac cavities, beyond the limits of Starling's law and the appearance of edema. It is clear that in these conditions the volume of circulating liquid must be reduced by using diuretics. One of the diuretics that can be used for this purpose is the above-mentioned *spironolactone*. Experimentally, the inhibition of aldosterone with the use of spironolactone has also been seen to prevent myocardial fibrosis and hypertrophy.

The above is a classical view of the RAS as an endocrine system involved in blood pressure regulation and body electrolyte balance. However, RAS is now considered a "ubiquitous" system that is expressed locally in various tissues and exerts multiple paracrine/autocrine effects involved in tissue pathophysiology and homeostasis. It is now clear that RAS plays key roles in apoptosis, differentiation, inflammation, migration, and proliferation, as well as on cellular growth and extracellular matrix remodeling. Anomalous tissue RAS expression is involved in multiple diseases including type 2 diabetes, renal fibrosis, atherosclerosis, and cardiac hypertrophy. Indeed, there is a RAS tissue-specific expression that may change under pathological conditions. The nature of maladaptive fibrosis and hypertrophy is multifactorial, but prolonged adrenergic and RAS actions on the heart are surely involved in these pathological processes (see also the BOX 13.2 on a new aspect of the RAS).

13.5 Apeline and the APJ Receptor

Apelin is a substance with interesting action on the cardiovascular system. Its action, in addition to regulating myocardial inotropism and vasomotor tone, also concerns post-conditioning myocardial protection (see Chapter 17.3).

Apelin is the ligand for the *APJ receptor*, previously considered an *orphan receptor* (receptors without an identified ligand). Produced in various tissues (endothelium, VSM, myocardium, etc.) of various species (man,

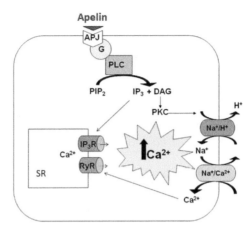

Figure 13.7   Increased contractility of myocardial fibers by apelin. APJ: apelin receptor; G: membrane protein; PLC: phospholipase C; PIP2: phosphoinositol diphosphate; IP3: inositol phosphate; DAG: diacylglycerol; PKC: protein kinase C; IP3R: IP3 receptor-channel; RyR: ryanodine receptor-channel; SR: sarcoplasmic reticulum. See the explanation in the text.

mouse, rat, cattle, etc.), it derives from the precursor pro-apelin, a peptide formed by 77 amino acids with the sequence active in the region COOH-terminal. The name apelin refers to the active fragments of pro-apelin, which are classified according to the number of amino acids that compose them.

The most studied active fragments are apelinin-13 and -36. Of the two fragments, the most active is apelin-13. APJ receptors are similar to A1 receptors for Ang II, although Ang II does not bind to APJ receptors and no form of apelin attaches to angiotensin receptors.

Apelin exerts an inotropic action on the heart and an action that can be constrictor or dilator on the vessels. The inotropic action takes place because in the cardiomyocytes the G membrane protein, to which the receptor is bound (Figure 13.7), activates a phospholipase C that cleaves the phosphoinositol-diphosphate (PIP2) in inositol-triphosphate (IP3) and diacylglycerol (DAG).

The DAG activates the Na^+/H^+ exchanger for the intervention of a protein kinase C. The consequent increase of Na^+ in the cytosol activates the Na^+/Ca^{2+} exchanger which allow the entry of Ca^{2+} into the cell. This Ca^{2+}, entering the cell from the interstitial fluid, also determines the opening of the ryanodic channels of the sarcoplasmic reticulum (CICR phenomenon, see Chapter 4.3), thus triggering a further increasing the concentration of intracellular Ca^{2+}. Also, IP3 allows the passage of Ca^{2+} in the cytosol acting on its receptors-channel of the SR.

On the myocardium, apelin mimics the ischemic postconditioning limiting the extension of the infarct area if administered immediately at the beginning of reperfusion, while it does not induce any protection if it is administered for preconditioning purpose before ischemia (for pre and postconditioning see Chapter 17).

Binding to APJ receptors, apelin can determine protection through a NO-dependent and a NO-independent way (Chapter 17.3). The NO-dependent pathway is induced by the activation of phosphoinositol-3-kinase, while the NO-independent pathway occurs by activation of the superoxide-dismutase that transforms the superoxide anion, which forms at the beginning of reperfusion, in the least toxic, hydrogen peroxide. In the apelin-induced cardioprotection, a role is played also by the epidermal growth factor and Src kinase, which regulate protein kinase B/Akt activity.

On the vessels, the response to apelin normally consists of vasodilation, while the vasoconstriction occurs in case of endothelial dysfunction when the VSM represents the only target for the action of the peptide. In the presence of intact endothelium, the vasodilator action is mediated by the endothelial production of NO which acts on the smooth muscle fibers, causing their relaxation. On the contrary, in the case of endothelial dysfunction, the vasoconstrictor action is exerted directly on the VSM using a cascade of signals similar to that which determines the inotropic effect on the heart.

Depending on how apelin stimulates NO production by endothelial cells, it has been suggested a Ca^{2+}-dependent or Ca^{2+}-independent vasodilation. In Ca^{2+}-dependent vasodilation, apelin binding with the G protein-coupled receptor may activate a phospholipase C which, through the release of IP3, mobilizes Ca^{2+} from intracellular deposits and causes e-NOS activation. In the case, instead, of the Ca^{2+}-independent vasodilation, the G protein may activate the phosphoinositol-3-kinase (PI3K) which in turn activates the e-NOS through the Akt.

Box 13.2 New Aspects of the RAS

A receptor binding renin and prorenin, its proenzyme, termed the *(pro)renin receptor* (PRR), was cloned in 2002. This receptor is ubiquitously expressed, and it is involved in several physiological and pathophysiology processes, including the cell cycle. Prorenin becomes enzymatically active after binding to PRR then it activates the ERK1/2, MAP kinases, and p38 pathways, leading to the upregulation of profibrotic and cyclooxygenase-2 genes. We still know very little about its cellular functions.

Angiotensin-converting enzyme 2 (ACE2) is a newly discovered, membrane-bound aminopeptidase responsible for the production of vasodilatory peptides such as angiotensin 1-7 (Ang 1-7), starting from angiotensin II or angiotensin I. Thus, in some tissues ACE2 and Ang 1-7 are important in counteracting the vasoconstrictor effects of Ang II. Also, a *testicular angiotensin-converting enzyme* (tACE) isozyme has been described, which is likely to play important functional roles in male reproduction.

Cardiac stress stimulates the expression of ACE and at the same time suppresses that of ACE2, resulting in a net production of Ang II that supports cardiac hypertrophy and fibrosis.

Four types of *angiotensin receptors* are known, of which the type 1 (AT1) and type 2 (AT2) receptors are the most represented and studied. Stimulation of the AT1 receptors leads to the classical processes attributed to Ang II, such as vasoconstriction, inflammation, and cell proliferation. These processes are involved in various cardiovascular diseases, including atherosclerosis, hypertension, and ventricular hypertrophy. In adults, the AT2 receptors are minimally expressed and their role is not well established, but their effects seem to oppose those of the AT1 receptors. Controversy exists about the existence of the AT3 receptor subtype. AT4 receptors are concentrated predominantly in the brain. They have been described also in the heart, kidney, adrenals, and blood vessels. They seem activated mainly by angiotensin IV and are involved in the memory processes. *MAS receptors* are candidate receptors for Ang (1–7). The main actions of Ang (1–7) through MAS seem to be the production of arachidonic acid and the activation of NOS. Of note, SARS-coronavirus 2 (SARS-CoV-2) uses the SARS-CoV receptor ACE2 for cell entry.

14

District Circulations

Each tissue has special functions and their vascular circulations have characteristics that allow these functions to be performed properly. Here, we will
describe some of these specialized districts to be able to analyze their main
characteristics and to be able to highlight which control mechanisms prevail
among those illustrated in the previous chapters.

14.1 Coronary Circulation

14.1.1 The Coronary Arteries and the Microcirculation

The myocardium receives blood from the two coronary arteries: the right and
left coronary arteries. Sometimes we refer to three coronary arteries as the
left coronary artery divides into two branches, the left anterior descending
and the circumflex arteries.

The right coronary artery arises from the right Valsalva sinus of the aorta
(Figure 14.1) and runs between the root of the pulmonary artery and that of
the aorta to move into the atrio ventricular groove, where it goes towards the
right face of the heart. It then reaches the posterior interventricular groove.
Several branches break off from the right coronary along its course. The most
important are: the branch of the pulmonary artery cone, the branch of the
right ventricle, the branch of the sinoatrial node, and the branch of the acute
margin or right marginal branch.

The left coronary artery arises from the left Valsalva sinus. The first
tract is the so-called common trunk, which, at the beginning of the anterior
interventricular groove, divides into the anterior descending branch, or left
anterior interventricular (LAV) branch, and into the circumflex branch.

The anterior descending artery runs along the anterior interventricular
groove to reach the posterior apex of the heart (Figure 14.1). The main
collaterals of the anterior descending branch are the adipose artery, which

313

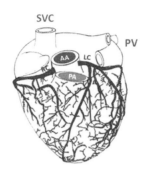

Figure 14.1   Distribution of coronary arteries. AA: artery aorta; PA: pulmonary artery; SVC: superior vena cava; PV: pulmonary veins; LC: left coronary artery; RC: right coronary artery. See the explanation in the text.

runs along the initial part of the pulmonary artery, the diagonal branches, and the septal branches, which penetrate the front part of the interventricular septum.

The circumflex artery runs through the atrioventricular groove heading to the left. It goes beyond the left margin, or obtuse margin, reaching the posterior part of the left ventricle. The main collaterals of the circumflex branch are the left anterior atrial branch, the oblique artery, the branch of the obtuse margin, the left posterior atrial branch, and various branches that are distributed on the wall posterior of the left ventricle.

The large coronary arterial vessels are divided into smaller and smaller branches until they give rise to vessels of resistance represented by arterioles and precapillary sphincters. The vessels of resistance are followed by the exchange vessels formed by the capillaries, in turn, followed then by venules and finally by veins.

Capillary density is much higher in the myocardium than in skeletal muscle (Figure 14.2). While in fact in the skeletal muscle about 400 capillaries cross the section of one mm^2, in the myocardium about 3000–5000 capillaries cross a section of one mm^2. Indeed, in basal conditions, while a skeletal muscle receives 2–4 mL/min, the myocardium receives 80–100 mL/min of blood per 100 g of tissue.

Myocardium and skeletal muscle are surrounded by 1 capillary for fibers, but the myocardial fibers are smaller. Therefore, the capillary density is greater and the diffusion distance for gases and nutrients is shorter in the myocardium (Figure 14.2). This large amount of capillaries creates a very large endothelial area of exchange and facilitates the delivery of oxygen and nutrients to the restless working cardiomyocytes.

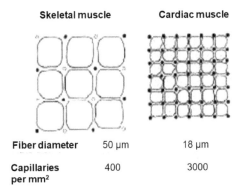

Figure 14.2 Capillary density in skeletal and cardiac muscle. Cardiac cells are smaller and diffusion distance is shorter in the heart muscle. The small black circles represent the perfused capillaries. The small white open circles in the skeletal muscles represent the non-perfused capillaries at rest. In the heart muscle, there are very few, if any, non-perfused capillaries even at rest.

It is believed that in the subject at rest, more than 80% of the coronary capillaries are patent. The remaining 20% opens in conditions of physical exercise.

14.1.2 The Venous Coronary Circulation

The coronary blood flows back to the cardiac cavities through the following venous systems: the large cardiac vein (*vena cardiaca magna*) ending in the coronary venous sinus, the posterior cardiac vein, the oblique vein, the mean cardiac vein, the small cardiac vein, the cardiac accessories veins and the veins of Thebesius.

The *large (magna) cardiac vein* receives blood from most of the left ventricle and the front of the septum and runs parallel to the descending artery for most of its path. The *posterior cardiac vein* opens into the coronary sinus venous vein where it carries blood from the posterior part of the left ventricle. The *oblique vein* runs along the back face of the left atrium to also open in the venous sinus. The *mean cardiac vein* originates near the apex of the heart on the posterior face. It receives blood from the posterior wall of the two ventricles near the interventricular sulcus, as well as from the posterior part of the septum. This vein also opens in the coronary venous sinus. The *small cardiac vein* runs along the acute edge of the heart from which it also has its origin. It opens in the venous sinus where it carries blood from the myocardium of the right ventricle.

Given the number of veins that flow into the *coronary venous sinus*, about 70% of the blood returns to the heart returns through this vein into the right atrium. The *accessory cardiac veins* run on the cardiac surface and open directly into the right atrium. The *thebesian veins*, also called minimal cardiac veins (or smallest cardiac veins), originate within the walls of the heart to open directly into the atrial and ventricular cardiac cavities. The amount of venous blood returning through these veins is scarce. *Sinusoidal veins* have been also described as thin vessels made up of endothelium and a minimal amount of subendothelial connective tissue. Therefore, a little fraction of the venous coronary blood drains into the left heart rather than into the right atrium. This is partly why the oxygen saturation of arterial blood is not quite 100%.

14.1.3 Anastomosis of the Coronary Circulation

A few anastomoses are present both between the arterial and venous coronary vessels. Arterial anastomoses can be intracoronary, that is, present between the branches of the same coronary artery, or intercoronary, that is, present between the branches of the two coronary arteries. Anastomoses ensure collateral circulation in the presence of stenosis or occlusion of the large vessels on which they rise. Anastomoses are scarce in young people (coronary artery are called end-arteries or terminal arteries) and may increase in number in the case of coronary heart diseases. Actually, in some individuals, anastomosis increases (angiogenesis) with age, in case of coronary artery disease or previous myocardial infarction (see Box 14.1: Some more details in healthy and cardiovascular diseases)

14.1.4 The Resistance of the Coronary Circulation

The coronary circulation as all other circulations presents *autoregulative or vascular resistance* and *viscous resistance*. To these two resistances important *compressive or extravascular resistances* are added during heart contraction. The compressive resistance is due to the cardiac contraction acting on the intramyocardial coronary vessels and reduces the flow in systole in the left coronary artery (see below).

14.1.4.1 The phasic coronary flow and the compressive resistance

The shape of the coronary flow curve recorded instantly by instant, *i.e.,* the phasic coronary flow varies concerning the contractile activity of the heart

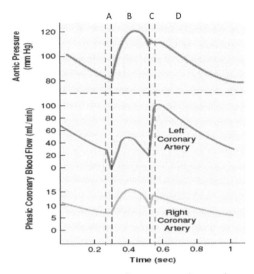

Figure 14.3 Right and left phasic coronary flows concerning aortic pressure and the cardiac cycle. A: isovolumic systole; B: Ejection period; C: isovolumic relaxation; D: Ventricular filling period. See the explanation in the text.

(Figure 14.3). The phasic coronary flow in the right and the left coronary artery has different characteristics depending on the different compressive resistance.

In the *right coronary artery*, the phasic flow curve is very similar to the shape of the aortic pressure curve (Figure 14.3). Indeed, in this territory, the flow instant by instant is determined prevalently by the pulsatile oscillations of the aortic pressure, thus resulting in higher in systole than in diastole. Because of the modest thickness of its walls, the contraction of the right ventricle has little influence on the intramyocardial coronary vessels and adds very little, negligible, compressive resistance.

The situation for the phasic flow in the *left coronary artery* is quite different (Figure 14.3). In this territory, the flow is considerably affected by the compressive effect that the thick ventricular walls exert on the intramyocardial vessels. Because of this compressive effect, the flow is lower in systole and higher in diastole when compression is absent. In Figure 14.3, we note that at the end of the diastole the flow, although decreasing with the aortic pressure, is however quite high. A moment before the systolic pressure rise occurs, a rapid fall begins which brings the flow to around zero, or even below (that is a backflow). The lowest point of the fall coincides with the beginning of the ventricular ejection revealed by the pressure rise in the pressure curve.

It is therefore understandable how the time in which the fall of coronary flow occurs coincides with the *isovolumic systole*, *i.e.* the period in which the myocardial tension increases and the pressure is at the lower level in the aorta and coronary arteries.

With the onset of the *systolic ejection*, there is an increase in pressure. This increase causes an elastic expansion of the large coronary branches located under the epicardium, vessels which, due to their location, are not subjected to the compressive action of contraction. The expansion of the subepicardial branches receives a small amount of blood, which in the phasic flow curve gives rise to a small but visible increase in the flow compared to the previous fall. During the systolic ventricular ejection, we see a parallel increase and decrease of both pressure and flow, at the end of systole, *i.e.* at the *protodiastole,* the flow returns to a value close to zero. Sometimes even this second fall of flow can be so pronounced as to result in a flow around zero, or even below (*backflow*), especially when aortic pressure is particularly low.

After protodiastole the ventricular myocardium relaxes isovolumically, ceasing to compress the vessels that supply it. During *isovolumic relaxation,* the flow rapidly rises to a value very close to the maximum value of the entire cycle. The maximum flow is reached at the moment of the *rapid filling phase* of the cardiac cycle when in the aorta the dicrotic pressure is recorded. Once the *slow filling phase* occurs, the coronary circulation is subject to aortic pressure only, and the flow gradually declines until the onset of the subsequent systole, in parallel to the aortic pressure decline.

In contrast to arterial coronary flow, venous flow is favored by myocardial contraction, so it increases in systole compared to diastole (Figure 14.4).

How myocardial contraction exerts extravascular resistance has been explained using three physical models: the water cascade model, the intramyocardial pump model, and the time-varying elastance model.

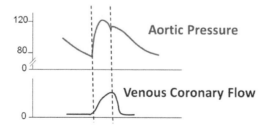

Figure 14.4 Venous coronary flow concerning aortic pressure. See the explanation in the text.

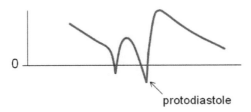

protodiastole

Figure 14.5 Left coronary flow with protodiastolic backflow. See the explanation in the text.

The *waterfall model* was discussed in Chapter 11.7. In coronary circulation, it can be applied in systole when myocardial contraction produces a sudden increase in chamber pressure. The *waterfall model* explains the fall till an eventual zero flow in systole but does not explain why sometimes, there is even a backflow. Backflow occurs especially when the aortic diastolic pressure is less than 60 mmHg in the protodiastole (Figure 14.5). It does not even explain why in systole the venous flow increases (Figure 14.5), nor why, if a large coronary branch is occluded, downstream of the occlusion, the blood pressure continues to pulsate, although progressively decreasing (Figure 14.6). Since the occlusion prevents the propagation of the pressure wave from the aorta, it is clear that the pulsation is generated by the contractile activity of the myocardium.

The waterfall model, however, explains the *zero flow pressure* which is that pressure, very low, but different from zero, for which in diastole the flow is stopped. In experimental preparations, the zero flow pressure was obtained with a prolonged stimulation of a vagal nerve able to stop the heart for a few seconds, during which the diastolic pressure is progressively reduced. It has been seen that when it reaches about 40 mmHg the flow stops. E ven the reduced myocardial tension in diastole acts as *chamber pressure* on the intramyocardial coronary vessels. The zero flow pressure is related to interstitial pressure (depending on the myocardial diastolic tension) and vasomotor tone. The lower is the vasomotor tone the lower is the zero flow pressure.

Based on the inadequacy of the waterfall model in explaining all aspects of the phasic coronary flow, Spaan et al. (1981) elaborated the *intramyocardial pump model* formulated in a capacitive version and a capacitive/resistive version.

In general, according to the *intramyocardial pump model*, in the coronary flow, a continuous component must be distinguished, due to the average pressure gradient between the arterial and venous ends. To this continuous flow a

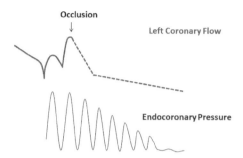

Figure 14.6   Persistence of pressure pulsations in an occluded coronary artery. After the occlusion, the flow drops to zero (dotted line). See the explanation in the text.

pulsatile component, due to cardiac contractions, must be added/subtracted. The cardiac muscle would exert a compression action capable of reducing the capacity of the intramyocardial vessels. The effect of this additional pump increases from the subepicardial to the subendocardial layers of the heart wall. The blood would be pushed towards the coronary veins and partly backward towards the aorta. The blood flow pushed backward would reduce the forward flow in systole or could even reverse its direction, thus producing a backflow. On the contrary, the flow of blood pushed forward would be added to the continuous component which would thus increase the venous flow during systole.

While the capacitive version of the model only considers the reduction of the capacity of the compressed vessels, the capacitive/resistive version also considers the increase in vessel resistance when the compression also reduces their radius.

The time-variable elastance model relates the amplitude of the systolic flow to the tension developed by the contracting myocardium rather than to the developed intracavitary pressure. Indeed, experiments performed on the isolated rabbit heart by Krams et al. (1989) pointed out that, with the same perfusion pressure and inotropic state, the systolic flow fall remains unchanged even if the increase in pressure in the ventricles is prevented by keeping their cavity empty. From these observations, it was concluded that the amplitude of the fall of the flow in systole depends on the tension developed by the myocardial walls concerning their contractility and not to the pressure developed in the ventricular cavity. The latter is shielded by rigid contracting cardiac muscle. However, these experiments were performed at low ventricular and perfusion pressures. We demonstrated that when the ventricular pressure is high enough a compressive component is added because the intracavitary pressure overcome the shielding effect due to the rigidity of the contracting muscle.

Given that during the development of tension, the cardiac walls see their elastance or rigidity to vary instant by instant, the model has been given the name of the *variable elastance model*. Indeed, during contraction, the ventricular walls can be so rigid that they do not allow the transmission of low levels of intracavitary pressure through their thickness. This is the shielding effect that according to the *variable elastance model* prevents coronary vessels from being compressed by the pressure that develops in the ventricles (when it is relatively low). It is the rigidity of the myocardium itself that occludes the vessels.

Different components considered by the three models likely contribute to determine the coronary flow pattern in the various experimental and clinical conditions. To make a long story easy, we can say that the first model considers a compressive component, the second model a variable capacity/resistance of vessels and the third one a variable rigidity (elastance) during the cardiac cycle that can affect the pattern of coronary flow, making it to oscillate and to be lower in systole and higher in diastole.

14.1.4.2 Coronary autoregulative resistance

As stated above, it is clear that coronary autoregulative resistance, although present throughout the cardiac cycle, cannot be easily calculated. In diastole, when the compressive influences are minimal an index of autoregulative resistance can be calculated by the ratio between mean diastolic pressure and mean diastolic flow (see Chapter 8.1).

The compressive strengths added by the systole do not concern the coronary vasomotor tone, which is responsible for autoregulative or vascular resistance. However, a cyclic variation of vessel capacity influences also the diastolic mean coronary flow. Therefore, as there is a capacitive component in the diastolic flow, the proposed ratio provides only an index of resistance and not the real resistance.

The mean diastolic pressure is the average value of all the infinite values that the pressure assumes during the aortic diastole, which in the relative pressure curve is included between the instant in which the notch or incisure appears until the beginning of the subsequent ejection. *The mean coronary flow* can be considered between the value recorded at the end of isovolumic relaxation and the beginning of the subsequent systolic fall. A ratio between the two values can give a good index of autoregulative vascular resistance (Figure 14.7). To obtain a better index of autoregulative

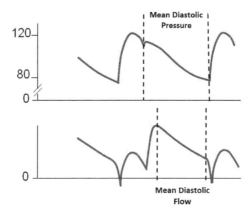

Figure 14.7 Coronary vascular resistance index as a ratio between mean diastolic aortic pressure and mean diastolic coronary flow. See the explanation in the text.

resistance the measurement must be limited to the slow filling phase. Indeed, autoregulative resistance can vary greatly, especially with a variation of heart metabolism, and to have a good index allows to better study the various regulatory mechanisms.

Resistance equals the pressure gradient divided by flow. For simplicity, we consider the ratio between mean diastolic pressures and mean diastolic flow during the slow filling phase. This ratio can give a piece of information on the variation of vascular resistance, whatever is the level of diastolic pressure or flow. For example, coronary vasodilation can be present even if the flow is unchanged or slightly decreased if the pressure is decreased. Similarly, vasoconstriction occurred if the pressure is increased and the flow remained constant, or if, having both increased, the diastolic pressure is increased more than the flow.

Sometimes, even in some physiology book, it is reported that the increase in heart rate would lead to a reduction in mean coronary flow during the entire cardiac cycle. The wrong reasoning at the basis of this opinion is the observation that during an increase in heart rate increases the number of systolic flow falls in the unit of time and the diastole duration will be shorter. The opinion is valid only for those pathological situations in which the coronary circulation has lost the ability to vasodilate (*i.e.* the vasodilator coronary reserve has been already exploited; see below). *Under normal conditions, however, an increase in heart rate leads to an increase in mean coronary flow. This increase is due to an increase in myocardial metabolism, which causes coronary vasodilation and a consequent increase in diastolic flow to overwhelm the effect of systolic reductions (more b.p.m.), thus increasing the mean coronary flow.*

14.1.4.3 Coronary viscous resistance

We call viscous resistance those vascular resistances that can be calculated in diastole after the coronary vascular bed has been dilated to the maximum. The difference between the value of the autoregulative resistances in basal and viscous conditions determines the *vasodilatory reserve*.

The autoregulation of coronary vascular resistance, or more generally the autoregulation of the coronary circulation, are those mechanisms that tend to adapt the perfusion of the myocardium to its work requirements. To clarify the meaning of the autoregulation it is necessary to keep in mind some data.

In a resting person, the coronary flow is 70–90 mL/100g/min. Under stress and in the *fight or flight reaction* it can increase by over 4–5 times. If we consider that the weight of the human heart can be 260-300 g, the total flow is between 180 and 270 mL/min at rest and almost 1250 mL/min under stress or heavy exercise. The consumption of oxygen, per 100 g of the myocardium, is 7-9 mL/100g/min in a subject at rest to rise to 35–45 mL/100g/min under stress or heavy exercise.

While the concentration of oxygen in arterial blood is around 19%, in coronary venous blood it is only 6%, so the arterial-venous difference is around 13%. Bearing in mind the hemoglobin dissociation curve, it can be seen that at an O_2 concentration of 19% corresponds a partial pressure of O_2 (pO_2) of about 100 mmHg and at a concentration of 6% corresponds a pO_2 of 18 mmHg.

The oxygen content of 6% is much lower than that of 14% present in the mixed venous blood of a resting person. From such a low value it can be inferred that not only basal flow is high but also basal oxygen extraction is very high. Moreover, we can understand that when the myocardium needs a greater supply of oxygen, it has the scarce possibility of obtaining it through greater O_2 extraction from the blood, already very high in resting conditions, but must rely on an increase in blood flow, which can be assured by vasodilation, *i.e.,* the vascular resistance can reduce from the autoregulative level to the vicious level.

14.1.5 Metabolic Mechanism

The main mechanism of autoregulation is metabolic. The metabolic mechanism is evoked by the workload that the heart can meet in the unit of time and therefore by the demand for oxygen. Mean coronary blood flow can increase in almost linear relation with myocardial metabolism and oxygen consumption. In heavy exercise, oxygen extraction can increase modestly so that *the extra oxygen required is supplied chiefly by the increase in blood flow*. An increase in workload in a minute can be due to multiple factors, the most important are the increase in heart rate and the increase in contractility. Also, an increase in cardiac output and systolic aortic pressure may increase the cardiac workload.

When myocardial metabolism increases, and therefore the oxygen demand increases, it seems that the extraction of this gas from the blood initially increases a little without an increase in blood flow. However, when the initial increase in the extraction begins to slightly reduce the concentration of oxygen in the venous blood of the coronary sinus (for example from 6 to 5.5%), vasodilation begins and flow increases and venous oxygen concentration returns to be identical to the rest conditions. As said, it is possible that, under certain circumstances of intense exercise or excitement, there is both vasodilation and an increase in extraction, so that the concentration of oxygen in the venous sinus can drop well below 5.5%.

Myocardial metabolism generates vasodilator agents. The nucleoside *adenosine* has been proposed as one of the most important vasodilator metabolites deriving from the myocardium. Any imbalance between myocardial intake and oxygen supply would cause cardiomyocyte release of adenosine. The classical scheme for adenosine linking metabolism to flow *via* a negative feedback pathway is likely incorrect. Feed-forward dilator mechanisms – likely involving CO_2 and H_2O_2 – are involved in coupling the flow to metabolism. However, it must be said that neither dilator can explain the full connection between metabolism and flow in physiological conditions. For sure, under physiological conditions adaptation of coronary flow to exercise involves also feed-forward neuronal and endothelium-dependent control.

Adenosine is formed by the action of the 5-nucleotidase enzyme on the intracellular AMP derived from the ATP catabolism. Adenosine has several receptors (A1, A2$_A$, A2$_B$, and A3). All four subtypes are known to regulate coronary flow. In general, A2$_A$ is the predominant receptor subtype responsible for coronary blood flow regulation, which dilates coronary arteries in both an endothelial-dependent and -independent manner.

By differential coupling to either Gs (A2$_A$ and A2$_B$) or Gi proteins (A1 and A3), along with variable tissue distribution of AR subtypes, adenosine elicits both relaxation (A2$_A$- and A2$_B$-mediated) and constriction (A1- and A3-mediated) of vessels. A3 adenosine receptors have been localized on endothelial cells, leading to contraction of VSM through cyclooxygenase I (COX I).

The predominant and classical vasodilator effect of adenosine is obtained by acting on $A2_A$ receptors located on VSMs. Via $A2_A$ receptors adenosine activates an adenylyl cyclase which, through the production of cyclic AMP and the activation of a PKA, stimulates the sarcolemmal pump for Ca^{2+} and SERCA2, reducing the intracellular content of Ca^{2+} and thus determining relaxation. It is also possible that the formation of cyclic GMP following adenosine binding with receptors participates in the intracellular depletion of Ca^{2+}.

After performing its vasodilating action, adenosine is inactivated to inosine by an adenosine deaminase present in red blood cells. The inosine is subsequently transformed into hypoxanthine by a nucleoside phosphorylase present both in red blood cells and in capillary endothelial cells.

As said, recently the mechanism of adenosine is considered important only in the regulation of the coronary circulation in conditions of metabolic demands of particular intensity and hypoxia, while the coronary vasodilation of the normal life would be due to H_2O_2. This is formed by the superoxide anion by the action of superoxide dismutase. Since H_2O_2 is a reactive oxygen species, it s vasoactivity is likely due to its actions as an oxidant, presumably on thiol groups which are sensitive to the actions of oxidants. The vasodilator action of H_2O_2 would be mediated by its ability to hyperpolarize smooth muscle fiber cells, presumably by opening potassium Kv channels, and/or large conductance MaxiK channels.

Among the metabolic mechanisms that mediate coronary vasodilation, there is also the fall in pO_2, the increase in carbon dioxide production, and the decrease in tissue pH. Indeed, under several different conditions, e.g. anemia, vessel stenosis, or high altitude exposure, cardiac oxygenation may become critical, especially during physical exercise. Under such conditions, the *fall of coronary pO_2* may directly result in the opening of oxygen-sensitive potassium or closure of calcium channels. Furthermore, the fall of pO_2 results in the production of vasoactive metabolites, *e.g.* adenosine, nitric oxide or prostaglandins, and proton accumulation.

Of note, in cardiac tissue, adenosine binds to A1 receptors, which are coupled to Gi-proteins. Activation of this pathway decreases cAMP, which inhibits L-type Ca^{2+} channels and therefore Ca^{2+} entry into the cell. Moreover, adenosine opens K^+ channels, which hyperpolarizes the cell. On SAN, adenosine inhibits the pacemaker current (If), which decreases the slope of diastolic depolarization of the pacemaker action potential thereby decreasing heart rate (negative chronotropic effect). Inhibition of L-type Ca^{2+} channels also slows conduction velocity (negative dromotropy). Finally, adenosine inhibits the release of norepinephrine by acting on presynaptic receptors located on sympathetic nerve terminals. From a finalistic point of view, the negative chronotropic effect has the meaning of slowing down the heart activity when the heart is jeopardized by hypoxia/ischemia. Of note, intracoronary infusion of adenosine decreases heart rate and reduces conduction velocity, especially at the AVN, which can produce AV blocks; however, when adenosine is infused into veins, heart rate increases because of baroreceptor reflexes caused by systemic vasodilation and hypotension.

$A2_B$ adenosine receptors have usually a very low sensitivity to adenosine. Their sensitivity is increased after transitory ischemia by a phosphorylation mechanism and may play an important role in the cardioprotection induced by adenosine (see Chapter 17). Also, the upregulation of $A2_B$ receptor gene expression has been observed in chronic ischemic mouse hearts.

14.1.6 Nervous Mechanisms

The nervous mechanisms perform their direct action, but they act prevalently by the intermediation of the metabolic mechanism.

Action of the sympathetic system

If sympathetic fibers are stimulated in an experimental model, a transient (a few seconds) small reductions in coronary flow is observed, then followed by a conspicuous increase in flow. The ensuing increase in HR and contractility raises the cardiac work and induces metabolic vasodilation which can last beyond the stimulation period.

The adrenergic receptors are of type α and β (see Chapter 12.4). Also on the coronary vessels, the receptors are divided into postsynaptic $\alpha1$ and $\alpha2$. These postsynaptic receptors are further distinguished in intrasynaptic $\alpha1$ and extrasynaptic $\alpha2$. These receptors mediate vasoconstriction. On the other hand, $\alpha2$-like receptors, located on endothelial cells, could induce NO production and thus vasodilation. Also, presynaptic $\alpha2$-receptors on nerve terminals have been described. These receptors may limit noradrenaline release from the corresponding nerve terminal.

Although presynaptic receptors reduce the release of noradrenaline and $\alpha2$-like ones are responsible for the release of NO, the overall stimulation effect of α receptors is a vasoconstrictor and is responsible for the reduction in the flow that occurs at the beginning of sympathetic stimulation.

Beta-receptors are divided into subtypes $\beta1$, $\beta2$, and $\beta3$. On the coronary circulation, the most represented receptors are $\beta2$. Nevertheless, small quantities of $\beta1$-receptors have also been found at the level of the membrane of smooth muscle cells in an extrasynaptic position. The binding of catecholamines with these receptors causes vasodilation. For the cascade of signals triggered by these receptors see Chapter 12.4. Although β-receptors are more sensitive to adrenaline than to noradrenaline, intrasynaptic $\beta1$-receptors can be activated by the nerve-induced release of noradrenaline. Because of their extrasynaptic position, the $\beta2$-receptors are subsequently activated by the diffusing noradrenaline.

It is necessary to reiterate that the initial and transient response to cardiac sympathetic stimulation is vasoconstriction, while vasodilation takes place after a few seconds and remains throughout the time of stimulation and, after this is completed, for the time necessary for noradrenaline removal.

An extremely important concept is that vasodilation must be attributed mainly to an indirect action of the sympathetic system mediated by the increase in cardiac metabolism. The vasodilator action of the cardiac sympathetic is due to the increase in HR and contractility that norepinephrine causes by binding to abundant myocardial $\beta1$-receptors both on SAN and on contracting myocardium. The increases in HR and contractility are indeed the most important factors in determining an increase in the demand for oxygen by the heart.

In summary, we can, therefore, reiterate the concept that the vasodilation induced by the sympathetic is above all of a metabolic nature. The metabolic "overriding" of alpha-mediated vasoconstriction ensures that any activation of the sympathetic system does not compromise blood flow to this vital organ in healthy subjects.

In patients with ischemic coronary disease, metabolic vasodilation may be compromised, as a result, stress and/or exercise may provoke angina and other signs of cardiac ischemia, including ECG alterations, such as ST-segment shifts. It is likely that in these pathological conditions α-mediated vasoconstriction prevails and may be responsible for coronary spasms. Moreover, endothelial dysfunction can concur with this picture. To reveal latent ischemia, the *cold pressor test* or more frequently the *exercise test* is used during ECG monitoring. In the case of latent coronary ischemia cold pressure test, obtained by immersing a hand in crushed ice, can trigger reflex sympathetic vasoconstriction which can cause angina. Also, anger and stress can induce a prolonged sympathetically mediated vasoconstriction in patients with diseased coronary vessels. In these patients, angina and ST-segment alterations occur.

Action of the vagus

The heart receives parasympathetic vagal fibers, which are distributed mainly to the atria and a small extent to the coronary circulation. Vagus stimulation in a regularly beating heart indirectly causes coronary vasoconstriction due to decreased metabolism caused by HR reduction. The poor distribution of the vagal fibers to the ventricles greatly attenuates the importance of possible reduced contractility.

The direct action of the parasympathetic on the coronary circulation is instead of a vasodilator type. This is an action that can be highlighted only if the heart does not undergo a reduction in the HR, and therefore of the metabolism, following the stimulation of the vagus. Experimentally, to prevent vagal stimulation from reducing the HR, ventricular fibrillation (under extracorporeal circulation) or artificial pacing can be induced. In these experimental conditions, the vagus produces coronary vasodilation not exceeding a 10% increase in flow. The vasodilatation from ACh is likely mediated by the release of NO and/or hyperpolarizing factors by the endothelium. Although there is a controversy on whether neural ACh can reach the endothelium, nevertheless, the ACh infused into coronary circulation (ACh test) is used to verify whether or not the endothelium is dysfunctional: ACh test induces vasoconstriction instead of vasodilation when the endothelium is dysfunctional.

Reflex nerve regulation (see also Chapters 8 and 12)

In the same way as the general circulation also the coronary circulation is subjected to the action of reflex mechanisms, such as the reflexes starting from baroreceptors and chemoreceptors.

Action of baroreceptors: if the distal tract of a common carotid artery is cannulated, a local increase in pressure can be produced at the carotid sinus. This procedure allows studying the isolated action of the baroreceptor reflex on the coronary circle. Since the stimulation of the baroreceptors reduces the activity of the sympathetic and increases that of the vagus, the heart undergoes a reduction in HR and contractility, conditions that reduce its metabolism resulting in coronary vasoconstriction.

If the increase in pressure is not local but systemic, the increase in ventricular afterload leads to a certain increase in cardiac work. In these conditions, the possible vasoconstriction due to the decrease in HR and contractility is counteracted by the metabolic vasodilator stimulus triggered by the increase in afterload. Since coronary perfusion pressure (the pushing pressure) is increased, the flow increases instead to be reduced.

Chemoreceptor action: the chemoreceptors of the aortic and carotid bodies or glomi are activated by the decrease in pO_2 and pH, as well as by the increase in pCO_2 in arterial blood. Since it evokes an increase in heart rate and total peripheral resistance, their stimulation induces metabolic coronary vasodilation as normally induced by sympathetic stimulation. Since chemoreceptors are activated also when pressure falls dangerously, the metabolic vasodilation does not imperil the coronary blood flow, regardless of the diffuse sympathetic tone which induces systemic vasoconstriction in an attempt to reverse the fall in pressure.

Action of reflexes of cardiac origin: cardiac reflexes capable of influencing coronary vascular resistance can respond to stimuli of a chemical or physical nature. A chemical stimulus may be due to *veratridine*. Injected into a coronary artery of an experimental preparation, this alkaloid causes vasodilation that disappears with the cutting of the vague or block of muscarinic receptors by atropine. A mechanical stimulus instead evokes the Bainbridge reflex characterized by an increase in heart rate in response to the distension of the atria (see Chapter 12.7). In this case, the vasodilation is mainly metabolic.

14.1.7 Humoral Mechanisms

Catecholamines of adrenal origin are among the humoral control factors of the coronary circulation. Adrenaline is prevalent compared to noradrenaline amount. Similarly, to coronary sympathetic nerves, the overall action of the adrenal catecholamines on the coronary circulation is of a vasodilator type.

Other endogenous substances active on coronary vascular resistance are arachidonic acid derivatives, angiotensin II, bradykinin and histamine. These humoral mechanisms very often intervene when the factors responsible for their formation are activated in circumstances that do not concern the regulation of the circulatory district, such as the widespread inflammatory response.

The humoral derivatives of arachidonic acid active on the coronary circulation are: prostaglandins, thromboxanes and leukotrienes. Their production is due to the action of cyclooxygenases on arachidonic acid. *Prostaglandins H2 and F2a* should have a vasoconstrictor action, while prostaglandins E1 and

E2 and prostaglandin I2, or prostacyclin, would be responsible for vasodilation. Despite the direct action on the vessels, the role of prostaglandins in the physiological regulation of coronary circulation is rather uncertain, as their production depends on the activation modalities of cyclooxygenase I (COX I) and, especially, of cyclooxygenase II (COX II). *Thromboxanes* are synthesized by COX especially in platelets. They have a vasoconstrictor action and favor platelet aggregation. For this reason, the inhibitors of their synthesis are used in some cardiovascular diseases. *Leukotrienes*, in addition to taking part in inflammatory responses, possess vasoconstrictor activity also on the coronary circulation.

Angiotensin II has a constricting effect on vessels, including coronary vessels. It also seems capable of determining the release of noradrenaline from the sympathetic terminals, so that its physiological action on the coronary circulation appears to be rather complex. Moreover, ACE is produced in the vascular endothelium of many tissues; thus, angiotensin II can be synthesized at a variety of sites, including the heart and the brain. Also, alternative enzymatic pathways may contribute to angiotensin II production. Human heart chymase may be one such enzyme but its physiological significance remains uncertain. Tissue angiotensin II can be responsible for maladaptive remodeling in the post-ischemic heart.

Bradykinin is responsible for coronary vasodilation. It is an effect mediated by both NO and above all by one of the epoxy-trienoic-acids (EETs) released by the endothelium after bradykinin has bound to endothelial B2 receptors and has stimulated cytochrome P-450, which acts on arachidonic acid.

The fact that the inhibition of cytochrome P-450 almost makes the bradykinin vasodilation disappear suggests that this pathway is prevalent compared to that of NO. Because bradykinin is released due to the decrease in tissue pH in hypoxia, it may represent a compensatory mechanism in the case of myocardial ischemia.

Histamine is also responsible for coronary vasodilation. This occurs directly due to the action of the substance on the vascular H1 receptors and, perhaps, even indirectly due to the action on myocardial H2 receptors that increase cardiac contractility.

14.1.8 Endothelial Mechanisms

As we have already seen in Chapter 13.2, the endothelial factors that regulate coronary vascular resistance are the endothelial-derived relaxing factor

(EDRF), the endothelial-derived hyperpolarizing factors (EDHFs), and the endothelial-derived contractile factors (EDCFs).

The EDRF is *nitric oxide* (NO). This performs its vasodilator action also on the coronary circulation and is responsible for the *flow-induced vasodilation* especially on the epicardial artery. NO production requires the activation of the e-NOS by chemical and mechanical factors (Chapter 13.2). In particular, in the case of coronary endothelium, acetylcholine, shear stress and pulse pressure are stimuli capable of inducing vasodilation in many physiological conditions. Bradykinin is formed and intervenes in the event of ischemia.

The importance of basal NO release in determining basal vasomotor tone is a matter of controversy. Its importance rang from null to 30–40% of basal tone lowering in different experimental studies. There is, however, no doubt on the role of stimulated (chemically or mechanically) NO release in regulating coronary vascular resistance.

The study of the coronary vasodilator role of NO was performed, *inter alia*, by the administration of acetylcholine (*ACh test*) and the induction of *reactive hyperemia* after transient occlusion of a large coronary branch. It has been seen that the inhibition of NO synthase significantly shortens the duration of vasodilation and suggests the need to mitigate the traditional distinction between endothelium-dependent and endothelium-independent vasodilation. While vasodilatation such as acetylcholine is certainly endothelium-dependent, vasodilation due to reactive hyperemia, although initiated by a metabolic mechanism, is also prolonged over time by the NO produced by the increased shear-stress which determines the so-called *flow-induced vasodilation* (see also below).

Among the hyperpolarizing endothelial factors, we include *prostacyclin, superoxide anion*, EETs (non-prostanoid derivatives of arachidonic acid) and *hydrogen peroxide* to which is attributed a significant coronary vasodilator action as a compensatory factor concerning the absence of NO and/or adenosine.

The redundancy of the above-described factors and vasodilator mechanisms may also explain why in experimental conditions the blockage of one or more mechanisms does not prevent the increase in coronary flow during exercise. We can certainly say that metabolic hyperemia is a certainty, but unfortunately, its fine mechanisms are still unknown.

Endothelins and prostaglandins H2 and F2a are classified among the *endothelial contractile factors*, which we have already considered among the humoral factors (Chapter 12). Acting on specific receptors, the endothelins cause transient vasodilation followed by much more lasting vasoconstriction. The increase in their concentration may also be due to turbulent shear stress. Low, in basal conditions, the concentration of endothelins increases in cases of chronic endothelial dysfunction, essential hypertension, pulmonary hypertension, heart failure, and myocardial infarction. Whether the increase in endothelin is a cause or an effect of these diseases is a controversial issue.

14.1.9 Myogenic Mechanism

The coronary circulation is also subject to myogenic autoregulation (see Chapter 11.6). This means that in a range of variations in mean transmural pressure between 70 and 120 mmHg, the flow remains almost constant. The vascular smooth muscle contracts (vasoconstriction) in the presence of increased pressure and releases (vasodilation) following a decrease in pressure (Figure 11.12).

Since the variations in transmural pressure are generally related to changes in the aortic pressure that represents the ventricular afterload, myogenic responses may be more or less masked by changes in cardiac work and metabolism. It follows that, when an increase in transmural pressure causes myogenic vasoconstriction, this can be partially attenuated by metabolic vasodilation.

If we look at Figure 11.12, we see that, outside the lower (70 mmHg) and upper (120 mmHg) autoregulation limits, the increase in flow is closely correlated with the increase in pressure. This correlation, however, is not strictly linear. Indeed, below 70 mmHg, the pressure-flow curve has a concavity towards the y-axis. This concavity tells us that the coronary vessels are elastically distensible so that they are made to expand by the increase in pressure.

14.1.10 Coronary Flow at Rest and Under Stress

We have already seen that in a normal heart physical exercise and states of excitement are accompanied by an increase in coronary flow up to 4–5 times. Figure 14.8 shows that during a physical exercise there is an increase in heart rate, stroke volume (SV), and aortic pressure. In Figure 14.8, we can see how, despite the increase in the number of systolic reductions, the overall coronary flow is enhanced by an increase in diastolic flow. Note the greater amplitude of the fall from the maximum diastolic value at the minimum systolic value. This can be seen as an index of increased contractility, which determines the fall in flow during the isovolumic systole. Also, an increase in

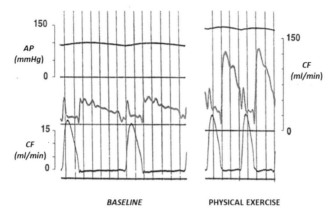

Figure 14.8 Increase in heart rate, systolic flow and mean aortic pressure in a dog during exercise with a consequent increase in mean coronary blood flow even though the number of systolic reductions increases in one minute. AP: arterial pressure (average); AF: aortic flow (systolic flow); CF: left coronary flow. Note that during exercise the area under the curve of coronary flow (an index of flow per beat) is greater and that the number of beats in a minute is increased. See the explanation in the text.

systolic pressure due to an increase in SV may be an indicator of increased contractility.

Since the diastolic flow increased much more than the pressure, it is clear that there has been a considerable reduction in coronary vascular resistance. This reduction was induced by the increase in metabolism brought about by the increase in HR, contractility, and the greater work due to the increase in systolic pressure. The increase in pressure may have stimulated myogenic vasoconstriction, which was completely overwhelmed by metabolic vasodilation. Moreover, if the pressure increases also in diastole, the flow increases even more. However very often during diastole the pulse pressure increase, because of the systolic pressure increases, and the diastolic pressure decreases. Despite this decrease, the flow in diastole increases because the pulse pressure, together with shear stress, triggers NO release and enhances the concomitant metabolic vasodilation. The concomitance of multiple mechanisms may explain the conspicuous vasodilation and increase in flow. Even the increase in the nervous and adrenal adrenergic discharge contributes to the vasodilation due to 1) *cardiac β1-receptor activation* which mediates metabolic vasodilation by increasing HR, contractility, and workload, and 2) *vascular β2-receptor activation*, especially by adrenaline (see also below for other mechanisms).

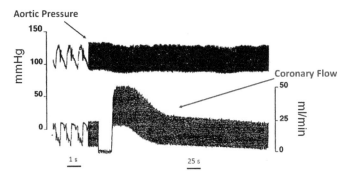

Figure 14.9   Coronary reactive hyperemia. The recording started at 25 mm/s, is subsequently slowed to 1 mm/s. See the explanation in the text.

14.1.11 Reactive Hyperemia

The study of reactive hyperemia is a way to analyze the contribution of several vasodilator mechanisms, namely metabolic, endothelial, and myogenic. In an experimental model, reactive hyperemia can be obtained occluding transiently (a few seconds) a large subepicardial arterial branch. When the occlusion is removed there is an increase in the blood flow which in a few seconds rises to a value much higher than the basal value. Once the maximum level is reached, the flow progressively descends towards the basal level (Figure 14.9).

Amplitude and duration of reactive hyperemia depend on the duration of the occlusion. For an occlusion of 20 sec, the maximum amplitude of the hyperemic flow could be 3–4 times higher than the basal/control flow, while the return to the control value can take a few minutes. The maximum amplitude and duration of the hyperemia increase with the increase of occlusion duration until a certain point. Too long occlusions (more than 10 min) can alter the myocardium and blood vessels and the characteristics of hyperemia.

Coronary reactive hyperemia is primarily due to a *metabolic mechanism* which in coronary circulation is the dominant control process because of the high oxygen and metabolic needs. In this case, the metabolic mechanism depends on the lack of oxygen supply to the myocardium during occlusion and is likely expressed through the increased production/accumulation of adenosine and protons (low pH).

The *myogenic mechanism* is the consequence of the pressure drop that occurs distally to the coronary occlusion during flow interruption. This pressure drop causes a relaxation of smooth muscle cells, which contributes to increasing in flow when the occlusion is removed. This consensual dual mechanism of vasodilation (metabolic/myogenic) causes the flow to increase considerably at the time of artery re-opening, causing a sharp increase in *shear stress*.

The *endothelial mechanism* is due to the *shear-stress*-induced release of NO and hyperpolarizing factors (namely, hydrogen peroxide, H_2O_2), which seem to play a predominant role. Since they are produced by *shear stress*, NO and H_2O_2 acts after the increase in flow have already begun, thus being responsible mainly for the duration of hyperemia. The vaso-relaxant action of NO is attributed to the generation of cGMP. The hyperpolarizing action of H_2O_2 is attributed to its ability to open the K^+ channels on smooth muscle. On the other hand, K^+ can be released by the endothelium and can act as a hyperpolarizing factor, *via* the opening of K^+ channels and/or upregulating Na^+/K^+ pump on smooth muscle cells.

Metabolic hyperemia and endothelial-dependent vasodilatation are strictly linked as demonstrated by several studies, including studies in which coronary microvessels were directly visualized in beating hearts. These studies have shown that metabolic vasodilation occurs principally in arterioles <100 μm in diameter. For example, adenosine and *dipyridamole* (a drug that blocks adenosine catabolism) produce vasodilator effects mainly in these small vessels. However, up to 40% of total coronary autoregulative resistance concerns vessels 100–400 μm in diameter, which are not under direct metabolic control. When the direct metabolic vasodilation of the small arterioles causes an increase in blood flow, the resultant increase in shear-stress will induce NO production and vasodilation of the bigger vessels. In this way, endothelial-dependent vasodilation also contributes to metabolic vasodilation. Therefore, physiologically, the vasomotor tone is directly, but also indirectly coupled to myocardial metabolic needs *via* endothelium-dependent flow-mediated NO synthesis.

In pathological conditions as well as in hyperlipidemia, shear-stress-mediated vasodilation may be impaired or lost. In this situation, adenosine or dipyridamole would be expected to produce a subnormal increase in blood flow, because these substances dilate only the small arterioles. The absence of flow-mediated vasodilation of the 100–400 μm vessels could impair the further lowering of vascular resistance because substantial resistance resides at the level of these vessels. The different sites of action of adenosine and NO are in line with the observation of Pagliaro *et al* (1999) of a synergistic vasodilator effect of adenosine and enhanced perfusion pulsatility, which induces NO release.

14.1.12 The Coronary Reserve

The coronary reserve can be studied as a reserve of flow or as a reserve of vasodilation. The *flow reserve* refers to the *maximum increase in flow,* at a given perfusion pressure, in comparison to basal flow. The *vasodilation reserve*, on the other hand, means the *maximum degree of fall of the autoregulative vascular resistance*, at a given perfusion pressure, in comparison to the basal coronary resistance.

At first glance, the distinction between the two types of reserves may seem completely obvious and meaningless, given that, for each perfusion pressure value, an increase in flow can only depend on corresponding vasodilation. However, this is true only if the coronary circulation has no stenosis along the large arterial branch. *If critical stenosis is present, the vasodilation reserve may be greater than the flow reserve.* Critical stenosis of a vessel means a narrowing that reduces the vessel diameter by at least 75%: e only stenosis of this magnitude can reduce blood flow at rest.

When the reduction in the diameter of a large branch exceeds 75%, the reduced initial supply of oxygen to the myocardium causes vasodilation of the resistance vessels placed distally to the stenosis. If the 75% limit is exceeded a little, distal vasodilation is sufficient to bring the flow back to the control value. If, on the other hand, the stenosis far exceeds the critical value and the flow is considerably reduced, distal vasodilation will not be able to compensate for the effect of the stenosis. It is clear that in this case, the vasodilation reserve does not correspond to an equally large flow reserve. Also, the Bernoulli effect, which may occur when distal vasodilation increases the flow velocity and reduces the lateral pressure at the site of stenosis, may favor an elastic recoil of a compliant (non-rigid) stenosis, which aggravates its restriction, thus limiting the flow increase. In other words, the vasodilation of the distal resistance vessels will cause an increase in blood flow and blood velocity (kinetic energy) within the stenotic segment. A proportionate decrease in lateral pressure, acting to distend the stenosis, occurs. Therefore, the arterial wall elastic recoil, as well as the vasomotor tone, will tend to collapse the stenosis and worsen the degree of narrowing. With severe stenosis, these effects could cause a paradoxical decrease in coronary blood flow in response to adenosine or dipyridamole, which dilates mainly the small ($<100 \mu$m) resistance vessels.

In Figure 14.10, a flow reserve is schematically shown. For each perfusion pressure value, it is given by *the difference between the autoregulated basal flow curve and the maximum flow recorded in the completely vasodilated coronary circulation*. At low perfusion pressure there is no reserve, while

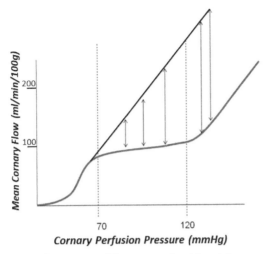

Figure 14.10 Coronary flow reserve. The autoregulated basal flow curve (in red) and the maximum flow recorded in the completely vasodilated coronary circulation (in black). The dotted lines delimit the autoregulation range. At lower pressure the two curves are identical; therefore, only the red curve can be seen. The arrows represent the extension of flow reserve at various perfusion pressure. See the explanation in the text.

in the range of auto-regulation the "amount" of reserve depends on pressure. Then exceeding the upper limit of the auto-regulation range, when the pressure/flow relationship becomes linear, the reserve remains constant.

In the absence of stenosis, the coronary flow reserve may decrease 1) due to a decrease in vasodilation reserve and 2) due to an increase in the basal flow. The first case occurs in the presence of small vessel disease, while the second case is typical of cardiac hypertrophy.

Both the flow reserve and the vasodilation reserve can be absolute or relative. A reserve is called absolute when it is referred in absolute terms to increase in flow or to decrease in resistance, while it is called relative when it is expressed as a percentage respectively of the maximum increase in flow or maximum fall of the resistance.

It must be said that the picture may be complicated by the presence of anastomosis which alters differently the distal perfusion and microvascular response. Moreover, mechanical interference during systole may be responsible for flow reduction when the heart rate increase in a completely or almost completely dilated distal microvasculature. Finally, the microvascular disease may alter the physiological responses and stimuli that usually induce vasodilation may induce vasoconstriction.

Box 14.1 Additional Features of Coronary Circulation in Healthy and Cardiovascular Diseases

Coronary flow and exercise

Exercise is a primary stimulus for increased myocardial oxygen demand. The ~6-fold increase in oxygen demand of the left ventricle during heavy exercise is met principally by augmenting coronary blood flow (~5-fold), as hemoglobin concentration and oxygen extraction (which is already ~70% at rest) increase only modestly in most species. As a result, coronary blood flow is tightly coupled to myocardial oxygen consumption over a wide range of physical activity. This tight coupling has been proposed to depend on periarteriolar oxygen tension, signals released from cardiomyocytes and the endothelium as well as neurohumoral influences, but the contribution of each of these regulatory pathways, and their interactions, to exercise hyperemia in the heart remain incompletely understood. In humans, nitric oxide, adenosine, and K_{ATP} channels each appear to contribute to resting coronary resistance vessel tone, and some other conditions, but evidence for a critical contribution to exercise hyperemia is lacking (*Journal of Molecular and Cellular Cardiology 52 (2012) 802–813*).

Coronary vascular α-adrenoreceptors

Activation of coronary *vascular α-adrenoreceptors* results in vasoconstriction which competes with metabolic vasodilation during sympathetic activation. Usually, metabolic vasodilation prevails. Epicardial conduit vessel constriction is largely mediated by α1-adrenoceptors, which might have a beneficial effect on redirecting transmural blood flow distribution towards endocardial tissue. The constriction of the resistive microcirculation largely mediated by α2-adrenoceptors, but also by α1-adrenoceptors. There is no firm evidence that α-adrenergic coronary vasoconstriction exerts a beneficial effect on transmural blood flow distribution. A-blockade in anesthetized and conscious dogs improves blood flow to all transmural layers, during both normoperfusion and hypoperfusion. Also, in patients with coronary artery disease, blockade of α1- and α2-adrenoceptors improves coronary blood flow, myocardial function, and metabolism (*Journal of Molecular and Cellular Cardiology 52 (2012) 832–839*).

Left ventricular hypertrophy
Two distinct types of left ventricular hypertrophy (LVH) have been described: the so-called "physiologic" hypertrophy, which is normally found in professional athletes, and "pathologic" LVH which is found in patients with inherited heart muscle disease such as hypertrophic cardiomyopathy (HCM) or patients with cardiac and systemic diseases characterized by pressure or volume overload. Patients with pathologic LVH have often symptoms and signs suggestive of myocardial ischemia despite normal coronary angiograms. Under these circumstances, ischemia is likely due to coronary microvascular dysfunction (CMD) (*Journal of Molecular and Cellular Cardiology 52 (2012) 857–864*).

Anastomosis
"Chronic" obstruction of coronary arteries may lead to an arteriogenic response. Pre-existent collateral networks enlarge, forming large conductance arteries with the capability to "partially" compensate for the loss of perfusion due to the occlusion. Interestingly, significant differences exist between patients regarding the capacity to develop such collateral circulation. This heterogeneity in arteriogenic response is also found between and even within animal species. It strongly suggests that next to environmental factors, innate genetic factors play a key role (*Journal of Molecular and Cellular Cardiology 52 (2011) 897–904*). The more or less rapid formation of collateral circles can make the difference between life and death.

14.2 Cerebral Circulation

The blood supply to the brain is ensured by the internal carotid arteries and vertebral arteries which form the *circle of Willis* (Figure 14.11).

The cerebral circle is placed in a rigid box that limits the possibilities of distension. It follows that, while due to the *bellows/windkessel effect* in the other districts at a given moment (but not in a minute!) the arterial flow can exceed the venous flow, in the cerebral circle arterial and venous flow is almost identical in each instant.

Although the brain is only 2% of the body mass, it receives about 15% of the cardiac output at rest. The mean cerebral flow is 52–55 mL/min per 100 g of tissue and does not change during physical exercise. The majority of blood goes to the grey matter. It is a flow that is very well regulated if

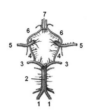

Figure 14.11   Circle of Willis. 1: vertebral arteries; 2: basilar artery; 3: posterior cerebral arteries; 4: posterior communicating arteries; 5: internal carotid arteries; 6: anterior cerebral arteries; 7: anterior communicating artery.

we consider that, while on the one hand, it remains constant in the presence of large variations in arterial pressure, on the other hand, it may undergo small local variations in response to the increased metabolism of certain areas. It is known that *positron emission tomography* reveals a greater supply of blood to the visual cortex (extrastriate area V4) during the color vision. To underline the importance of autoregulation, it should be borne in mind that a few seconds of interruption of cerebral blood flow leads to the loss of consciousness and fainting.

14.2.1 Autoregulation and Metabolic Regulation

Cerebral blood flow does not undergo appreciable variations in the presence of mean arterial pressure variations between 60 and 160 mmHg *circa*. This autoregulation is attributed mainly to a *myogenic mechanism* for which the increase in transmural pressure causes vasoconstriction while the decrease causes vasodilation. Below 60 mmHg there is a reduction in the flow capable of inducing loss of the senses (syncope), while above 160 there may be impairment of the blood-brain barrier and possible appearance of cerebral *edema*.

In addition to the myogenic mechanism, the *metabolic mechanism* is also a very important factor in the regulation of cerebral circulation. This type of control can intervene in the circulation of particular areas concerning their activity. The cerebral circulation is remarkably sensitive to *carbonic dioxide* (CO_2) and pCO_2 in arterial blood: a PCO_2 increase (hypercapnia) can cause vasodilation and a decrease can induce vasoconstriction. If the arterial pCO_2 is increased with the breathing of a gaseous mixture rich in CO_2, strong cerebral vasodilation occurs, whereas if the arterial pCO_2 is decreased with hyperventilation vasoconstriction takes place.

It is believed that CO_2 diffuses into the smooth vascular muscles and interstitial fluid resulting in a decrease in pH, which would be the real cause of the decrease in vascular resistance. The passage in the interstitial fluid is possible because the CO_2 can cross the blood-brain barrier.

Like hypercapnia, *hypoxia* can also cause vasodilation, albeit to a lesser extent. It appears that the action of diminished pO_2 is mediated by an increased concentration of K^+ in the interstitium and by the production of adenosine. While the increase in K^+ concentration has a limited duration that may be less than that of hypoxia, adenosine production lasts as long as hypoxia.

Studying the brain flow inverse variations of the pCO_2 and pO_2 can be observed. If pressure and cerebral flow start to decrease at the cerebral level there is an increase in the pCO_2 and a reduction in the pO_2 with consequent vasodilation and recovery of flow, while the opposite occurs if the variations determine an increased in flow.

At unchanged pressure, vasodilation can occur if an increase in cerebral metabolism leads to a greater demand for oxygen and greater production of carbon dioxide. This mechanism can also affect individual parts of the brain, making it responsible for local variations in the flow. Although controversial, it has been proposed that much of the flow increase during neuronal activity is generated first by dilation of capillaries, rather than of arterioles (which dilate later). Capillary dilatation is achieved mainly by a relaxation of contractile pericytes present at intervals along the walls of capillaries and venules. Perycites are important for blood vessel formation, and for the regulation and maintenance of the blood-brain barrier. Pericytes also play a key role in pathology: in ischemia pericytes constrict capillaries, trapping blood cells, which prevents microcirculatory reperfusion after artery reopening in stroke.

14.2.2 Nervous Regulation

For a long time, the role of innervation on cerebral microcirculation was considered to be of little or no importance. However, in 1998, Roatta et al. performing the *cold pressor test* in humans, observed that the arterial pressure increases but the cerebral flow – measured in the middle cerebral artery using the echo-Doppler – does not change thanks to an immediate vasoconstriction. Since in the cold pressor test the increase in pressure is due to a generalized sympathetic stimulation, which can occur also in other circumstances of real life, it is likely that the sympathetic stimulation also affects cerebral circulation increasing vascular resistance to prevent the flow from increasing along with pressure. In this case, the reflex nervous mechanism is evidenced by the fact that the increase in resistance begins simultaneously with the increase in pressure. Other mechanisms, such as metabolic or myogenic have a longer latency period which generally lasts a few seconds.

In the experiments of Roatta and colleagues, the reduction of the flow pulsatility index was also observed. This fact has suggested that nerve regulation does not only concern resistance vessels but also conductance vessels characterized by a greater diameter. The abundant innervation of conductance vessels is a classical view of nervous cerebral regulation. This abundant innervation is important for the "protection" of cerebral blood flow from the reflex control operating in other vascular districts, in many circumstances. The "supremacy" of the cerebral circle is the basis of the *Cushing*-like response (see Chapters 8 e 12), in which the perfusion of other organs (except the heart) is sacrificed (vasoconstricted) to preserve cerebral perfusion, when necessary. The sympathetic cerebral vasoconstriction seems mainly due to the vasoconstriction of the arteries, richly innervated, rather than to arterioles, poorly innervated. Therefore, generalized sympathetic activation may not limit brain flow.

Data in favor of a vasoconstrictor sympathetic innervation had been reported by Molnar and Szanto since 1964. These authors state that in animals, the stimulation of *bulbar pressure area* (the *rostro-ventrolateral-medulla*) reduced cerebral blood flow. Yet, Heistad and Kontos (1983) suggested that sympathetic stimulation only reduces flow when arterial pressure is abnormally high (more than 150 mmHg), whereas for lower pressures the flow is similar to and without sympathetic stimulation.

In brief, the cerebral circulation is organized to maintain oxygen supply to hypoxia-intolerant grey matter and to adapt local perfusion to local activity. Structural adaptations (circle of Wills, high capillary density, and tight endothelial junctions responsible for blood-brain barrier) guarantee stable support of nutrients and stable composition of intestinal fluid. A peculiar problem for this circulation is postural hypoperfusion and possibility to lose consciousness if baroreflex is impaired; normally strong sympathetic vasoconstriction may involve all districts except brain and heart circulation, which have a predominant metabolic control.

Space occupying lesion may compress the vessels and lead to bulbar ischemia, which may trigger the *Cushing phenomenon* (increased blood pressure, bradycardia and irregular breathing).

14.3 Splanchnic Circulation

The splanchnic circle consists of three districts: splenic, mesenteric and hepatic.

14.3.1 The Splenic Circulation and the Spleen

The *splenic artery* originates from the celiac trunk along with the hepatic artery and the left gastric artery. The splenic artery supplies the pancreas and spleen. Through the short gastric arteries, it also carries blood to the bottom and body of the stomach. This latter is also perfused from the right and left gastric arteries, and from the gastro duodenal artery.

In addition to haemocataretic function (blood cell recycle) and immune function, *the spleen* is also a blood reservoir. In basal conditions, it contains about 200 mL of blood, which can sometimes rise to 2000 or drop to 50 mL. The spleen can widely vary its capacity as its capsule is rich in smooth muscle cells. It contracts in response to sympathetic stimulation and to the action of catecholamines released by the adrenal medullary, situations that occur during exercise, and allow blood to circulate through the vein of the leg. In this way, there is an increase in the cardiac output that adapts to the body's greatest demands. It appears that the blood pushed into the bloodstream by the spleen is particularly rich in red blood cells since the organ's vascular structure causes it to contain a high percentage of these cells under basal conditions. When it reaches the spleen, the splenic artery divides into small arteries and then into capillaries. From these, the blood passes into the venous sinuses and therefore into the trabecular veins which, in turn, give rise to the venous sinuses. Normally a certain amount of blood does not pass directly from the capillaries to the venous sinuses, but escapes from the capillaries and diffuses into the pulp to enter then venous sinuses. During this passage the pulp retains a good number of red blood cells, leaving to return mainly plasma into the venous sinuses. This explains the spleen's ability to store red blood cells in the pulp and to re-circulate them when required.

14.3.2 The Mesenteric Circulation and the Intestinal Villi

The mesenteric circulation consists of the celiac artery and the superior and inferior mesenteric arteries. The venous blood coming from the mesenteric veins and the splenic vein flows into the portal vein destined for the liver, which also receives blood from the hepatic artery. The blood flow of this vascular district was calculated to be about 10–20% of cardiac output.

In the intestinal wall, the arterioles of the mesenteric circle form a submucosal arteriolar plexus. From this plexus, smaller vessels originate which partly distribute to the muscular layers and partly penetrate the intestinal villi.

In the *intestinal villi,* the arterioles extend up to the apex where they are subdivided into capillaries that flow into a venule (Figure 14.12). This runs

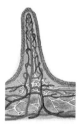

Figure 14.12   Circulation in the intestinal villus. See the explanation in the text.

from the apex to the base of the villus parallel to the arteriole and with a flow opposite to this. At the base of the villus, the venule ends in the submucosal venular plexus. The passage of blood from the arteriole to the venule does not occur only through the capillaries present at the apex of the villus, but also through many systems of capillaries parallel to each other which, from the base to the apex, connect the two vessels, among which therefore there is a counter-current blood exchange system.

In addition to being carried by the blood flowing through the various capillary systems in parallel, oxygen can pass directly from the arteriole to the venule, spreading from the first to the second through the gap in the villus and thus generating a counter-current oxygen exchange system. This exchange is more efficient when the speed of blood is reduced. Since the oxygen removed from the blood of the arteriole from the counter-current system does not reach the mucous cells of the most distal parts of the villus, it sometimes happens that a noticeable slowing of blood flow causes more or less widespread necrosis of the intestinal villi.

The vascularization of the mucosa and submucosa of the intestine is rich and the vasomotor tone of the mesenteric vascular bed is very low already in basal conditions. Yet, the flow increases after the ingestion of food and during the absorption of digested food. After ingestion of food, the secretion of two hormones, the gastrin produced by the mucous membrane of the stomach and the secretin produced by the mucosa of the duodenum contributes to the vasodilation and the increase in intestinal perfusion.

Even bowel movements can induce changes in the flow, either by increasing it due to the greater metabolism of the smooth muscles or by reducing it by compression of the vessels. The *metabolic mechanism* is important in the autoregulation of the mesenteric circulation, as demonstrated by the vasodilator action of a decrease in pO_2 and an increase in pCO_2 in the interstitium and capillary blood. Also, bradykinin may induce vasodilation, mainly through the endothelial release of hydrogen peroxide.

The mesenteric vessels possess a rich *sympathetic innervation* which however under basal conditions exerts a rather modest activity so that the denervation increases the blood flow by no more than 25%. On the other hand, it is interesting what happens if the large splanchnic nerve, which contains the sympathetic fibers directed to these vessels, is stimulated. Under these circumstances, there is a noticeable reduction in the mesenteric flow accompanied by an increase in arterial pressure. This latter effect is mainly due to an increase in cardiac output rather than to the increase in total peripheral resistance. Even if it is big, the mesenteric circle is located in parallel with all the other district circles of the organism, so variations in its resistance cannot modify the total peripheral resistance significantly. The increase in cardiac output is instead possible because the sympathetic stimulation extends its vasoconstrictor action also to the venous section of the mesenteric circle, thus causing an increase in the venous return to the heart. This situation has an important implication in the *fight or flight reaction*, or during exercise when the sympathetic vasoconstrictor activity on the intestine allows the redistribution of blood flow in favor of the skeletal muscles.

In the regulation of the mesenteric circulation, it is possible the participation of a myogenic mechanism, even if not all the researches in this regard have given certain results.

14.3.3 The Hepatic Circulation

The liver is perfused with about 1500 mL of blood per minute, of whom 1200 mL arrive with the *portal vein* and only 300 mL with the *hepatic artery*. The blood carried by the portal vein comes from the stomach, intestine, pancreas, and spleen. After nourishing the liver, blood from the two systems reaches the *inferior vena cava* through the *hepatic veins*.

The hepatic parenchyma cells or hepatocytes present a lobular organization. The lobules, polygonal in shape, are constituted by columns of hepatocytes arranged in a radial pattern around a vein which, due to its position at the center of the lobule: is called *centrolobular vein*. The blood that reaches this vein comes from the *liver sinusoids* surrounded by the columns of hepatocytes. Sinusoids are considered true capillaries. Along the sinusoid wall, there are also *Kupfer cells* belonging to the phagocytic-mononuclear system.

The portal vein gives rise to the *interlobular venules* located outside the lobule. From these veins, the blood passes into the sinusoids to then pour into the centrolobular venule. The sinusoids also receive blood from the interlobular branches of the hepatic artery. In sinusoids then portal blood mixes with arterial blood. Since sinusoids receive more blood from the portal vein than from the hepatic artery, the capillary system that they form is considered a "mirable" venous network. A "mirable" network is a system of capillaries interposed not between an arteriole and a venule, but between two vessels of the same type.

More centrolobular veins drain into an *intercalated vein*. Subsequent confluences give rise to the interlobular veins from which the hepatic veins arise.

While the blood pressure in the portal vein is about 10 mmHg, in the hepatic artery it has an average value of about 90–100 mmHg. The arterioles that carry blood to the sinusoids must have a resistance much higher than that of the other arterioles of the systemic circulation so that the distal arterial blood pressure drops and does not prevent the portal blood from accessing the sinusoids.

Although the blood in the hepatic artery contains approximately 19% of oxygen, as in all systemic arteries, the liver is mainly oxygenated by the high portal blood flow (venous blood) which still contains about 16–17% of the oxygen that can be partially transferred to the liver.

The hepatic arteriole system is subject to metabolic autoregulation of resistance. The branches of the portal vein are not subject to autoregulation. If the pressure in the vein increases, the resistance does not vary or may even decrease due to the passive expansion of the vessel walls. Nevertheless, the regulation mechanisms are less important than in the other organs, as the liver can cope with an increased metabolism by a greater extraction of oxygen from both the arterial and oxygen-rich portal blood.

The term *ascites* is of Greek origin (*askos*) and means a bag or sac. In the clinic, the *ascites* is the accumulation of fluid in the peritoneal cavity. In western countries, the two most common causes of ascites are liver cirrhosis and congestive heart failure. Several other pathological conditions, including tumors and tuberculosis, can be the cause of ascites. A particular form of diffuse edema and ascites is due to the lack of proteins in the plasma. This can occur in *nephropathies with proteinuria* and in the so-called *Kwashiorkor*. This is a condition, observed mainly in the child of developing poor countries, characterized by severe protein malnutrition and accompanied by an enlarged liver with fatty infiltrates, edema, and ascites.

Hypoproteinemia favors diffuse filtration at the capillary level, as the oncotic plasmatic pressure is reduced in these conditions (see Chapter 11.9). In liver cirrhosis, the fibrotic liver is an obstacle to the portal flow. The pressure then increases in the portal vein and, backward, in the capillaries of the entire splanchnic circle, from which plasmatic water exudes in the cavity.

In right heart failure, on the other hand, the fluid exudes mainly from the liver vessels, in which the increase in pressure present in the inferior vena cava is propagated backward. When in liver disease, portal hypertension and hypoproteinemia occur, ascites is present.

14.4 Circulation in the Skeletal Muscle

Blood flow in skeletal muscle circulation has different values depending on the muscles considered. In the muscles with pale-fibers the flow varies between 2–3 mL/100g/min at rest at 60–70 mL/100g/ min under stress; in red-fiber muscles, the variation is between 25–30 and about 150 mL/100 g/min. Intermediate values occur in the muscles composed of both pale- and red-fibers. As will be seen in Chapter 16.1, in the presence of maximal exercise there is a conspicuous increase in cardiac output.

If we consider a muscle with pale-fibers, we see that the increase in exercise flow occurs in two stages. During the preparatory phase and in the initial phase of the exercise it goes from 2–3 to 20–30 mL/100g/min, while only in a second phase it reaches the maximum value of 60–70 mL/100g/min.

Muscle contraction, especially the isometric one, can be a limiting factor in blood flow. In the presence of isotonic and rhythmic muscular contractions as in the race, the mean muscular flow increases, though the compressive effect of contractions reduces the flow similarly to a systolic reduction of coronary flow. It follows that the compressive effect overlaps with the increase in mean flow during the activity. Of note, at the end of the exercise, the increase in the mean flow lasts for a while, though the compressive rhythmic oscillations completely disappear, the effects of accumulated catabolites are still there. Among metabolites, *lactate* formation is an index of the deficit of oxygen supply, and this "oxygen debt" is repaid by the flow that continues after the exercise, which gradually removes lactate and other metabolites.

The massaging effect of rhythmic muscle contraction helps muscle perfusion and increases venous return (see the skeletal muscle pump in Chapter 6.2). In a standing person, the venous pressure in the calf can reach 80–90 mmHg, during walking the muscle pump lowers the venous pressure to about 30-35 mmHg.

To understand the regulation of circulation in skeletal muscles it is useful to remember the organization of their microcirculation. As can be seen in Figure 14.13, an arteriole does not immediately give rise to a capillary network but continues in a *metarteriole* characterized by having a reduced number of smooth muscle cells in its wall. The metarteriole is the first part of the *preferential channel* and is continued, forming a loop and presenting

arteriola

venula

Figure 14.13   *Circulation in skeletal muscle.* The arteriole does not immediately give rise to a capillary network but continues in metarterioles which have a reduced number of smooth muscle fibro-cells indicated by thickened lines in black color. The metarteriole is the first part of the preferential channel and continues, forming loops with progressively decreasing number of smooth muscles, in the proximal tracts, and therefore in the distal tract (highlighted in blue color) of the preferential channel. The preferential channel ends in a venule that runs parallel to the arteriole. The presence of an anastomosis is possible between the arteriole and the venule. See the explanation in the text.

a progressively decreasing number of smooth muscles, in the proximal tract and therefore in the distal tract of the preferential channel. The preferential channel ends in a venule that runs parallel to the arteriole. The presence of an anastomosis is possible between the arteriole and the venule.

The arterioles are innervated by sympathetic fibers. These exert a modest vasoconstrictor tone which increases in standing position and has the function of preventing the hydrostatic pressure due to the upright position from excessively increasing the pressure in the microcirculation, especially of the lower limbs.

The capillaries arise from the metarteriole and the proximal tract of the preferential channel and flow into the distal tract after forming a network between them. At the origin of each capillary, there is a ring of smooth muscles that form the *precapillary sphincter*. In resting conditions, most capillaries are closed (see Figure 14.2). They are recruited during exercise.

At the beginning of the exercise, or even in the preparatory phase, there is a dilation of the arteriole and the preferential part of the channel with smooth muscles. This dilation allows an increase in the flow along the preferential channel, but does not affect the precapillary sphincters and does not modify the perfusion of the capillaries. In these circumstances the flow rises from the resting value up to 20–30 mL/100g/min without guaranteeing a significant increase in the oxygen supply to the tissue.

It is believed that the initial increase in blood flow to skeletal muscles has a hemodynamic role. The first response of the cardiovascular system to exercise is an increase of up to three times the cardiac output (see Chapter 16.1). The expansion of the preferential canal in such a large district may end up reducing the total peripheral resistance, although it is arranged in parallel with the other districts. The result of this decrease in resistance prevents the increase in arterial pressure concerning the increase in cardiac output.

What is the mechanism of the first phase (hemodynamic phase) of exercise vasodilation? A reduction of the sympathetic vasoconstrictor tone is unlikely. Indeed, it is likely that sympathetic tone, instead of decreasing, increases. The hypothesis of the activation of the *cholinergic sympathetic* has in the past found many supporters. Its presence, however, ascertained in some animal species such as the cat, has recently been questioned, although not entirely denied, in humans. In humans, the presence of *β2-receptors* on smooth muscle cells of skeletal muscle is considered the main factor for the first phase of exercise vasodilation. The *β2*-receptors can be stimulated either by noradrenaline released by common sympathetic terminations or by circulating catecholamines. Nevertheless, it is interesting that an increase in sympathetic activity is already present in the emotional state that precedes the beginning of the exercise and that this induces vasodilation.

After the exercise has begun, the *precapillary sphincters* are released by metabolites. The blood flow, therefore, rises to its highest value, while the skeletal muscle fibers are guaranteed to have adequate oxygenation to the effort. In working conditions, the possible sympathetic vasoconstrictor effect is overwhelmed by the vasodilation due to metabolism. That is "metabolic override" or "functional sympatholysis".

The *metabolic regulation* of this second phase of muscular vasodilation is due to the release of adenosine, CO_2, H^+, and K^+. It seems that K^+ accumulation is important in the early phases and that adenosine intervenes in later phases. Adenosine also inhibits norepinephrine release from sympathetic terminals. The intervention of *nitric oxide* produced by the action of the increased shear stress on vascular endothelium is also likely, since the first phase (hemodynamic phase) of increased in flow.

We have seen how in the red-fiber muscles the blood flow has values much higher than those of the pale-fiber muscles, being able to oscillate from 20–30 to 150 mL/100 g/min. This increased flow is made possible by the increased density of vessels in this type of muscle. The functional significance of this greater density lies in the fact that the red-fiber muscles, having a tonic and postural function, are rarely at complete rest. Their ability to receive up to 150 mL/100g /min of blood under maximum stress (exercise) means that they, like the myocardium, do not experience oxygen lack/debt and fatigue (two aspects present in pale muscle).

An interesting feature of the circulation in skeletal muscles is that the venous blood coming from the microcirculation can have a pO_2 higher than that present in the muscle tissue with which the exchanges took place at the level of the capillaries. The fact seems to be paradoxical in that, the blood present in the veins must have the same respiratory gas tensions, or only slightly different, compared to those of the tissue that has nourished. The explanation lies in the parallel path of arterial and venous vessels and the flow of blood in the opposite direction concerning one another. Through the walls of these vessels, a certain amount of O_2 passes directly from the arterioles to the venules before the blood reaches the capillaries. This is a real counter-current exchange, confirmed by the fact that in arterioles the pO_2 decreases as the blood proceeds towards the capillaries, while in the venules it increases as it moves away.

14.5 Renal Circulation

The renal circulation is closely connected with the function of the kidneys. In this paragraph, only some regulatory aspects of renal circulation are treated. For a more complete discussion, readers should refer to texts on kidney physiology.

Blood flow through the two kidneys is approximately 1250 mL/min. Given the anatomic-functional aspects of the kidney, the blood flow is distributed more to the cortical than to the medullary regions of the kidney.

The renal blood flow (RBF) can be assessed starting from the clearance of para-aminohippuric acid (PAH) which allows knowing the renal plasma flow (RPF), which is 650–700 mL/min. If we know the hematocrit (Htc), usually 0.45 in men and 0. 42 in females, the blood flow can be easily calculated by the formula:

$$RBF = RPF/1 - Htc.$$

The vascularization of the kidneys is particularly abundant and complex (Figure 14.14).

From the *renal artery* derive the *interlobar arteries* and then the *arcuate arteries*, located at the border of the renal cortex and renal medulla. There are *interlobular arteries, afferent arterioles,* and *true straight arterioles* (arteria recta). The afferent arteriole is subdivided into the *glomerular capillaries* which, in turn, flow into the *efferent arteriole*. Besides glomerular capillaries, there are *peritubular capillaries* and the *vasa recta*. These microvessels drain into *straight veins* and the *intrarenal veins* that drain into the *renal vein*. The

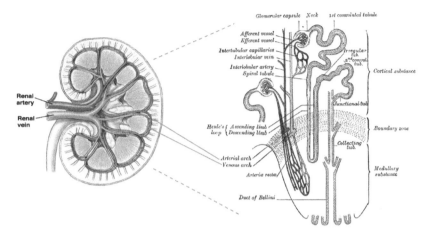

Figure 14.14 Renal vascularization. On the left we see the interlobar arteries raising from the renal artery and giving origin to arterial arch (arcuate arteries); arteries are accompanied by correspondent veins. On the right, we see the details of microcirculation, including interlobular arteries, afferent arterioles in the cortex, and true straight arterioles (arteria recta) in the medulla.

afferent arterioles and the efferent arterioles are the vessels in which there is maximum resistance (Figure 14.15 and 14.16). The glomerular capillaries, being located between two arterioles, form a *"mirable arterial network"*, in which the pressure is high (the highest capillary pressure of our body).

The regulation of renal circulation can be extrinsic or intrinsic. *Extrinsic regulation* is nervous. The renal vessels are innervated exclusively by *vaso-constrictor sympathetic fibers* originating from the T4-L4 spinal segments. Under basal conditions, these fibers produce a very modest vasoconstrictor tone. This tone may increase in the transition from lying to standing when the stimulation of the baroreceptors is reduced.

Sympathetic discharge and related vasoconstriction may increase to reduce blood flow in a kidney to 200 mL/min, as, for example, in the *fight or flight reaction*, in physical exercise, and after hemorrhage. This strong reduction in flow allows a redistribution of the blood to the advantage of the vital organs and the skeletal muscles during exercise. The reduced blood flow may determine a transitory reduction in glomerular filtration making it possible to retain liquid after there has been a reduction in the circulating volume. In the renal circulation vasoconstriction of an extrinsic nature with a reduction in the glomerular filtration rate can also be due to endogenous substances such as adrenal catecholamines, ADH, serotonin, and angiotensin II.

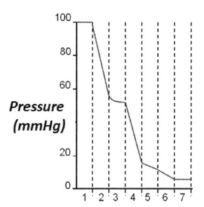

Figure 14.15   Pressure changes in the renal circulation. 1: renal artery, 2. afferent arteriole, 3: glomerular capillaries, 4: efferent arteriole; 5: peritubular capillaries and vasa recta; 6. intrarenal veins; 7: renal vein. Note the *high pressure in the glomerular capillaries* (favors filtration) due to high post-capillary resistance in the efferent arteriole. For the same reason, we have very *low pressure in the peritubular capillaries* (favors reabsorption). See the explanation in the text.

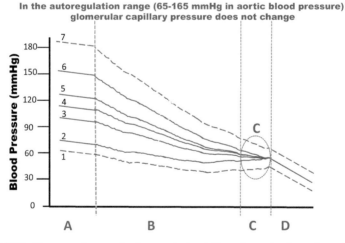

Figure 14.16   Pressure changes in the renal circulation. A: systemic and renal artery, B. afferent arteriole, C: glomerular capillaries, D: efferent arterioles, and peritubular capillaries. Note that the pressure in the glomerular capillaries does not change (it favors constant filtration) regardless of the five (from 2 to 5) different levels of systemic blood pressure in the range of autoregulation (levels 1 and 7 are out of the range of autoregulation).

Intrinsic regulation, or autoregulation, of the renal circulation, can be highlighted in the denervated kidney where blood flow can remain unchanged despite extremely large changes in blood pressure. This type of regulation is based on a myogenic mechanism, on plasma-red blood cell separation and tubulo-glomerular feedback. These mechanisms limit the variation of pressure in the glomerular capillaries (Figure 14.16).

According to the *myogenic mechanism*, the stretching produced by the pressure on the vessel wall causes vasoconstriction (see Chapter 12.6). The opposite occurs in case of pressure reduction. This phenomenon in kidney regards mainly the afferent arteriole.

The *plasma-red blood cell separation*, also known as *plasma skimming* (*Fåhræus-Linqvist effect*, see Chapter 11.3), depends on the tendency of the red blood cells to locate in a central position, and the plasma in a peripheral position in a vessel in which there is laminar blood flow. The higher the blood velocity is, the greater the separation between plasma and red blood cells. So that the red blood cells are concentrated in the main vessel. Since an increase in hematocrit above normal values leads to a noticeable increase in blood viscosity, while a decrease only results in a modest reduction, it results that an increase in blood velocity determines an increase in flow resistance. Nevertheless, this mechanism is considered marginal to the effects of the regulation of renal circulation.

The *tubule-glomerular feedback* concerns the ability of the *macula dense* to promote adenosine, as well as renin formation. If, due to an increase in systemic pressure, the blood flow and the glomerular filtration rate increase, the filtered load of NaCl increases, and with it the concentration of this salt in the liquid that reaches the macula dense. The NaCl then enters in greater quantities in the cells of the macula dense leading to the production of adenosine, which causes constriction acting on A1 and A3 adenosine receptors (*see above*) of the afferent arteriole and returns to the norm the blood flow as well as the velocity of glomerular filtration. If, on the contrary, the systemic pressure, the blood flow, and the glomerular filtration rate are reduced, the macula dense signals lead to a reduction of adenosine production, allowing dilation of the afferent arteriole and a greater filtration. The production of NO by the juxtaglomerular apparatus also participates in the vasodilation of the afferent arteriole. At the same time, due to the reduced concentration of NaCl in the tubular liquid, the juxtaglomerular apparatus secretes renin which, through the production of angiotensin II, and possibly aldosterone, brings arterial pressure back to the norm. Angiotensin II acts also on efferent arteriole inducing an increase in glomerular capillary pressure and favoring filtration.

14.6 Cutaneous Circulation

The *cutaneous or skin circulation*, in addition to supplying the skin with oxygen and nutrients and removing the catabolites, has a fundamental role in allowing *thermoregulation*. This is possible when the blood conveys to the surface of the body both the heat to be dispersed and the water to be excreted by the sweat glands. According to the needs of the general thermoregulatory function of the organism, the blood flow is more than 20 times greater than that which would be necessary for the thermoregulation of the skin alone and can undergo variations ranging from 20–50 mL/min to 2–3 L/min for the entire body surface. The skin of adult weights 2–3 kg; therefore, the skin flow ranges from 1 mL/100g/min (in a cold environment) to 150 mL/100g/min (in a cold environment), with a mean value of 15 mL /100g/min in a thermoneutral environment (27°C). Transiently in some territory, it can reach 200 mL/100g/min.

The cutaneous circulation has a rich vasoconstrictor sympathetic innervation. The sympathetic tone is already active in basal conditions so that the cutting of these nerve fibers causes evident vasodilation. Subsequently, however, the tone of the resistance vessels progressively returns to be the one preceding denervation. This resistance recovery likely depends on the increased sensitivity of the vessels to circulating catecholamines.

The sympathetic tone increases in cold exposure limiting the dispersion of heat and decreases when greater dispersion is necessary. The thermoregulatory nervous control is exerted by the hypothalamic defense centers against the cold and against the heat on which impulses coming from the skin thermoreceptors act. Although there may be flow regulation due to local regulatory factors, the nervous control mechanism is prevalent. In addition to the hypothalamic centers of thermoregulation, the sympathetic directed to the vessels of the skin may be also influenced by other higher centers that would be responsible for the blushing and blanching of the face in case of emotions.

In some parts of the body, such as hands, feet, and ears, instead of vasoconstriction, the cold can cause vasodilation. This response allows a greater supply of blood, and therefore of heat, at extremities that at very low temperatures could easily be frozen. The opening of *arteriolar-venular anastomoses* present in the skin of these sites contributes to an increase in blood flow to the hands, feet, and ears. These anastomoses closed in normal conditions and lacking in sympathetic innervation, open by an axonic reflex (see Chapter 12.6), evoked when the skin temperature drops below 10°C. It is noteworthy that below this temperature value a painful sensation can occur.

Other *arteriolar-venular anastomoses* are controlled by the sympathetic tone and are fundamental for thermoregulation control; they constrict with cold to conserve heat and dilate when the core temperature is high so that also skin temperature rises and heat loss increases.

The skin also contains fibers belonging to the cholinergic sympathetic system. These fibers do not innervate the resistance vessels but the sweat glands. Nevertheless, when these are activated there is also vasodilation.

15

Pulmonary Circulation

15.1 The Characteristics of Pulmonary Circulation

The pulmonary circulation supplies the pulmonary capillaries where gas exchange takes place between the air present in the alveoli and the blood. Here the mixed venous blood takes on oxygen and gets rid of excess carbon dioxide. This circulation differs substantially from the systemic districts: it receives the entire cardiac output of the right ventricle which vastly exceeds the nutritional/metabolic needs of the lungs and, therefore, metabolic factors exert no appreciable influence on lung circulation. The autoregulation is absent and during exercise, the vascular system is expanded passively by a small increase in lateral pressure. This circulation has a very high capillary density, very low vascular resistance, and low blood pressure. Due to the low pressure, the flow is greatly affected by gravity, which determines a vertical distribution of blood flow in a standing subject (the lung apex is under-perfused and the bases hyper-perfused). A unique characteristic of this circulation is the *vasoconstrictor response* to hypoxia and hypercapnia. The metabolic needs of the bronchi are satisfied by the bronchial circulation, which is a systemic district.

After having undergone gaseous exchanges, the mixed venous blood is transformed into *arterialized blood* (see respiratory physiology). Here we recall that the *arterialized blood* is the blood present at the venular-end of an alveolar-capillary, which has the pO_2 and the pCO_2 identical to alveolar air. A part of the *deoxygenated* venous blood of the bronchial circulation drains into the pulmonary veins. Therefore, mixing with venous *deoxygenated* blood the arterialized blood starts to become *arterial blood*. This blood has a slightly lower pO_2 than the arterialized blood. The pO_2 decreases a little bit more in the left atria and ventricle where a little amount of *deoxygenated* blood is added by the *thebesian* and *sinusoidal coronary vessels*. Due to these *anatomical shunts* (bronchial and coronary shunts), the pO_2 of arterial blood is always lower than arterialized/alveolar pO_2 by at least 5–10 mmHg, even in a healthy person with normal ventilation and perfusion, that is the so-called *alveolar-arterial gradient*. Of note, these anatomical shunts are responsible for the fact that the left ventricle pumps a little more blood than the right ventricle in the unit of time (a few mL/min).

The pulmonary circulation also plays a filter role and stops the emboli which, having reached with systemic venous blood could block arteries of the pulmonary circulation. Given the large number of parallel circuits that characterize the pulmonary circulation, these emboli do not generally cause significant obstructions. In the case of a blood clot (thrombus), these can be digested by fibrinolytic enzymes released by the endothelium. Also, air emboli, not too large, can be eliminated by the gas exchange in the alveoli. On the other hand, fatty emboli that form after bone fractures or orthopedic surgery are more difficult to eliminate. Generally, to determine a serious obstruction in the pulmonary circulation, the emboli must have a size such as to interrupt the flow in a large branch of the pulmonary artery, thus causing a pulmonary infarction. The risk of developing a pulmonary embolism increases in some circumstances, such as the presence of a tumor or a prolonged and forced stay in bed, secondary to various problems such as trauma, fractures, or advanced age. The symptoms of pulmonary embolism may be absent or of different magnitude depending on the extent of the embolism. These include dyspnea, chest pain during inspiration, and palpitations. In the case of extensive embolism, there may be low oxygen saturation in the blood, cyanosis, tachypnea, and tachycardia. Severe cases can lead to low systemic pressure, shock, and sudden death.

As already said in other Chapters (*e.g.* Chapters 7, 13, and 14), the surface of the endothelium of the pulmonary circle is rich in ACE which transforms Ang I into Ang II. ACE also catabolizes bradykinin. ACE inhibitory are drugs used in the treatment of systemic hypertension and heart failure. The accumulation of bradykinin (a potent vasodilator) contributes to the lowering of the blood pressure. In some patients, bradykinin sensitizes somatosensory fibers and may be responsible for the dry cough that is a disturbing side effect of ACE inhibitors. Lungs and respirator tract are also rich in ACE2.

The vessels make up about 40% of the lung and contain about 500 mL of blood, *i.e.* 1/10 of the blood contained in the circulation. The arteries of the pulmonary circulation are almost 7 times more distensible than those of the systemic circulation. This greater distensibility is explained by the scarcity of smooth muscle cells and the thinness of the walls.

As said, *the pulmonary circulation is at low pressure*. In fact, in the pulmonary artery, the systolic and diastolic pressures are about 20–25 and 10–15 mmHg, respectively. The cause of the low pressure is the *low pre- and post-capillary resistance*. If only the precapillary resistances were low, the pressure in the alveolar capillaries, which is normally 5–6 mmHg, would be higher due to easy entry and a difficult exit of blood from the capillaries. Such high pressure would cause plasma liquid to go out from the capillaries resulting in pulmonary edema. As we saw in Chapter 1.3 and Figure 1.5, the presence of low resistance favors the pulsatility of pressure and flow in the alveolar capillaries. Since the post-capillary resistances are also low, the pressure oscillations generated in the left atrium also go backward in the capillaries. The capillary pressure and thus the atrial pressure can be measured as *wedge pressure* (Box 1.1).

Figure 15.1 illustrates how the pulsatility present in the alveolar capillaries can be recognized. This procedure allows also to calculate blood velocity in the pulmonary capillaries.

In a person in an orthostatic position, the various parts of the lung are not equally perfused. West distinguished the lung in three zones; these are the *West's zones of the lung*. *Zone 1* is the slightly perfused upper/apical zone; *zone 2* is the normally perfused middle zone and *zone 3* is a high perfused basal/low zone (see Figure 15.2). Perfusion depends on the difference between the air pressure in the alveoli and the blood pressure in the capillaries. The second varies in the three different zones according to the *hydrostatic pressure* whose value, in a subject in orthostatic position, is given by the level of the zone compared to the level of the heart. Therefore, the *capillary blood pressure* decreases from the base to the apex of the lung, while the pressure of the air within the alveoli (PA) does not change. The mean pressure in the capillaries, moreover, decreases from the arteriolar value, Pa, to the venular value, Pv, present in the venules (in the three zones Pa>Pv).

In the basal/low zone 3, located below the level of the heart, the hydrostatic pressure due to the difference between the level of the heart and that of the considered vessels is added to the capillary pressure. It follows that Pa> Pv> PA (Figure 15.2), or that the blood pressure exceeds the pressure of the alveolar air along the entire length of the capillaries. In this zone, therefore, the capillaries are distended and the flow is high.

In the middle zone 2, placed at the same level as the heart, the hydrostatic pressure is not added to the capillary pressure. It follows that Pa > PA > Pv, *i.e.* the blood pressure exceeds that of the alveolar air only from the arteriolar side, while the opposite occurs from the venular side. Due to the

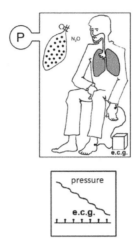

Figure 15.1 Determination of flow pulsatility in alveolar capillaries using the whole-body plethysmograph. P: pressure in the plethysmograph chamber; N_2O: nitrous oxide. The subject is in a sealed chamber that serves as a body plethysmograph. A pressure gauge P allows the recording of air pressure in the chamber (lower box). As long as the patient breathes ambient air, there are no pressure variations, because air moves from the chamber atmosphere to the lungs without changing the overall volume and pressure inside the plethysmograph. The subject is then invited to inhale from a balloon a gaseous mixture consisting of 20% O_2 and 80% N_2O. Thereafter, the patient holds his breath for a few moments. Since the balloon is contained in the body plethysmograph, during the inhalation there will be no variation in either volume or pressure chamber. A decrease in both occurs instead when the subject holds his breath because of a loss in a volume that occurs when N_2O diffuses into the blood. The pressure decrease recorded by the pressure gauge P does not occur continuously, but in steps synchronous with cardiac activity, as shown by the simultaneous recording of the ECG. This procedure, in addition to highlighting the pulsatility of the pressure in the capillaries, allows assessing the instantaneous speed of the blood inside them. Knowing the *solubility coefficient* of N_2O and its alveolar concentration it is possible to calculate the blood velocity, which is about 10 L/min in systole and 2–3 L/min in diastole.

prevalence of PA pressure, from the venular side, the capillary is subjected to a certain degree of compression according to the *waterfall model* (Chapter 11.7). Because of this compression, the blood flow is lower compared to the basal zone. This zone extends from where Pa = PA to where PA = Pv.

In the upper/apical zone 1, located above the level of the heart, the hydrostatic pressure, due to the level difference between the capillaries considered and that of the heart is subtracted from the capillary pressure. It follows that in this case PA > Pa > Pv so that the capillaries are restricted throughout their length by the waterfall effect. This is theoretically a *dead space* (ventilated

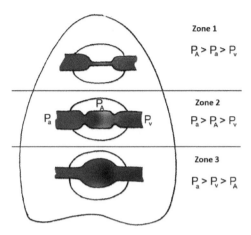

Figure 15.2 Zone of West in the lung. PA: alveolar air pressure; Pa: arteriolar pressure, Pv: venular pressure. See the explanation in the text.

but not perfused). However, little gas exchange takes place here because the blood in systole (especially in the inspiration phase) is high enough to open the capillaries.

The increased pressure within capillaries of the basal zone explains why in the case of acute left ventricular failure the pulmonary edema begins in the basal part of the lungs. The modest perfusion of the upper zone is considered responsible for the high prevalence of nodular tuberculosis at the apical site.

Box 15.1 How to Determine the Different Blood Flow in the Various Levels of the Lung

To highlight and measure the perfusion at different levels of the lung, two scintillation counters are placed near the upper part and the basal part of the chest, respectively, in a person in an upright position (Figure 15.3). The subject is then invited to inhale pure radioactive carbon dioxide ($C^{15}O_2$) and then to hold the breath for 15 seconds.

During this short period, carbon dioxide having a higher pCO_2 in the alveoli than in the venous blood, passes into the capillaries to be removed by the blood stream. The removal is accompanied by a decrease in the radioactivity detected by the counter. It can then be seen that the decrease in radioactivity is more rapid at the base of the thorax where the flow is high than in the upper part where the flow is lower.

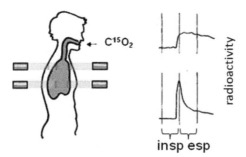

Figure 15.3 Radio-labelled carbon dioxide and external scintillation counter to measure the flow in the upper and lower zone of the lung. See the explanation in the text.

15.2 Variations of Volume of the Lung and the Resistance to the Blood Flow

Pulmonary vascular resistance is not constant over time but varies cyclically with changes in lung volume due to the alternation of inspiration and expiration. The vascular resistance is low at the end of a normal exhalation when the lung contains the volume of air corresponding to the so-called *functional residual capacity*, but increases during inspiration (Figure 15.4). In addition to inspiration, pulmonary vascular resistance increases also in forced expiration, when in addition to the tidal volume the *expiratory reserve volume* is also expelled and the lung contains only the *residual volume*. In other words, vascular resistance is low when the thorax-pulmonary system is in the rest position (*i.e.* at the end-expiration position or functional residual capacity) and increases both in forced inspiration and expiration.

The reasons for these variations in resistance during breaths reside in the action of changes in the volume of the lung on the lumen of the vessels. As can be seen in Figure 15.4, *in forced inspiration* the increase in the radius of the alveoli causes a *compression of the alveolar vessels*. In terms of increased resistance, the effect of this compression exceeds the effect of the *distension of extra-alveolar vessels* induced by the increase in lung volume; *in forced expiration*, on the other hand, the *extra-alveolar vessels are less distended* (almost wrinkled or convolute) than the normal (at functional residual capacity) and, for this reason, the resistance to blood flow increases.

Given the distensibility of the pulmonary vascular bed and the limited presence of smooth muscle cells, even variations in cardiac output (CO) can vary the resistance. In basal conditions, the pulmonary vascular resistance is already quite low, although some capillaries are not perfused. When the CO

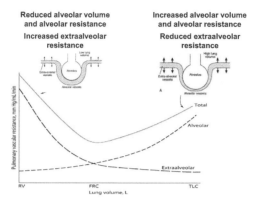

Figure 15.4 Variation in resistance to the pulmonary circulation during breathing. RV: residual volume; FRC: functional residual capacity; TLC: total lung capacity. See the explanation in the text.

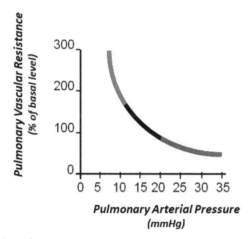

Figure 15.5 Variation of pulmonary vascular resistance with changes in pulmonary arterial pressure. It should be noted that the increases in pressure are caused by increases in cardiac output (normal in black color, reduced in gray color, and increased in red color) which causes pulmonary vascular resistance to vary. See the explanation in the text. (modified by Rhoades R.A: and Tanner G.A. Medical Physiology. EdiSES 1996).

increases, the low resistance allows a limited increase of the pressure which determines, however, the opening of the previously occluded capillaries and the distension of those already open: the result is a decrease in the total resistance, while the opposite occurs when the CO decreases (Figure 15.5).

When the CO has typical values of the rest condition (5 L/min), the transit time of the blood in the alveolar capillaries is about 0.9 s. If CO increases, the transit time can be reduced to 0.25–0.3 s. This corresponds to a 3 to 4-fold increase in blood velocity (v), but a 5-6-fold increase in lung flow (i.e. $CO = 25$–30 L/min), as expansion and recruitment of new vessels enlarged the section (S^2) of the vascular system. Remember $CO = v \times S^2$.

15.3 The Chemical and Nervous Regulation of the Pulmonary Circulation

Hypoxia causes vasoconstriction of the pulmonary circulation. Hypoxia can be local or generalized. It is local when it is due to a part of poorly ventilated alveoli, while it is generalized when all the pulmonary circulation is involved: for example, when the mixed venous blood has an oxygen content lower than the normal 75% or at the high altitude. While the local increase in resistance leads to a modest, though present, increase in pressure in the pulmonary artery, the generalized vasoconstriction induces a noticeable increase in pressure. Regional *hypoxic-vasoconstriction* reduces blood flow to poorly ventilated alveoli and "redirects" blood towards well-oxygenated alveoli.

Hypoxia causes constriction of the pulmonary vessels through the closure of the K^+ channels (Kv channels) of the smooth muscle cells. The consequent reduction of the outward K^+ current reduces the amplitude of the resting membrane potential with the consequent opening of the voltage-dependent channels for Ca^{2+} and increases in the intracellular concentration of the ion which determines vasoconstriction.

Like hypoxia, *hypercapnia* can also cause pulmonary vasoconstriction, especially in general anesthesia. Vasoconstriction due to hypercapnia is exacerbated in the case of concomitant metabolic acidosis. Although the increase in alveolar pCO_2 in poorly ventilated alveoli will be considerably less than the fall in pO_2 and cannot go beyond the value found in mixed venous blood, the pulmonary vasculature is very sensitive to changes in pCO_2, and even a rise of a few mmHg will induce *hypercapnic-vasoconstriction*. In an analogous fashion but the opposite direction, small increases in pCO_2 will cause vasodilation in the systemic vasculature (especially in the cerebral circulation), and increases in ventilation. Thus, hypercapnia may be as important as hypoxia in initiating pulmonary perfusion redistribution to minimize the extent of resultant hypoxemia and hypercapnia in acute asthma, for example. Moreover, blood pO_2 decreases, and pCO_2 rises to determine a rise in lung vascular resistance through a *chemoreceptor reflex sympathetic activation*.

In addition to hypoxia and hypercapnia, various endogenous substances are also active on the pulmonary vessels. *Catecholamines* exert a constrictive action through adrenergic α-receptors. Also, *neuropeptides, endothelins,* most *prostaglandins, serotonin, histamine,* and *angiotensin II* are vasoconstrictors. On the other hand, *adrenaline* exerts a dilating action through the β-receptors. Also, *nitric oxide, bradykinin, dopamine,* and *prostacyclin,* as well as *hyperoxia* can induce vasodilation.

The pulmonary vessels are innervated by *postganglionic sympathetic fibers* that originate in the stellate ganglion. When these fibers are stimulated exert a vasoconstrictor action, with the increase in pulmonary pressure. In nature the sympathetic discharge on the pulmonary vessels is regulated by the same mechanisms (*e.g.,* chemoreflex and baroreflex) seen in the systemic circulation: the stimulations of high-pressure baroreceptors, and low-pressure baroreceptors (within pulmonary artery), can inhibit the sympathetic discharge to the pulmonary vessels, reducing pulmonary resistance (see Chapter 12.6).

15.4 The Formation of the Pulmonary Edema

The mean pressure in the alveolar capillaries is only 5 to 6 mmHg. In the interstitial spaces of the lungs, the pressure is between 5 and 8 mmHg (lower than atmospheric pressure). Starting from these values the filtration pressure can have a maximum value of 14 (6 + 8) mmHg, *i.e.* a pressure lower than the oncotic pressure of plasma proteins which is 24–25 mmHg. The resulting pressure balance should, therefore, prevent the leakage of fluid from the capillaries and the formation of edema. Normally the lung capillaries have a 0.75 reflection coefficient, $\int$, that can fall to 0.4 or lower during lung inflammation, thus favoring edema. However, the alveolar capillaries have *high permeability* (higher than those present in many systemic districts). This feature together with other anatomical-functional aspects (see lung physiology) can allow leakage of proteins. In these conditions can be sufficient a slight increase in capillary pressure to have the formation of edema. Due to the low value of post-capillary resistance, the pressure in the alveolar capillaries may increase in the acute failure of the left ventricle due to the backward propagation of the pressure increase. Usually, the edema begins in the lower parts of the lungs where the blood pressure is higher. In the event of pulmonary edema, a reduction of the venous return to the right heart is recommendable. This can be achieved by administering fast-acting diuretics (*e.g.* furosemide) or morphine. In the absence of these drugs, bloodletting or phlebotomy once represented the only effective treatment. Three tourniquets on the three limbs can be a less invasive way to reduce venous return (three limbs at a time, in turn, so as not to subject the limbs to prolonged ischemia). Even inotropic drugs could be used to improve the contractility of the left ventricle. Following treatment or after spontaneous resolution the fluid present in the alveoli may be reabsorbed, while that present in the interstice is drained by the lymphatic vessels, which are highly active in the lungs. Pulmonary edema can also occur without increased pressure in the capillaries when the wall of these vessels is injured due to inflammatory processes or inhalation of irritating gases.

16

Coordinated Cardiovascular Adaptations

Th is chapter aims to analyze how the various components we have seen in the previous chapters adapt in a coordinated manner during different challenging situations.

16.1 Physical Exercise

Muscle exercise can be mild, moderate, intense, and very intense. While mild, moderate, and intense exercises are classified as submaximal, the very intense exercise is considered supramaximal.

Exercise is characterized by a series of responses of the cardiovascular parameters that may support a 13-fold increase in oxygen consumption during an intense exercise. Figure 16.1 summarizes the principal adaptation observed during exercise. In these cases, the responses include a 5-fold increase in CO due to a 3-fold increase in HR and a 1.5-fold rise in SV, accompanied by a 3-fold increase in the artero-venous difference in O_2 concentration (a-vO_2), and a 15 to 20-fold increase in blood flow in the contracting skeletal muscle vessels. Of course in some other vascular districts, the blood flow can decrease.

While in the submaximal exercise the increase in CO and district flow, together with increased respiratory activity, ensure that the active muscles receive the oxygen they need, in supramaximal exercise the supply of oxygen may not be any longer adequate to the effort. In this case, the muscles undergo ATP exhaustion and lactic acid accumulation due to the prevalence of anaerobic glycolysis.

In the transition from rest to intense exercise, CO can increase from 5 L/min at rest to 20 L/min in normal/sedentary subjects and 25–30 L/min in Olympic athletes. With a CO of 20 L/min, the exercise can be supramaximal for a sedentary but not for an athlete. The following discussion will concern the circulation adaptation during intense but not supramaximal exercise.

367

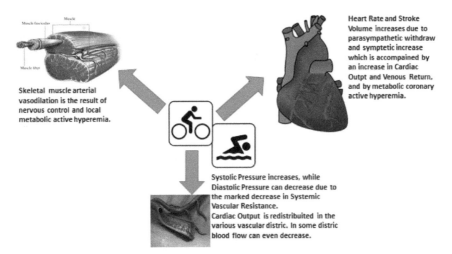

Figure 16.1   Principal adaptation during exercise (see explanation in the text).

In healthy subjects, arterial blood pressure has a moderate increase in systolic value, whereas diastolic blood pressure varies little or may even decrease, and mean pressure may undergo little variation.

The fact that substantial increases in cardiac output are not accompanied by proportional increases in arterial pressure tells us that the total peripheral resistance as a whole must be diminished. This decrease depends on the considerable dilation of the muscle vessels in which most of the increase in CO is diverted.

Since the muscle that contracts have a greater demand, and therefore extraction, of oxygen, the a-vO$_2$ can triple, going from 5 to 15%.

Three factors contribute to increase CO: an increase in HR, SV, and venous return (VR). These factors are closely integrated to allow an increase in CO. In this integration, the increase in sympathetic activity plays a leading role.

In physical exercise, the heart, in addition to an inhibition of vagal activity, which explains mainly the increase in HR, is subjected to a greater sympathetic activity which increases cardiac contractility. Deeper contraction contributes to increasing the SV, exploiting part of the blood which normally constitutes the systolic residue. At the same time, the increase in sympathetic activity reduces the venous capacity contributing to an increase in VR. The massage on the veins by the active muscles also contributes to the increase in VR. Therefore, the increase in CO is accompanied by a parallel increase in

VR, and, of course, these two parameters are identical when a steady-state is reached.

As already seen in Chapter 14.4, in the pale skeletal muscle fibers the blood flow from 2–3 mL/100 g/min at rest rises to 70–80 mL/100 g/min during intense activity, while in the red fiber muscles it rises from 20–30 mL/100 g/min at rest to 150 mL/100 g/min during intense activity. In the case of red fiber muscles, the terms of rest and activity must be taken with caution, since they refer to muscles that are always partially active as they perform a postural function and are rarely subjected to a maximal effort comparable to that of pale fiber muscles.

In Chapter 14.4 we saw the two stages of increased flow in skeletal muscles when they enter into activity. Initially, the vasodilation is of nervous and humoral nature, while subsequently it is regulated mainly by the increase in muscle metabolism. Metabolic products are also responsible for local sympatholysis, thus potentiating vasodilatation.

In the pale-fiber muscles, the nervous and humoral vasodilation begins already in the preparatory phase of the exercise. At this stage, an increase in sympathetic activity begins to cause an increase in CO which in turn would determine a parallel increase in arterial pressure if vasodilation did not occur in the muscles involved in the exercise. This situation extends from the preparatory phase to the initial phase of the exercise. Preparatory vasodilation has been attributed both to an activation of the sympathetic cholinergic vasodilator system (see Chapters 8.5 and 12.3) and to an action of the circulating adrenaline on the $\beta2$-adrenoreceptors located on the smooth muscle of the resistance vessels, and subsequently to the reduction of the sympathetic vasomotor tone of α adrenoreceptors (sympatholysis). Recently, in humans, there is a tendency to exclude a cholinergic innervation directed to the vascular smooth muscle and the first phase of vasodilation is attributed mainly to the $\beta2$-adrenoreceptors.

As stated in Chapter 14.4, vasodilation due to cholinergic sympathetic vasodilator and/$\beta2$-adrenoreceptors involves precapillary arterioles, arteriolo-venular shunts, metarterioles and the proximal tract of the preferential route of muscle microcirculation, without extending to the precapillary sphincters. In this way, the initial increase in blood flow in the pale muscles does not concern the exchange vessels and does not reach the maximum value but settles at around 30 mL/100 g/min. Only when the effect of the increased metabolism intervenes (directly or for sympatholysis), the precapillary sphincters dilate and flow rises to 70–80 mL/100 g/min. If in resting conditions only a quarter of the capillaries are patent, once the flow has reached its maximum value, they are all patent.

The first phase of vasodilation, which does not lead to greater perfusion of the exchange vessels, but simply to a greater flow through the preferential path and the arterial-venular shunts, has a hemodynamic role. It prevents the immediate increase in CO from being accompanied by a corresponding increase in pressure.

Once the exercise is over, the arterial pressure can transiently fall below the normal values due to the persistence in the muscles of a certain degree of metabolic vasodilation, and of the reduction of the volume of circulating fluid by both sweating and hyperventilation. Only when the metabolites have been removed, the situation recovers to control conditions, thus providing the restoration of the volume of circulating liquid with the adequate introduction of water and salts.

The increase in CO matches the increase in oxygen consumption. However, how the CO is increased to match the oxygen consumption rate is not completely understood. Likely, a combination of *central command* from the nervous system and reflex from peripheral receptors, including *metaboreceptors* play important roles in the active muscle. The central command is somewhat like the alerting response accountable for the *fight or flight reaction* (see below) and may start from the cortex with the beginning of the activation of the motor program. The metaboreflex originates from metaboreceptors stimulated by metabolite accumulation in the working muscle. It is responsible for the so-called *pressor response* which can be observed even after exercise had been stopped if we inflate a cuff around a limb to block washout of metabolites by blood flow.

The *circulatory adaptations in fitness training* are many. They include an enlargement of the cardiac cavities accompanied by wall thickening of the ventricles. This is the main reason for an increased SV even at rest. This increase in SV may be responsible for increased pulse pressure as well as an increased stimulation of baroreceptors, which may explain the typical bradycardia (*e.g.* 40 b.p.m.) of athletes. Being SV increased and HR reduced the CO may be unchanged at rest. During exercise a greater reserve for HR can be exploited, so that a 4 to 5-fold increase in HR and a 1.5 increase in SV may explain a CO as high as 35 L/min in Olympic athletes. In some vascular districts, especially in trained skeletal muscle and myocardium, and increased vascularization with an increased capillary density and greater endothelial responsiveness to mechanical stimuli have been described.

16.2 Alerting Response

In previous chapters, we saw how the hypothalamus can regulate the so-called *fight or flight reaction*. This is an alerting response to unusual stimuli and is known also as *alarm or defense reaction*.

The alerting response consists in putting into action responses that allow an animal to escape or attack in the face of danger and is characterized by increased HR and CO, as well as by increased pressure and blood flow to the skeletal muscles. It is generally accompanied by other signs of generalized sympathetic activation, such as *mydriasis and piloerection*. In humans, all these responses are more or less present in stressful and emotional situations. These are responses quite similar to those present also in physical exercise. In all these conditions, vasodilation in the skeletal muscles may be mediated by cholinergic sympathetic fibers (see above).

Cats and felines, in general, are among the animals in which the cholinergic sympathetic system has been most frequently described. The cardiovascular variations typical of the alarm or defense reaction were experimentally induced in the anesthetized cat with stimulation of the posterior hypothalamus. The answer was a simultaneous increase in blood pressure in its systolic, diastolic, and pulsatile components, as well as in heart rate.

An increase in pressure, due to an intense sympathetic activation with the release of neural and adrenal catecholamines, should be accompanied by a decrease in HR by stimulation of arterial baroreceptors. The hypothesis has been then formulated that the stimulation of the hypothalamus also causes a vagal inhibition. It has also been observed that in these conditions the *perifornical region of the hypothalamus* can inhibit the nucleus of the solitary tract. This is responsible for the so-called *central resetting of baroreflex,* which explains the absence of bradycardia and the persistence of the increase in pressure during the *fight or flight reaction.*

At the end of the emotional state, an induced alerting response can be observed *post-excitatory bradycardia*. The ceased inhibition of the solitary tract nucleus allows the recovery of baroreceptor reflexes. For some time the pressure remains high due to the slow catabolism of catecholamines, the stimulation of the baroreceptors may now reduce the heart rate. It is interesting how in these conditions it is possible to develop ventricular arrhythmias (*extrasystoles*) induced by the excess of adrenergic action on the heart and favored by the long interval that separates two consecutive sinus beats due to post-excitatory bradycardia. If the extrasystoles arise in rapid succession they can give rise to ventricular tachycardia which could become ventricular fibrillation. With this mechanism are explained the sudden death that sometimes follows strong emotions.

Apart from the cases of genetic predisposition (see Boxes 10.3, 10.9, and 10.12) there is strong support for a triggering role of *emotional stress in sudden death* even in healthy subjects. It is assumed that the sympathetic and parasympathetic nervous system plays an important role. Sympathetic activation has been associated with arrhythmias preceding sudden death and the reduction of sympathetic tone has decreased the incidence of arrhythmias during behavioral stress. Most of the literature indicates a protective influence of vagal tone over the electrical stability of heart tissue, although some studies suggest that vagal stimulation can lead to arrhythmias. Atropine has in some cases proven to be able to block stress-induced arrhythmias. A captivating hypothesis is that there must be an interaction between sympathetic and vagal activity due to which the threshold for ventricular arrhythmias can be lowered.

The alert reaction, besides cardiovascular responses, displays also an increase in pulmonary ventilation and a decreased sensitivity to pain. All these responses seem to be coordinated by *orexins or hypocretins*, peptides produced by some hypothalamic neurons located mainly in the *perifornic area,* and in the *lateral hypothalamus area* (see Chapter 13.3).

In some animals, such as in the opossum or rabbit, defense against aggression by stronger animals is characterized by a reduction in HR and respiratory activity, so that the animal pretends to be dead and the aggressor moves away. Blood pressure is also decreased. Evidently in these circumstances, the activity of the *depressor area* of the *anterior hypothalamus* prevails resulting in activation of the vagal tone and inhibition of the sympathetic tone. This response is called *"playing dead response"* and originates in the *cingulate gyrus* of the *limbic system*. In humans a similar situation to the "playing dead response" is the passing out in response to stressing situations, already described by Dante Alighieri in the *Divina Commedia*: "I came as I died; and fell like a dead body falls" (Canto V, 142).

16.3 Responses to Hemorrhage

It is possible that, due to a conspicuous hemorrhage, but not life-threatening, a patient presents a not excessively reduced arterial pressure accompanied by a very high HR. The picture can be explained as follows: a decrease in the circulating volume and the arterial pressure determine a lower stimulation of the baroreceptors (baroreflex withdraw) with less "buffering" impulses towards the nucleus of the solitary tract. The minor activity of this nucleus will result in reduced stimulation of the *vagal nuclei* as well as the *caudal ventrolateral medulla*, which acting as a brake on the *rostral ventrolateral medulla* will inhibit the sympathetic discharge. The resulting lower vagal activity and increased sympathetic activity will increase myocardial contractility and vasomotor tone, and, especially, the HR with attenuation of arterial pressure drop.

When pressure drop is not limited by the above described compensatory mechanisms an acute hemorrhagic shock characterized by a series of cardiovascular responses occurs. Besides hemorrhage, shock can be determined by other causes of hypovolemia (e.g. heat stroke), as well as by marked vasodilation (in septicemia, anaphylaxis, or heatstroke), or heart failure (*e.g.* in a heart attack or valve rupture).

If we consider hemorrhage as an emblematic case of *hypovolemic shock*, it is possible to understand how the nervous and humoral mechanisms try to work in concert to restore an adequate blood pressure level for organ perfusion: acute hemorrhage cause hypovolemia, central venous pressure reduction which, in turn, determines the reduction of SV (Frank-Starling mechanism) and, therefore, fall in CO and arterial pressure. To this point the above described nervous mechanisms, which are mainly due to the withdrawal of baroreflex, are accompanied by the kidney intervention with the production of renin which leads to angiotensin II production. The latter stimulates thirst and the production of vasopressin/ADH and, aldosterone. Due to the increased sympathetic activity, the skin becomes pale and sweaty (due to vasoconstriction and stimulation of the sweat glands). The renin-angiotensin-aldosterone system enhances the vasoconstriction and promotes a slow restoration of extracellular volume and, therefore, blood, retaining NaCl and liquids and increasing the feeling of thirst.

If bleeding is high (30% blood loss or more of the blood volume) a phase of irreversible decompensation can occur due to the loss of red blood cells and hemoglobin leading to paradoxical vasodilatation. Indeed, this loss of hemoglobin, along with the excessive vasoconstriction, can lead to hypoxygenation and hypoperfusion of tissues and, therefore, a metabolic vasodilator response can counteract the nervous and humoral vasoconstriction. This paradoxical vasodilation decreases blood pressure, leading to hypoperfusion of other tissues. The subject may have a *sympathetic storm* in the unlikely attempt to increase CO and may, therefore, enter a vicious circle (hypoperfusion/hypoxygenation-vasodilation-hypotension-vasoconstrictor-hypoperfusion/hypo-oxygenation) that is difficult to reverse without the intervention of a physician and a blood transfusion.

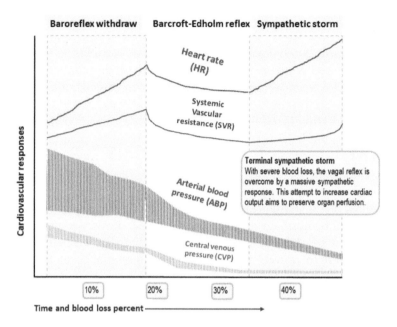

Figure 16.2   Variation in heart rate, systemic vascular resistance, arterial pressure, and central venous pressure during a progressive loss of blood. Note that during the Barcroft-Edholm reflex, the reduction in SVR is not accompanied by tachycardia probably due to an increase in vagal tone induced by low CVP.

Pressure collapse can be accompanied by acute renal failure, with anuria and, especially in patients with pre-existing coronary problems, myocardial hypoperfusion with acute heart failure. Because of the alteration of coagulation factors could be induced by the so-called *disseminated intravascular coagulation*, which aggravates the hemorrhagic picture. Figure 16.2 summarizes the cardiovascular responses during hemorrhage.

A *Barcroft-Edholm reflex* can occur at a certain critical low value of *central venous pressure* (CVP). This reflex determines skeletal muscle vasodilation and arterial pressure drop whose meaning is avoiding perfusion increases in the site of the wound respons e of blood loss. This reflex appears paradoxical because it aggravates and does not correct the hypotension associated with blood loss. For some authors, this reflex is an indirect proof of neurogenic sympathetic cholinergic vasodilation in skeletal muscle. Whether or not this reflex is present when the blood loss is severe (more than 30%

blood loss) the above described sympathetic storm can then occur. Again, blood transfusion is the only way to reverse this condition.

Barcroft-Edholm reflex can also be involved in the post-exercise collapse. In this condition, the optimum management is the *Trendelenburg maneuvers* consisting of the elevation of the feet and pelvis of collapsed athletes above the level of the right atrium. This maneuver results in a rapid reversal of hypotension and symptoms of dizziness. According to some authors, these beneficial effects would not result simply by increasing venous return and hence SV and CO, but also by increasing the CVP and reversing the skeletal muscle vasodilation induced by the Barcroft/Edholm reflex.

17

Myocardial Protection Against Ischemia-Reperfusion Injury

17.1 Ischemia and Reperfusion Injury

Myocardial infarction (MI), colloquially known as "heart attack," is due to a decreased or complete cessation of coronary blood flow to a portion of the myocardium (regional ischemia). Myocardial infarction very often leads to acute hemodynamic deterioration and sudden death, but it may be occasionally "silent" and go undetected. Death is often due to the onset of ventricular fibrillation before myocardial necrosis has had time to appear. To have cell death and necrosis, ischemia must last at least 30 minutes. Most MI is due to underlying coronary artery disease, the leading cause of death in Western countries. With coronary artery occlusion, the myocardium is deprived of oxygen. Patients can present with chest pain or discomfort, often described as squeezing pressure, fullness, tightness, or a heavy weight in the center of the chest. The pain can radiate to the neck, jaw, shoulder, or arm. In addition to the history and physical exam, myocardial ischemia may be associated with elevated biochemical markers such as cardiac troponins and typical ECG changes (see Box 10.10).

The damage caused by ischemia is aggravated in the first few minutes of any subsequent reperfusion. It follows that, if the opening of a coronary branch is necessary to avoid the extension of the infarcted myocardial mass, it paradoxically represents *per se* a cause of aggravation of the injury. To emphasize the role of reperfusion in infarct damage, we talk of *ischemia-reperfusion injury* (IRI).

During ischemia and reperfusion, *calcium overload, oxidative stress*, and *lowering of ATP levels* may reach critical levels at which the ability of the cardiac cells to remain viable is compromised, and the cells undergo uncontrolled death through processes of oncosis and necrosis. Moreover, even before this step, programmed cell death pathways may be initiated including apoptosis, necroptosis, and pyroptosis. Each of these cell death types contributes to infarct size and may activate different levels of inflammatory response.

IRI affects both the cardiomyocytes and the coronary vessels. Some mechanisms underlying myocardial damage, such as necrosis and apoptosis, start during the ischemia period and worsen during the first minutes of reperfusion. Other mechanisms of myocardial damage and the mechanisms of vascular damage occur only in the first few minutes of reperfusion. Vascular damage consists mainly of *acute endothelial dysfunction* and its consequences (see below).

The overall result of IRI is represented not only by cell death but also by hypo-contractility called *myocardial stunning*. This is partly due to the presence of dead tissue and partly to the fact that vital tissue does not contract or contract with less vigor. Stunning can have a variable duration, from minutes to days, depending on the severity of the ischemia. Stunning is a serious problem in patients recovering from *global* ischemia during cardiac surgery. These patients need external support to survive until recovers from stunning. We can say that "if the heart doesn't pump anything else matters".

During ischemia, the lack of oxygen limits or prevents the re-synthesis of ATP by the processes of oxidative phosphorylation. There is, therefore, accumulation of its catabolites (nucleotides, nucleosides and purine bases) and inactivation of ATP-dependent ion pumps. The inactivation of the Ca^{2+} pumps, both sarcolemmal and sarcoplasmic reticulum pump 2 (SERCA2), determines a cytoplasmic and mitochondrial overload of Ca^{2+}. Calcium overload, in addition to inhibiting mitochondrial respiration, contributes, together with the accumulation of nucleotides and purine bases, to an increase in the osmolality of the cells, which, attracting water from the interstice, undergo *swelling*. Ischemia is also accompanied by loss of phospholipids of the membrane, which therefore becomes more vulnerable especially in the presence of cellular swelling.

A nucleoside that is formed in abundance during ischemia is *adenosine*. As we saw in Chapter 14.1, adenosine is transformed into inosine by the action of the enzyme adenosine deaminase. In turn, adenosine can be transformed into hypoxanthine by a nucleoside phosphorylase.

Most of the myocardial reperfusion injury is due to the opening of the *mitochondrial permeability transition pores* (mPTP) and the *explosive swelling* of the cells. The mPTPs are complex pores present in the membranes of mitochondria. They allow the passage of molecules weighing less than

1,500 Daltons. Their opening leads to swelling of mitochondria and cell death.

At the beginning of reperfusion, the opening of mPTPs is the consequence of the aggravation of the Ca^{2+} overload and the production of *reactive oxygen species* (ROS). The first form of these ROSs, the superoxide anion (O_2^-), is produced by many enzymes. The action of the xanthine oxidase enzyme on hypoxanthine which was formed during ischemia is considered one of the principal sources of ROS during IRI. Xanthine oxidase is expressed mainly by granulocytes and endothelial cells adhering to each other. This adhesion is favored by the lack of nitric oxide (NO) that occurs in *endothelial dysfunction* (see below). At the same time, there is also the activation of the enzyme NADPH-oxidase which, in the presence of the large quantity of molecular oxygen brought by reperfusion, determines the production of O_2^- and hydrogen peroxide (H_2O_2). The mPTP is also central in the so-called *ROS-induced ROS release*. A *vicious cycle* enhances ROS production and damage.

The cellular overload of Ca^{2+} that occurs in ischemia and reperfusion is also responsible for the incomplete diastolic relaxation of the myocardium which results in an increase in ventricular diastolic pressure, which can limit ventricular filling.

In reperfusion, the *explosive swelling* of cardiomyocytes is the consequence of the further increase of cellular osmolarity due to the aggravation of Ca^{2+} overload at the time when the sarcolemma has been made more vulnerable by the loss of phospholipids and the so-called pH paradox occurs.

The *pH paradox* is the consequence of changes in cell pH in ischemia and subsequent reperfusion. During ischemia, the cellular pH drops, deactivating those hydrolytic enzymes such as phospholipases and proteases, which otherwise can "attack" the sarcolemma. In this sense, the acidosis situation plays a protective role. On arrival instead of reperfusion, the pH returns to normal because of H^+ washout reaching the pH an optimal level for the action of enzymes which can thus damage the cell membranes causing them to break.

The *reperfusion-induced vascular damage* is due to the lack of NO production by the endothelium during ischemia/reperfusion. As a result of the lack of NO, vasoconstriction, platelet aggregation, and activation of cellular adhesion molecules (CAM) can occur, which promotes interaction between leukocytes and endothelium. The main CAM molecules involved in this mechanism are selectins (P, L, E), integrins (CD11, CD18) and immunoglobulins such as the intracellular adhesion molecule-1 (ICAM-1), the vascular

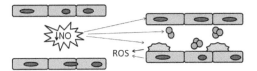

Figure 17.1   The reduction of the NO concentration at the beginning of reperfusion results in vasoconstriction and activation of the adhesion molecules. It follows that the vasoconstriction is added to the vascular obstruction by aggregation of platelets (in blue) and adhesion of neutrophils (in green) to the endothelium, so the flow is reduced or prevented (*no-reflow*). Note that after adhesion neutrophils and endothelial cells produce reactive oxygen species (ROS).

cell adhesion molecule-1 (VCAM -1) and platelet endothelial cell adhesion molecule-1 (PECAM-1).

Platelet aggregation and adhesion of leukocytes to the endothelium may result in a reduction in the cross-sectional area of small vessels with increased resistance to flow, which is however more accentuated by vasoconstriction due to NO deficiency (Figure 17.1). The concomitance of these factors can be among the causes of the so-called *no-reflow phenomenon, i.e.* a reduced or absent perfusion of a previously reperfused territory. This phenomenon may be at the base of a further aggravation of myocardial damage.

Since endothelium in reperfusion are also responsible for the production of ROS that damage myocardial fibers, this component of myocardial reperfusion injury is also due to the initial endothelial damage.

An important aspect of ischemia/reperfusion is the appearance of *arrhythmias* accentuated especially in reperfusion (see Box 10.9: arrhythmia mechanisms, in particular, the early and delayed after depolarization). These arrhythmias can result in ventricular fibrillation with a lethal outcome.

17.2 Pre-Conditioning and Post-Conditioning

A good part of the ischemia-reperfusion damage occurs in the first minute from the beginning of reperfusion due to the opening of the mPTP. The mPTP opening is prevented by low pH, therefore their opening is part of the pH paradox. To prevent their opening is considered a main myocardial protection mechanism to achieve. Indeed, the limitation of pH recovery has been considered as a possible therapy to limit reperfusion injury. However, arrhythmias may be aggravated by acidosis.

Ischemic preconditioning (IPC) is the protection against ischemia-reperfusion damage that can be achieved with one or more short occlusions of a large coronary branch (*e.g.,* 5 min ischemia $\times$ 4 times separated by 5 min reperfusion has been the first IPC protocol tested in 1986 by Murry CE, Jennings RB, and Reimer KA). If the same coronary branch is occluded for a sufficiently long time (generally 30–60 min) to cause a MI, the size of the infarcted area is reduced concerning the **area at risk** (the territory perfused by culprit artery). Ischemic preconditioning also induces a better recovery of contractility after ischemia and a lower incidence of arrhythmias. In some animal models the infarct area, which generally affects 50–60% of the area at risk, is reduced to 25–30% of this area. It is now clear that in addition to delaying ischemic cell death, the IPC protects against reperfusion injury.

The brief preconditioning ischemia ranges from 2 to 10 min. The protective effect of IPC occurs in two phases or two windows. The *first window of protection*, lasting from 1 to 3 hours depending on the duration of preconditioning ischemia, is followed by 20–24 hours of no protection. Thereafter, there is the *second window of protection* that persists for 70–90 hours. It was also observed that while in the first window the limitation of the infarct area prevails, in the second the limitation of the stunning prevails.

Post-transcriptional modifications are among the main mechanisms responsible for the first preconditioning window. It has been suggested a fundamental role for adenosine, ROS, and NO as triggers of this post-transcriptional modifications.

Adenosine released in the short preconditioning ischemia(s) may act on the adenosine G-protein coupled receptors (GPCR) A1, A2a, A2b, and A3. The prevalent receptor depends on many factors, including the animal species and the phase of the IPC phenomenon considered. It is considered prevalent the fact that adenosine would activate a phospholipase C (PLC), which in turn would break down phosphoinositol-diphosphate (PIP2) into inositol-triphosphate (IP3) and diacylglycerol (DAG) (Figure 17.2). In these processes, especially for the intervention of IP3 and the DAG the isoforms α, δ, and ε of the *protein kinase C* (PKC) are activated. These kinases would, in turn, determine the opening of the *ATP-dependent mitochondrial K^+ channels* (mito-KATP). In particular, IP3 would activate the isoform α through a transient increase in the intracellular Ca^{2+} concentration due to the opening of the ryanodine ion channels.

The opening of mito-KATP channels would be followed by the formation of a small amount of ROS by mitochondria. ROS signaling, through a series of survival kinases, would prevent the opening of mPTPs. It is important to note the apparent paradox that ROS, which is normally

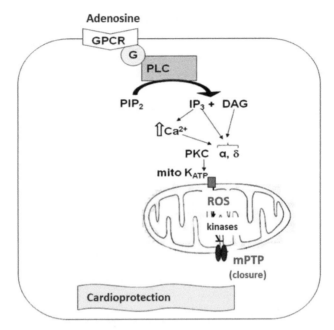

Figure 17.2   Pathways of myocardial protection triggered by adenosine. GPCR: G-protein coupled receptor; G: G protein isoform; PLC: phospholipase C; PIP2: phosphoinositol-diphosphate; IP3: inositol phosphate; DAG: diacylglycerol; PKC α, δ: α and δ isoforms of the protein kinase C; mitoK$_{ATP}$: ATP-dependent mitochondrial K$^+$ channels; ROS: reactive oxygen species; mPTP: mitochondrial permeability transition pores. See the explanation in the text.

considered detrimental factors, in this particular case, at low concentration and appropriate timing, would participate in myocardial protection.

Several survival kinases have been involved in IPC, these include P38 mitogen-activated protein kinase (P38-MAPK), c-Jun N-terminal kinases (JNK), the extracellular signal-regulated kinase 1/2 (ERK 1/2) and protein kinase B/Akt (PKB/Akt). PKC ε and PKB/Akt may be mPTP inhibitors. Also, activators of other kinases such as tyrosine kinase (TK) are involved in the IPC phenomenon.

Adenosine acting on receptors A1, A2b, and A3 (Figure 17.3) may lead to the activation of e-NOS both *via* the phospho-inositol-3-kinase (PI3K) and PKB/Akt pathways, and via the mitogen-activated extracellular regulated kinase 1/2 kinase pathway (MEK 1/2). The nitric oxide thus formed leads to the protection through the activation of guanylyl cyclase-cGMP-PKG cascade, the reduction in the intracellular Ca^{2+}concentration, the activation

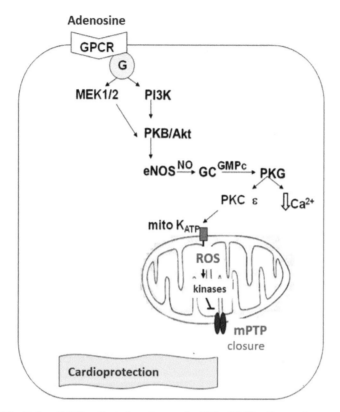

Figure 17.3 Role of NO-activated pathways in IPC. GPCR: G-protein coupled receptor; G: G protein isoform; MEK1/2: mitogen-activated extracellular regulated kinase 1/2; PI3K: phospho-inositol-3-kinase; PKB/Akt: protein kinase B/Akt; eNOS: NO endothelial synthase; GC: guanylyl cyclase; PKC: protein kinase G; PKCε: ε isoform of protein kinase C; mitoK$_{ATP}$: ATP-dependent mitochondrial K$^+$ channels; ROS: reactive oxygen species; mPTP: mitochondrial permeability transition pores. GMPc: cGMP or cyclic guanosine monophosphate See the explanation in the text.

of PKCε and therefore in the opening of the mito-KATP channels which lead to the ROS signaling described above. The action of adenosine on A2b receptors seems particularly efficient in the reperfusion phase.

Short ischemia to induce IPC can also produce NO through an adenosine-independent pathway. In particular, during ischemia, the reduction of the pH in the vessel wall activates a plasma kininogenase which leads to the formation of bradykinin, which acts on bradykinin B2 endothelial receptors (a GPCR), which in turn activate the e-NOS.

As we said before, after the first window of protection, we observe a period of 20-24 hours characterized by the absence of protection. After this period, the second window of protection starts.

Since superoxide anion and nitric oxide are formed in reperfusion with preconditioning ischemia, peroxynitrite may form according to the following reaction:

$$O_2^- + NO => ONOO^-.$$

The *peroxynitrite* activates the ε-isoform of the PKC which, activating TK in turn, initiates a cascade of signals, which, with the involvement of the NF-κB, determines the formation of the inducible-NOS (iNOS). Therefore, the NO produced by eNOS ensures the trigger, while the NO produced by iNOS mediates the late protection. The time required for the appearance of the second window is due to the complexity of the cascade as well as the time required for the expression of the iNOS and other enzymes. In the transition from early to late protection, the expression of some heat shock proteins (HSPs) was also observed. However, it is not clear whether HSPs constitute an obligatory passage or are the simple markers of late protection.

Ischemic postconditioning is the protection that is obtained immediately after the end of ischemia by performing some very short occlusions (a few seconds; the exact duration varies among species: 1–2 minutes have been used in the humans). It has been proposed that after infarction and a few seconds of reperfusion, the artery studied is re-closed two or three times for a few seconds, separating the re-occlusion with the full reopening of a few seconds. The protection against infarct size is quite similar to that obtained with IPC, which underlines the importance of reperfusion in determining the total damage caused by ischemia/reperfusion.

The fact that post-conditioning protection is suppressed by inhibition of NOS and guanylyl cyclase shows that the path to protection passes through NO. Once the NO is formed, the next steps are the ones we saw for preconditioning.

It was the group of D.E. Yellon of University College London that proposed the same pathway for pre- and postconditioning and they called it the *Reperfusion Injury Salvage Kinases* (RISK). As can be seen in Figure 17.4, both pre- and postconditioning would lead to the formation of adenosine and bradykinin. Through the respective GPCR receptors, these substances would initiate a cascade that subsequently involves TK, PI3K, and PKB/Akt. The cascade would end with the activation of the e-NOS.

Preconditioning or Postconditioning

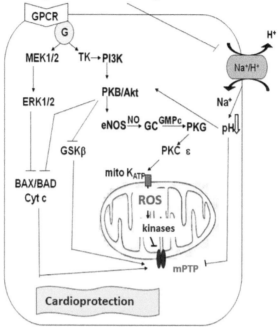

Figure 17.4 RISK cascade to myocardial protection. GPCR: G-protein coupled receptor; G: membrane G protein; MEK 1/2: mitogen-activated extracellular regulated kinase 1/2; ERK 1/2: extracellular signal-regulated kinase 1/2; BAX/BAD: pro-apoptotic proteins; Cyt c: cytochrome c; TK: tyrosine-kinase; PI3K: phospho-inositol-3-kinase; PKB/Akt: protein kinase B/Akt; GSKβ: glycogen synthase kinase β; Na$^+$/H$^+$: exchanger Na$^+$/H$^+$; eNOS: endothelial NO synthase; GC: guanylyl cyclase; PKC: protein kinase G; PKC and: isoform and protein kinase C; mitoK$_{ATP}$: ATP-dependent mitochondrial K$^+$ channels; ROS: reactive oxygen species; mPTP: mitochondrial permeability transition pores. GMPc: cGMP or cyclic guanosine monophosphate See the explanation in the text.

Another cascade, parallel to the main RISK pathway, inhibits the opening of mPTP through the inhibition of pro- apoptotic BAX/BAD proteins and cytochrome *c* by MEK 1/2 and ERK 1/2 kinases. Other pathways would be activated together to the RISK pathway and would also be common for pre- and post-conditioning. One of these pathways sees the inhibition of GSKβ by the PKB/Akt. Since GSKβ tends to open the mPTP, its inhibition is a protective factor. An important role is also played by the NO/cGMP/PKG pathway.

The pathway called *survivor activating factor enhancement* (SAFE) has been also proposed in pre- and post-conditioning. This pathway starts with the activation of the tumor necrosis factor α (TNFα), which binds to the specific membrane receptor TNF-R resulting in the phosphorylation of Janus kinase (JAK). The phosphorylated JAK, in turn, phosphorylates the signal transducer and activator of transcription-3 (STAT-3) which prevents the opening of mPTPs. In these protective pathways, a role for low ROS has been proposed.

Finally, ischemic postconditioning, by reducing the activity of the Na$^+$/H$^+$ exchanger, lowers the cytosolic pH which is responsible for both PKB/Akt activation and mPTP opening inhibition.

Postconditioning has a much greater applicative significance than pre-conditioning, as it can be used immediately after a coronary vessel has been opened by angioplasty in an infarcted patient, while the time of use of a preconditioning maneuver cannot be programmed. This does not mean that the preconditioning has no clinical implication.

Indeed, protection by IPC can be induced locally in the heart, as above described, but protection can also be induced by ischemic episodes in tissues remote from the heart, namely, *remote ischemic conditioning* (RIC). Brief controlled episodes of intermittent ischemia of the limbs may confer powerful systemic protection against prolonged ischemia in a distant organ. It is a feasible clinical tool that can be performed prior (remote preconditioning or pre-RIC), during (remote preconditioning or per-RIC), or after (remote postconditioning or post-RIC) a prolonged infarction myocardial ischemia.

A sort of *similarity exists between RIC and exercise*. Whether exercise training can be used as a sort of pre-RIC is not clear yet. We know that subjects who regularly practice sports or, paradoxically, experience angina attacks, have reduced lesions if they are hit by a subsequent myocardial infarction. In fact, both in angina attacks and in the practice of sport, the myocardium can undergo episodes of more or less marked stresses and ends up being constantly in one of the protection windows. Indeed, a sedentary lifestyle is an established risk factor for cardiovascular diseases and regular exercise may be cardioprotective. However, the exact mechanisms through which regular exercise and RIC confer cardiovascular protection are not yet well understood. Exercise, as well as RIC, probably acts at multiple levels. One possibility is that regular intermittent stimuli during exercise modify genes expression, thereby changing the production of bioactive molecules such as proteins and enzymes. Moreover, exercise can affect endothelial and platelet functions, thus reducing the risks of atherosclerosis and consequently the risk of infarction and stroke. The practice of sport, through intermittent increases in shear stress and pulse pressure, may be the cause of a greater expression of messenger RNA for eNOS. A similar response is not present in RIC.

17.3 Pharmacological Pre-Conditioning and Post-Conditioning

Numerous studies have been conducted to obtain myocardial protection with pharmacological compounds. *Diazoxide*, an opener of mito-KATP channels, has been seen to trigger both pre and postconditioning. The preconditioning was also obtained with the *phorbol esters* and with the *monophosphoryl-lipid A* (MPLA). The mechanism of action of the phorbol esters, which triggers both the first and the second windows of protection, comprises the activation of the PKCs, while that of the MPLA, responsible only for the late protection, passes through the expression of the inducible heat shock protein 70 (HSP70).

Recently, the protective effect of endogenous peptides has been studied, such as vasostatin-1 (VS-1), catestatin, and apelin. Of these substances, the VS-1 was seen to protect the isolated rat heart in experiments in which it was infused before 30 min of ischemia, while in the same model the catestatin and the apelin have given good results if given at the beginning of reperfusion. For all these peptides the protective effect is mediated by many mechanisms, including NO production. In particular, Apelin is a ligand of the orphan receptor APJ, which is a GPCR. Apelin like many other substances exerts post-conditioning-like protection against IRI through activation of PI3K-PKB/Akt-NO signaling cascade. The pathway connecting APJ to PI3K may occur through transactivation of the epidermal growth factor receptor (EGFR) *via* a Src kinase.

It must be said that the attempts to translate cardioprotection from experimental evidence to clinical outcome benefit for patients undergoing cardiovascular surgery or suffering acute myocardial infarction have been disappointing so far. The causes for such lack of clinical translation are likely the confounding roles of many factors including the *comorbidities, aging, gender, and/or comedications*. Attempts to analyze these problems have been performed in clinical and experimental contexts. For example, G protein-coupled estrogen receptor (GPER) is an estrogen receptor expressed in the cardiovascular system. G1, a selective GPER ligand, exerts cardiovascular effects through activation of the PI3K-PKB/Akt pathway and *Notch signaling* in normotensive animals but also in female spontaneous hypertensive rat hearts. Experiments like this open the hope to overcome the blunting due to comorbidities. However, further studies are necessary. For example, *platelet inhibitors* (antagonists of the $P2Y_{12}$ receptor) recruit cardioprotection *per se* thus limiting the possibility to obtain further protection by other means, thus this is one of the best pieces of evidence for the interference of comedications. Moreover, some substances, like *propofol anesthetic*, abrogate protection from RIC in cardiovascular surgery. Several comorbidities may limit endogenous cardioprotective potential. For instance, hyperglycemia and diabetes may attenuate the cardioprotective effects of pre-, post-conditioning, and RIC. Yet, diabetes mcllitus patients are on anti-hyperglycemic agents some of which are known to be per se cardioprotective (*e.g.,* GLP-1 agonists, DPP-4 inhibitors, and SGLT2 inhibitors) or can interfere with conditioning (*e.g.,* glibenclamide). Nevertheless, the data available until now are limited to overcome the lack of clinical translation and further studies are required.

18

Lymphatic Circulation

18.1 Formation of the Linfa

In the study of microcirculation (see Chapter 11.9), we have seen that filtration through the capillary wall occurs thanks to four pressures known as *Starling forces*:

Pc = blood pressure in the capillary (or capillary hydrostatic pressure);

Pi = interstitial fluid pressure (or interstitial hydrostatic pressure);

πi = oncotic pressure of interstitial fluid proteins (or interstitial oncotic pressure);

πp = oncotic pressure of plasma proteins (or capillary oncotic pressure).

From the gradients of the mean values of all these forces it is possible to calculate the quantity J_v of liquid filtered in the unit of time:

$$J_v = K_f([P_c - P_i] - \sigma[\pi_c - \pi_i]).$$

Being K_f a *filtration coefficient* typical of the vascular district in a particular moment. σ is a unit-less factor called *reflection coefficient*, which is equal to 1 when no proteins cross the vessel wall, and it is lower to 1 when proteins cross the wall. It has a mean value of 0.85-0.90 in our body.

If with a certain approximation, the values of Pi and πi are considered equal to zero and σ is considered equal to 1, the formula can be so simplified:

$$J_v = K_f(Pc - \pi p).$$

From this simplification, we see that in many vascular districts, the values of these pressures favor the liquid exit along the entire blood capillary (see Chapter 11.9). Nevertheless, σ is rarely 1 and proteins cross the vessel wall. The mean values of σ in many districts are between 0.8 and 0.9.

When arterioles tone increases, capillary pressure decreases, and the Starling forces may favor transient reabsorption on the venular side of the capillaries. Whatever be the case, *in many districts capillary filtration prevails over reabsorption so that a large part of the filtered liquid would remain in the interstice and enter the lymphatic circulation.* Therefore, we understand the importance of the lymphatic system for liquid reabsorption and to maintain the balance of fluids between the blood and tissues.

We can say that the *lymph* has a composition very similar to plasma. It is formed by the liquid resulting from outward filtration at the arteriolar side of the capillaries that is not followed by inward reabsorption at the venular side of the capillaries. Given its origin, apart from the different protein content, the lymph, before entering the lymph-nodes, has a composition very similar to plasma. Compared to plasma, the lymph contains a greater quantity of catabolites produced by the tissues. Since it contains coagulation factors, it can coagulate if placed *in vitro*.

However, the composition of the lymph is not uniform, as it may contain a different concentration of proteins depending on the tissue considered and its richness in lymph-nodes. The proteins are almost absent in the lymph of the eye's ciliary body and can present a concentration of 6 g per 100 mL in that of the liver. Furthermore, the lymph coming from the small intestine during the absorption processes, also known as *chyle*, has the milky characteristic as it contains a considerable amount of free fatty acids (FFAs) and triglycerides, derived from the lipid digestion in the small intestine which cannot be absorbed as they are but they are assembled with proteins forming *chylomicrons*. These lipoproteins cannot be directly taken up by capillaries because of their dimension.

18.2 Lymphatic Vessels

After having formed in the interstitial spaces, the lymph passes into the blind-ended lymphatic capillaries present in these spaces. In particular, the lymph, rich in fats that are produced in the *intestinal villi,* is drained in that particular capillaries called *lacteals*. These are connected with the *submucosal lymphatic network* at the base of the villi that have abundant interconnections.

Figure 18.1 illustrates the organization of the lymphatic network. The *lymphatic capillaries*, in addition to anastomosing each other, converge to give rise to the *lymphatic ducts*, equipped with *"swallow-nest"-semilunar valves.*

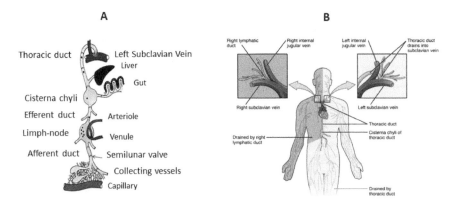

Figure 18.1 **A**: scheme of lymphatic circulation. **B**: in yellow the territory drained by *thoracic duct* and in violet that drained by *right lymphatic duct*.

The initial lymphatic networks contain blind-ended vessels that lack smooth muscle cells. These capillaries are non-contractile and lack real valves. Since they are non-contractile, their filling relies on other extrinsic forces and structural features; these include peristalsis, duct contraction (acting like a Pasteur pipette), vessels connection to local connective tissue, and unique cell-cell junctions overlapping that form *flaps* that serve a specialized function as unidirectional valves during the filling of the initial lymphatic network.

Downstream of the initial lymphatics are the collecting lymphatics ducts, which serve to transport lymph away from the initial lymphatic network. Wall ducts contain smooth muscle and semilunar valves and are not attached to local connective tissue. Along the course of these ducts lie the *lymph-nodes*, which have an immune function and are the site of lymphocyte production. They retain corpuscles and bacteria, as well as cancer cells, becoming, in this case, a site of metastasis. The vessels that carry the lymph to the lymph-nodes are called afferent, while those that carry it out of the lymph nodes are called efferent.

Some parts of the body are particularly rich in lymph-nodes; these are the lymph nodes of the head and neck, those of the thorax distributed in the mediastinum, at the hilum of the lungs, around the aorta and pulmonary artery, etc. The lymph nodes of the axilla and those of the upper limb and lymph-nodes of the groin and lower limb are particularly involved in the development of post-surgery *lymphedema* (see below).

The lymphatic vessels converge into increasingly large ducts, such as the *subclavian lymphatic duct*, the *jugular, bronchial, intestinal, and lumbar ducts*. The subsequent confluence of these ducts gives rise on the right side to the *right lymphatic duct* and, on the left, to the *thoracic duct*. These two ducts carry the lymph into the bloodstream opening, from each side to the confluence of the internal jugular vein with the subclavian vein (Figure 18.1).

In this way, the fluid that has been filtered from the blood capillaries into the tissues is returned to the bloodstream. Some fluid returns to the blood already in the lymph-nodes.

The amount of lymph that is formed every day and returns to the cardiovascular system is estimated to be about 4 liters. As said, lymphatic capillaries, like blood capillaries, are formed by endothelial cells located on a basement membrane. The lymphatic vessels, on the other hand, also have smooth muscle cells. Along the afferent and efferent ducts as well as the lymphatic ducts are present the *semilunar valves* that prevent the retrograde flow of the lymph. Therefore, the lymphatic flow is unidirectional because of the valves.

18.3 Lymphatic Flow

The unidirectional flow of lymph is made possible for the entry of interstitial fluid *via* the flaps of lymphatic capillaries as well as for the presence of the *semilunar valves* in the lymphatic ducts and the rhythmic contraction of the smooth muscles of the larger vessels. Smooth muscle cells do not contract when lymph does not come from the capillaries. When instead it arrives in lymphatic vessels and stretches the smooth muscles, these contract by a myogenic response. In addition to the myogenic mechanism, the musculature of the lymphatic vessels can also respond, by contracting, to sympathetic stimulation.

Normally, the water of the lymph represents 2–3% of the total water of the human organism. *Lymphedema* is a disease condition characterized by swelling due to the accumulation of lymph in the tissues. It can affect arms (after surgery and/or radiation therapy for breast cancer with involvement of the axillar lymph-nodes) or legs (for example after surgery and/or radiotherapy to the inguinal lymph-nodes for gynecological tumors) and sometimes other parts of the body if the (regional) lymph-nodes have been surgically removed or subjected to radiotherapy, or they are obstructed by a tumor. If the lymph nodes or lymphatic vessels are compromised or obstructed, the lymph cannot flow out. As a result, it stagnates in the tissues and causes swelling, which can result in the so-called *elephantiasis*. This is a monstrous enlargement characterized by lymphedema, hypertrophy, and fibrosis of the skin and subcutaneous tissues (hyperkeratosis). It typically affects the limbs or scrotum due to obstruction to the lymphatic flow due to the occlusion of the axillar or inguinal lymph nodes by nematodes (*filariasis*). Three known species of the filarial nematode are known that can cause *lymphatic filariasis*: *Wuchereria bancrofti, Brugia malayi, and Brugia timori.*

19

Functional Imaging of the Cardiovascular System: How to Study Human Physiology *In Vivo*

19.1 Functional Imaging vs Anatomical Imaging

Recent developments in biotechnology and bioengineering allow the evaluation of various aspects of human physiology in an increasingly less invasive (and therefore ethically acceptable), precise, and patient-specific manner, and this information often constitutes the basis for guiding and selecting appropriate therapies.

In this context, imaging methods have rapidly revolutionized clinical practice, allowing the visualization of cardiovascular structures and the measurement of multiple operating parameters in a non-invasive way.

The study of physiology is closely related and complementary to the study of normal anatomy. Indeed, the anatomical imaging methods, which provide morphological and structural data on the organs and systems, were the first to be used in the clinical setting. From these methods, functional imaging methods were gradually developed, in parallel with the technological development of radiological and echographic equipment.

An intriguing feature of *functional imaging methods* is the ability to measure, in a non-invasive way, the physiological processes that occur within the organism. They are based in most cases on the use of contrast media or tracers which, in the body, have a distribution proportional to the physiological phenomenon under study. The measurement of the tracer concentration, therefore, allows measuring the biological marker related to it.

There are several methods of functional imaging that allow the measurement of cardiovascular parameters *in vivo*. The purpose of this chapter is to provide, together with classical procedures, a brief overview of the different techniques, and the main functional parameters that they allow measuring cardiovascular function.

19.2 Functional Methods of Analysis

19.2.1 Cardiac Catheterization

Cardiac catheterization is one of the most successful methods of cardiovascular investigation in clinical practice. The procedure consists of inserting, usually percutaneously, a catheter into a cardiac chamber or a vessel. *Coronary angiography* is the most common application of anatomical imaging of cardiac catheterization. However, by using different types of catheters it is possible to measure *in vivo* a series of physiological parameters, such as blood pressure in the heart chambers, blood pressure and flow in the coronary arteries, or to map the electrical activation of the myocardium in a cardiac chamber. Although these mini-invasive procedures of physiological study are not strictly to be considered imaging methods, they are based on the possibility of visualizing in real-time the advancement of catheters inside the vessels and heart chambers (*radioscopy*). For this reason and due to the historical importance of cardiac catheterization, these methods will be briefly discussed in this chapter. In many cases, the percutaneous access route also allows the execution of minimally invasive therapies during the same session (coronary angioplasty, ablation of arrhythmic foci, valve dilations, etc.).

19.2.1.1 Measurement of cardiac output

Measurement of cardiac output is possible during a cardiac catheterization procedure and can be based on the *Fick principle* or the *dilution method.*

• **Fick principle.** Cardiac output (CO) is calculated as a function of the body's oxygen consumption (VO_2; measured by analysis of respiratory gases) and of the arterio-venous difference in oxygen contents (CA-CV), according to the following formula

$$CO = (VO_2/(CAO_2 - CVO_2)) \times 100.$$

The Fick method (Figure 19.1) has long been considered the most accurate method for measuring cardiac output. However, it is invasive, requiring accurate measurement of the oxygen content in arterial and mixed venous blood collected with catheters and an assessment of oxygen content in respiratory gases with spirometry and a Douglas bag to collect expired air.

• **Dilution method.** The dilution method was initially described for the venous injection of an inert indicator or dye. The same principle is currently in clinical use by injecting small amounts of water at the ice melting temperature (*thermodilution*). In the dilution method, CO is calculated as a function

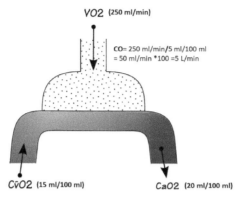

Figure 19.1 Fick principle. The cardiac output (CO) is calculated as the ratio between oxygen uptake (VO_2) and the difference of the arterial (CaO_2) and mixed venous (CvO_2) oxygen content. In other words, if 5 mL of O_2 are taken from the air and added to 100 mL of blood, it means that while 5000 mL/min of blood is pumped into the lung circulation, in one minute 250 mL of O_2 are taken from the alveolar air, added to the blood and then consumed by our body, in resting condition.

of the *mass (m, mg)* of an indicator (dye) injected into the bloodstream and of the *mean concentration (Ct, mg/mL)* of this indicator in the collected blood after a single passage through the heart. In other words, the CO is equal to the quantity of indicator injected divided by the integral of the dilution curve measured downstream and the reported to the unit of time (usually a minute or 60 s). In practice, a small bolus of dye is injected into a vein or right ventricles and in a systemic artery the blood samples are collected over time and dye concertation quantified. From the mass (m, mg) of the dye, the average measured concentrations (Ct, mg/mL), the volume (V, mL) in which the dye has been distributed is calculated ($V= m/Ct = mg/mg/mL = mL$). The time required for the first passage of this volume of blood at the sampling point is obtained by extrapolation and used to calculate the cardiac output (Figure 19.2). In other words, a quantity of dye (m) is dissolved in a volume (V) of blood, whose concertation is Ct. This V is pumped in a certain time (t; 35 s in Figure 19.2). To calculate CO, the V pumped in 35 s must be reported to 60 s.

19.2.1.2 Fractional flow reserve (FFR)

The *fractional flow reserve* (FFR) is a technique used during cardiac catheterization to measure the difference in blood pressure through coronary artery stenosis, to determine the likelihood that this lesion may limit the blood supply to the myocardium. The technique requires miniaturized pressure

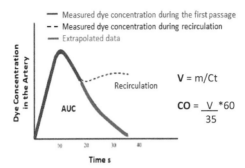

Figure 19.2   Dilution method. The cardiac output (CO) can be calculated if we know the quantity (mass, m) of dye injected into the vein and the mean concentration (Ct) of the dye collected by an artery. The area under the curve (AUC) is used to calculate the Ct of collected dye. This dye is collected in the theoretical time (t) individuated by extrapolation (*e.g.,* 35 s in this figure). This analysis tells us that if there was no blood recirculation, the concentration of the dye in the artery would gradually go from zero (when ventricular blood was not yet reached by the venous dye and the pumped blood was "clean") to a maximum (when the dye reached the maximum concentration into blood pumped by the ventricle) and then would return to zero (when all the injected dye had passed through the heart).

sensors inserted on top of a metal wire that is inserted inside the coronary vessels. FFR, measured during the administration of vasodilator drugs (adenosine), is defined as the ratio between blood pressure after coronary injury and immediately before it. The FFR is expressed as an absolute number. An FFR of 0.5 means that the examined stenosis can cause a 50% drop in blood pressure. Referred to the flow, this means that the lesion creates a 50% reduction of the maximum hypothetically expected flow in the vessel in the absence of stenosis and for the same coronary perfusion pressure.

It is currently believed that an FFR less than 0.75 or 0.8 (depending on the center or the clinical study considered) indicates hemodynamically *significant coronary stenosis*. A stenosis is worthy of acute treatment.

The FFR is currently considered the minimally invasive reference method for the diagnosis of hemodynamically significant coronary stenosis.

19.2.1.3 Electrophysiological mapping

This technique allows the minimally invasive study of the conduction system of the heart. Using special catheters, it is possible to record intracavitary electrocardiographic traces. Electrical or pharmacological stimulations allow to identify areas of abnormal generation or conduction of the electrical stimulus and, if necessary, to proceed with the ablation of the tissue. In recent years, the availability of three-dimensional navigation systems has allowed the visualization of the instantaneous position of the catheters within the heart, the reconstruction of anatomical-functional maps of the analyzed cardiac cavities, minimizing the number of X-ray photons delivered to the patient and the operator during the procedure.

19.2.1.4 Positron emission tomography (PET)

PET is a nuclear medicine technique that allows three-dimensional visualization of physiological processes within the organism. The device records pairs of gamma rays emitted by a radionuclide injected into the body. The radionuclides in use have a relatively short half-life, such as ^{11}C (about 20 min half-life), ^{13}N (about 10 min), ^{15}O (about 2 min) and ^{18}F (about 110 min half-life). They are incorporated into a biologically active molecule, such as glucose, water, or ammonium ion, or into molecules that bind receptors or other biological structures within the body. Such labeled compounds are known as *radioactive tracers* or simply tracers. The pharmacokinetics of the biologically active molecule determines the location and concentration of the radioisotope conjugated to the molecule.

PET is used in the study of the cardiovascular system for the measurement of cardiac metabolism by fluoro-deoxy-glucose (FDG) and the measurement of cardiac perfusion through the use of ammonia labeled with ^{13}N, water labeled with ^{15}O or ^{82}Rb. The tracer is taken up by the tissue concerning myocardial perfusion.

Due to the short half-life that characterizes most positrons radioisotopes, PET tracers are generally produced near the PET diagnostic center using a cyclotron. The half-life of the ^{18}F is long enough to allow the preparation of the radioisotope in another structure and therefore the delivery to the diagnostic center. Recently, ^{82}Rb portable generators have been introduced to the market. Such devices contain ^{82}Sr which decays into ^{82}Rb.

PET is an exclusively functional image method. Therefore, the localization of the biological phenomena measured inside the organism structures requires the fusion of the PET images with computerized axial tomography (CT) images acquired in the same session. To this end, modern PET equipment is a PET-TAC hybrid device.

19.2.2 Single-Photon Emission Tomography (SPECT)

SPECT is also a nuclear medicine method, based on the identification of single gamma rays emitted by a gamma-emitting radionuclide. The radionuclide can be injected as such or as part of low molecular weight molecules or as part of a biologically active molecule. In this case, the pharmacodynamic and pharmacokinetic properties of the tracer are dictated by the type of molecule to which the radioisotope is conjugated.

In the cardiovascular field, SPECT has found widespread use in the diagnosis of coronary heart disease, by the injection of ^{99m}Technzio-tetrofosmina, ^{99m}Tecnezio-sestamibi or ^{201}Tallio during exercise or during pharmacological stimulation with vasodilators, such as *adenosine* or *dipyridamole*, or chronotropic and positive inotropic substance such as *dobutamine*. SPECT allows visualization of the distribution of coronary perfusion flow to the myocardium. Perfusion abnormalities are diagnosed by comparing the results under stress with images acquired at rest, usually 1 to 7 days later (although the use of the ^{201}Tallio allows for the acquisition of images at rest during the same day).

19.2.3 Echocardiography

Echocardiography is an imaging method based on the recording of ultrasonic waves (2–18 kHz). The creation of images takes place in three steps: the production of ultrasonic waves, the recording of ultrasonic echoes, and the interpretation of signals.

Ultrasonic waves, generated by a piezoelectric crystal embedded in a surface probe, travel inside the body. The ultrasonic wave is partially reflected in the passage between one tissue and another and in correspondence of each variation of density (different impedance) within a tissue. The portion of ultrasonic energy reflected (echo) returns to the probe, within which it generates an electrical signal. The electrical signals thus formed are processed and transformed into digital images by the device.

The ***pulsed Doppler method*** is used to measure blood velocity. It is based on the fact that the reflected ultrasound energy changes concerning red cell velocity. The echocardiography allows measuring the diameter of the studied vessel and thus the cross-sectional area that must be multiplied for the mean blood velocity to obtain blood flow. If the vessel is the ascending aorta the calculated flow is the cardiac output (menus the coronary flow).

19.2.4 Computerized Axial Tomography (CAT or CT)

CT is an imaging technique that allows the acquisition of tomographic images of the human body. Starting from a large number of two-dimensional radiographic projections obtained by rotating the X-ray source around the patient, it is possible to reconstruct a three-dimensional image of the body.

Visualization of cardiovascular structures requires the injection of a radio-opaque iodine-based contrast medium, as the tissues of the cardiovascular system do not in themselves offer sufficient contrast to X-rays.

CT is still considered primarily an anatomical imaging method. However, in recent years the availability of new scanners capable of effectively minimizing the radiation dose delivered to the patient has opened up the possibility of using the CT scan also for function, perfusion, and myocardial vitality studies. In the context of pathologies with altered cardiac structure, CT plays an important role in planning interventions.

19.2.5 Cardiovascular Magnetic Resonance

Cardiovascular magnetic resonance is an imaging technique that allows non-invasive assessment of both the anatomy and the function of the cardiovascular system (fMRI, Functional Magnetic Resonance Imaging). The patient is exposed to an intense static magnetic field to which the protons of the nucleus of atoms align. In medical applications, the recorded signal comes from the protons of the nucleus of the *hydrogen* atoms that form the water and triglyceride molecules in the body. The systematic application of a radiofrequency pulse allows altering the alignment of the protons at regular intervals. When the radiofrequency pulse is interrupted, the protons tend to realign with the static magnetic field (relaxation), in turn emitting a radiofrequency signal. This signal (free induction decay) forms the basis for generating images. The speed with which the protons of the hydrogen atoms recover the initial alignment to the static magnetic field depends on the chemical bonds within the molecules and the physical interactions of these molecules with the external environment. On this basis, magnetic resonance allows obtaining a spontaneous contrast in the images between different tissues. In the case of some specific applications, however, the administration of a paramagnetic contrast medium containing Gadolinium is necessary. Paramagnetic contrast agents contain magnetic centers that create magnetic fields approximately one thousand times stronger than those corresponding to water protons. These magnetic centers interact with water protons in the same way as the neighboring protons, but with much stronger magnetic fields, and therefore, have a much greater impact on relaxation/recovery rates.

MRI does not require patient exposure to ionizing radiation (unlike CT, radioscopy, nuclear methods).

The electrocardiogram synchronizes the acquisition of images with the cardiac cycle (electrocardiographic trigger). Several acquisition protocols (sequences) are available, each optimized to obtain a specific type of image and therefore physiological information.

Cardiac magnetic resonance allows the acquisition of moving images for the measurement of contractile function and blood flows. Static images are used for the measurement of the amount of edema in the myocardial tissue or the redistribution of paramagnetic contrast agents (Gadolinium) and therefore the presence and extension of fibrous scar tissue in the myocardium. Also, other specific techniques allow the visualization and measurement of intramyocardial oxygenation levels (BOLD technique) and myocardial perfusion. The BOLD (blood oxygen level-dependent) technique is based on the fact that the change from diamagnetic oxyhemoglobin to paramagnetic deoxyhemoglobin that takes place with metabolic activation results in decreased signal intensity on MRI.

In recent years, MRI procedures have improved considerably and are now indispensable aids for evaluating chronic and acute Coronary Artery Disease (CAD) patients. In the most recent guidelines cardiac MRI is recommended to evaluate left ventricle function and volumes, to study tissues (in particular microvascular obstructions, scars, necrosis, edema and hemorrhage) and to evaluate the extent of ischemia. It allows accurate diagnosing of CAD in patients with intermediate pre-test probability. Thanks to the larger diffusion of this technique, data are accumulating that indicate a potential role of high resolution cardiac MRI to predict malignant arrhythmias and sudden cardiac death in patients with myocardial infarct.

19.3 Physiological Parameters in Clinical Practice (Table 19.1)

19.3.1 Contractile Function

One of the main applications of cardiovascular imaging methods is the measurement of end-diastolic and end-systolic ventricular diameters and the calculation of the *ejection fraction*. This parameter has assumed a key role in the clinical evaluation of cardiac patients, not only because it represents a reliable prognostic indicator, but also because it is easy to calculate with several techniques.

Echocardiography and magnetic resonance are considered the reference methods able to provide a measure of contractile function.

Echocardiography allows the evaluation of a cardiac function at the patient's bedside and a relatively low cost. Echocardiography is an essential

Table 19.1 Physiological parameters in clinical practice. $+++$: reference method in the clinical field; $++$ method currently in clinical use; $+$ method available but not frequently used in the clinical setting; -method not available; * method used in research. FFR: fractional flow reserve; EM: electroanatomical mapping

	Cardiac Catheterization	PET	SPECT	Eco-cardiography	TAC	Magnetic Resonance
Contractile function	$+++$	$+$	$++$	$+++$	$+$	$+++/++$
Myocardial perfusion	$+++$ (FFR)	$+++$	$++$	*/+	*/+	$+++$
Myocardial vitality	*	$++$	$++$	$++$	$+$	$+++$
Blod flow and Cardiac Output	$+++$	$-$	$-$	$++$	$-$	$+++$
Myocardial Oxygenation	$-$	*	$-$	$-$	$-$	$+++$
Electrical activity	$+++$ (EM)	$-$	$-$	$-$	$-$	$-$
Phasic Blood Flow	$+++$	$-$	$-$	*	$-$	*

method in cardiology departments and intensive or semi-intensive care units. Modern echocardiographs produce images of good quality and in real-time, allowing the assessment of the global parameters of cardiac chamber functioning and the identification of segmental abnormalities of contraction, at rest, or during pharmacological stimulation. Furthermore, the availability of Doppler methods also allows the evaluation of diastolic relaxation. The introduction of *strain imaging echocardiography* offers an imaging tool for a more objective study of myocardial dynamics and allows for early detection of subclinical ventricular dysfunction. Strain imaging is based on the recognized helical arrangement of left ventricle fibers which undergo characteristic systolic deformations and diastolic recovery. Three main echocardiographic modalities are currently available to perform strain imaging assessment of myocardial function. These include tissue Doppler imaging and 2-D or 3-D speckle tracking imaging.

Cardiac magnetic resonance is currently considered the reference method for the evaluation of cardiac function. The excellent image quality, superior in most cases to echocardiography and the clinical robustness of the method give it superior reproducibility to other techniques. Moreover, the availability of multiple resonance techniques applicable during the same session allows a complete evaluation of the heart and the large vessels. MRI is to be considered superior to echocardiography also for visualization and measurement of the right ventricle. The high reproducibility candidate, the magnetic resonance with the method of choice also in clinical studies, allows an improvement of the statistical power or a reduction in the number of subjects to enroll to obtain significant results. For the higher cost and the relatively reduced availability however, magnetic resonance is for the moment a second level method for the evaluation of cardiac function to be reserved for selected patients.

During cardiac catheterization it is possible the acquisition of *ventriculographic images*, which is a sort of film in radioscopy during the intraventricular injection of iodinated contrast medium. Ventriculography allows a precise and reproducible evaluation of left ventricular function but its use is limited to patients undergoing cardiac catheterization and is limited by the use of X-rays.

SPECT can provide high quality and reproducibility information on left ventricular function if the signal acquisition is synchronized with the electrocardiogram (gated SPECT). In the same way, even PET can provide information on the volumes of the left ventricle and the ejection fraction, although this application is mostly restricted to research protocols.

Finally, worthy of mention even if limited to research protocols, it is the possibility of acquiring functional images using the CT method. This method, made possible by the progressive reduction of X-ray doses administered during the CT scan, has the potential for wider development in the future.

19.3.2 Myocardial Perfusion

The identification of abnormalities of perfusion of the ventricular myocardium represents a fundamental step for the diagnosis of coronary artery disease. There is now abundant experimental evidence in the literature that demonstrates the superiority of functional tests on anatomical imaging (e.g., angiography) for the identification of patients who are candidates for revascularization therapies, percutaneous or surgical.

PET is still considered the reference method if you want to measure myocardial perfusion quantitatively, even if due to the high cost it is only used in a few medical centers and mainly as part of research protocols. The situation is different for SPECT, which represents a cornerstone among cardiology diagnostic methods, with thousands of centers around the world equipped for performing *stress studies*. In this application, SPECT is supported by a

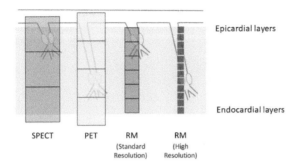

Figure 19.3 Importance of spatial resolution for the identification of subendocardial ischemic phenomena. The spatial resolution of SPECT, PET, magnetic resonance (RM) at standard resolution, and high resolution is represented concerning the thickness of the left ventricular myocardial wall. The superior spatial resolution allowed by RM allows the early identification of subendocardial ischemic phenomena without incurring in *partial volume phenomena* (the loss of information in small regions because of the limited resolution of the imaging system) that limit the diagnostic accuracy of PET and especially of SPECT.

considerable amount of scientific evidence to support diagnostic accuracy and its prognostic value.

The availability of *high field MRI* equipment at high performance allows the acquisition of high-quality perfusion images and diagnostic accuracy. The main advantage of resonance perfusion studies lies in the superior spatial and temporal resolution allowed by the method. The images, acquired in real-time, offer a simultaneous and global coverage of all the sectors of the left ventricle, and each voxel (the elementary unit that constitutes the image) represents a volume of 15–90 mm^3, depending on the equipment in use. In comparison, the PET allows images with voxels of about 125 mm^3, the SPECT of 350 mm^3 (Figure 19.3). A better spatial resolution allows the independent visualization of the layers of myocardium within the ventricular wall (endocardial layers) where, as a consequence of the systolic contraction and reduced vasodilator reserve, perfusion anomalies occur more early and more severely (Figure 19.4). Moreover, the visualization with temporal resolution in the order of a minute, according to the perfusion phenomena, allows easy identification of artifact possibilities and, ultimately, a very high diagnostic accuracy. We should also remember the absence of harmful ionizing radiation for patients in MRI.

Of note, myocardial perfusion functional data during cardiac catheterization can be obtained by the FFR method (see above). Although FFR is not to be considered strictly a functional imaging method, it is based on the possibility of the radioscopic guide of the catheters inside the coronary

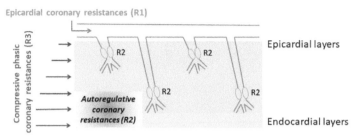

Figure 19.4 A schematic representation of the left ventricular wall and coronary vessels. Coronary resistance can be divided schematically into *epicardial coronary resistance* (R1), arteriolar or *autoregulative resistance* (R2), and *phasic compressive myocardial resistance* (R3), due to the systolic contraction of the myocardium and to the simultaneous increase in endo-ventricular pressure. Phasic compressive resistance has higher value proceeding from the epicardial layers to the endocardial layers and together to reduced vasodilator reserve are responsible for the greater severity and early onset of myocardial ischemia in the endocardial layers of the left ventricle.

vessels and is currently considered by many to be the reference method for the evaluation of myocardial perfusion abnormalities. Its prognostic usefulness in identifying patients who are candidates for revascularization therapies should be remembered. When determining FFR it is recommended to think about the Bernoulli effect and anastomosis effect described in Box 6.1 and Chapter 14.1 which can affect flow and pressure distal to a stenosis.

Finally, mention must be made of the ongoing attempts to visualize and measure abnormalities of myocardial perfusion through the use of echocardiographic methods (with the use of intravascular contrast agents) and using CT technology. The latter in particular promises a broad development in the years to come.

19.3.3 Myocardial Vitality

The identification of areas of fibrous scarring in the myocardial structure represents an important diagnostic and prognostic indicator. Currently, the reference method for ascertaining the presence of areas of fibrous scarring within the myocardium is cardiac fMRI. Through the late impregnation technique (*late gadolinium enhancement*) it is possible to create images in which areas containing a high concentration of contrast medium appear with a more intense signal (Figure 19.5). The gadolinium-based contrast agents commonly used in clinical practice share the property of permeating the capillary walls and of spreading in tissues in a quantity proportional to the blood concentration and the percentage of extracellular space. The cardiac muscle is a functional syncytium in which the cells adhere to each other,

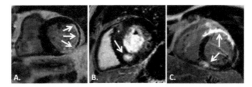

Figure 19.5 Example of late impregnation images (short axis of the left ventricle) acquired by magnetic resonance. Areas of fibrous substitution in the myocardium appear as areas of hyper-intensity (arrows), the vital myocardium is instead represented by low-intensity tissue. A: Thin subendocardial scar, chronic myocardial infarction. B: Intramyocardial fibrosis at the insertion point of the free wall of the right ventricle on the left ventricle in a case of hypertrophic cardiomyopathy. C: Epicardial fibrosis in a patient with cardiac sarcoidosis.

leaving very little extracellular space. In contrast, areas of fibrous scarring contain few fibroblasts surrounded by collagen and other matrix proteins and therefore have a high extracellular volume. Magnetic resonance imaging, using adequate image acquisition programs (sequences), allows highlighting the difference in concentration of the contrast medium that is created between scar and vital myocardium.

Even PET and SPECT allow the identification of areas of vital myocardium from scar areas. Similar to what was discussed in the previous paragraph; however, cardiac magnetic resonance allows a much higher spatial resolution capable of identifying subendocardial infarct areas with high sensitivity. The late impregnation by magnetic resonance also allows the differential diagnosis between different types of cardiomyopathies, thus representing a unique tool in cardiology practice.

Echocardiographic methods allow the non-invasive assessment of myocardial vitality. The wide availability of echocardiographs in cardiology departments and the intrinsic flexibility of the echocardiographic method are among the fundamental reasons that explain the success of the method. The echocardiographic study of vitality is based on the acquisition of functional images during adrenergic stimulation with *dobutamine*. The increased inotropy induced by this drug can be easily appreciated in qualitative terms, but some quantification is also proposed. Indeed, the global longitudinal strain is used to compare heart function in various conditions.

19.3.4 Assessment of Blood Flow in Large Vessels and Cardiac Output

Measurement of blood flow in a vessel or assessment of cardiac output plays a key role in diagnostic evaluation and therapeutic planning, especially in patients with congenital heart disease.

For a long time, cardiac catheterization (Fick method and thermodilution method) has been considered the reference method, although there have been many studies aimed at developing methods for non-invasive measurement of flows and cardiac output by echocardiography.

Currently, in some hospitals, the reference method is cardiac magnetic resonance. It offers a reproducible, completely non-invasive flow measurement (no injection of any contrast medium is required), independent of the geometry of the anatomical structure under examination, and does not require radiation.

Echocardiography and Echo Doppler, on the other hand, remain the reference imaging methods for alterations in blood flow found in valvular diseases (stenosis or insufficiency).

19.3.5 Myocardial Oxygenation

The level of myocardial oxygenation can be measured by coronary sinus catheterization and measurement of oxygen content in coronary venous blood. Alternatively, new developing PET tracers for clinical applications allow selective marking of tissue areas where the partial oxygen pressure is reduced. These tracers initially developed to visualize hypoxic areas within solid tumors, could soon find an application also in the cardiovascular field.

Finally, as said above there are the possibility of measuring the oxygenation status of the myocardial wall by using BOLD (blood oxygen level-dependent) methods in magnetic resonance.

19.3.6 Electric Activation

At present, cardiac catheterization and electrophysiological mapping are the only alternatives for the study of myocardial electrical activation. New-generation image methods, such as electrophysiological navigation systems, have recently enabled the generation of three-dimensional maps of cavities without the need to use X-rays.

At the end of this chapter and of this book, it is worth spending a few words on the lowest technological methods for evaluating the cardiovascular function, even if only qualitatively. These are the *pulse examination, auscultation* and *blood pressure measurement* with a *sphygmomanometer* (see Chapter 8.12). Examination of the peripheral pulse by palpation with the tip of the finger can give an idea of heart rate, pulse pressure, stroke volume concerning artery compliance. These simple procedures can help the medical doctor decide whether to use the most advanced technologies we have described above or the older but still indispensable ECG (see Chapter 10) Of course, not only these simple procedures can be used, but they cost nothing and are also useful for feeling empathically close to the patients.

About the Authors

Pasquale Pagliaro, *M.D., Ph.D* was born in Rossano, Italy, in 1961. He is a full professor of Physiology at University of Turin (Italy), Department of Clinical and Biological Sciences. He is also member of the National Institute for Cardiovascular Researches (Bologna, Italy).

Degrees awarded: MD, University of Turin (Italy), Thesis topic: Coronary Pathophysiology, 1988. PhD, University of Turin (Italy), Thesis topic: Endothelial Physiology, 1994. Research Fellowship in Medicine-Cardiovascular at the Johns Hopkins University Baltimore (USA); Research topic: Coronary and Endothelial Physiology, 1997–99.

Research experience/other activities: Prof. Pagliaro studies cardioprotection, coronary physiology and pathophysiology. The recent research activities concern endothelial factors and other endogenous substances in triggering cardioprotective signaling pathways. Prof Pagliaro lab focus also on redox-signaling and mitochondrial function.

He is Ordinary member of Italian Society of Physiology, The Physiological Society (London),

Italian Society of Cardiology, European Society of Cardiology, Italian Society of Cardiovascular Research (SIRC). Prof Pagliaro is Past-President of SIRC and the Coordinator of the PhD program in Experimental Medicine and Therapy of the University of Turin.He is author of over 200 articles on topics of Physiology and Molecular Pathophysiology in the cardiovascular field.

Claudia Penna, *BSc.D., Ph.D* was born in Asti, Italy, in 1967. She is an associate professor of Physiology at University of Turin (Italy), Department of Clinical and Biological Sciences. She is also member of the National Institute for Cardiovascular Researches (Bologna, Italy).

Degrees awarded: BSc, University of Turin (Italy), Thesis topic: Effect of venom in the isolated heart, 1991. Specialist in Clinical Pathology, University

411

of Turin (Italy), thesis topic: Modulation of cardiac current by Nitric Oxide, 1995; PhD, University of Turin (Italy), thesis topic: Hyperaemic response and Ischemic Preconditioning, 2000.

Research experience/other activities: coronary circulation pathophysiology and cardioprotection.

She is member of Italian Society of Physiology, Italian Society of Cardiology, European Society of Cardiology, Italian Society of Cardiovascular Research. She is in the Editorial Board of international journals and is Associate Editor of Oxidative Medicine and Cellular Longevity.

She published over 100 publications on impacted journals, including invited scientific reviews, and she is co-author of several book chapters.

Raffaella Rastaldo, *MSc, PhD* was born in Cigliano (VC), in 1970. She is an associate professor of Physiology at University of Turin, (Italy), School of Medicine.

Degrees awarded: Degree in Biological Sciences (MSc equivalent), University of Turin (Italy) in 1998. Thesis topic: Endothelial Physiology. PhD in Physiology, School of Medicine, University of Turin (Italy) in 2003. Thesis: Coronary Pathophysiology. Post-doc fellow at the Medicine Department - Cardiovascular Research Institute – New York Medical College, Valhalla, NY (USA).

Research experience/other activities: her research activity is focused on cardiac regeneration using innovative nanomaterial as well as on cardioprotection against the ischemia-reperfusion injury in particular using endogenous peptide.

She is a member of the Italian Society for Cardiovascular Research since 2007 and of the Italian Society of Physiology since 2008. She is a reviewer of various prestigious international journal and a member of the Editorial Board of BioMed Research International. She published over 40 publications on PubMed indexed journals, several scientific reviews and book chapters.